RIGHTS TO HEALTH CARE

# Philosophy and Medicine

VOLUME 38



*The titles published in this series are listed at the end of this volume.*

# RIGHTS TO HEALTH CARE


*Edited by*

THOMAS J. BOLE, III

*University of Oklahoma Health Sciences Center*
*Oklahoma City, Oklahoma, U.S.A.*

WILLIAM B. BONDESON

*School of Medicine, University of Missouri*
*Columbia, Missouri, U.S.A.*



SPRINGER-SCIENCE+BUSINESS MEDIA, B.V.

**Library of Congress Cataloging-in-Publication Data**

Rights to health care / edited by Thomas J. Bole, III, William B. Bondeson.
p. cm. -- (Philosophy and medicine ; v. 38)
Includes index.
ISBN 978-94-017-4092-0 ISBN 978-0-585-28295-4 (eBook)
DOI 10.1007/978-0-585-28295-4
1. Right to health care--Congresses. 2. Medical policy--Congresses. I. Bole, Thomas J. II. Bondeson, William B., 1938- . III. Series.
RA394.R56 1991
362.1'042--dc20 91-157

ISBN 978-94-017-4092-0

*Printed on acid-free paper*

# TABLE OF CONTENTS

## SECTION III / A QUALIFIED RIGHT TO HEALTH CARE: TOWARD A NOTION OF A DECENT MINIMUM



## SECTION IV / EQUALITY, FREE MARKETS, AND THE ELDERLY



## SECTION V / HEALTH CARE AS A COMMODITY

# EDITORIAL PREFACE

Human existence is marked by pain, limitation, disability, disease, suffering, and death. These facts of life and of death give ample grounds for characterizing much of the human condition as unfortunate. A core philosophical question is whether the circumstances are in addition unfair or unjust in the sense of justifying claims on the resources, time, and abilities of others.

The temptation to use the languages of rights and of justice is understandable. Faced with pain, disability, and death, it seems natural to complain that "someone should do something", "this is unfair", or "it just isn't right that people should suffer this way". Yet it is one thing to complain about the unfairness of another's actions, and another thing to complain about the unfairness of biological or physical processes. If no one is to blame for one's illness, disability, or death, in what sense are one's unfortunate circumstances unfair or unjust? How can claims against others for aid and support arise if no one has caused the unfortunate state of affairs? To justify the languages of rights to health care or justice in health care requires showing why particular unfortunate circumstances are also unfair, in the sense of demanding the labors of others. It requires understanding as well the limits of property claims. After all, claims regarding justice in health care or about rights to health care limit the property rights of those whose resources will be used to provide care. The languages of rights to health care or justice in health care, if secured, lead to others having duties to give aid and to relinquish claims over their own time, money, and resources. The languages of rights to health care and of justice in health care construe much of what is unfortunate as also unfair in the sense of supporting moral claims for particular allocations of health care.

This volume addresses the problem of grounding duties to provide care, rights to health care, and claims to justice in health care. As the volume shows, sense can be made out of this web of claims only if one goes to a foundational level to determine the scope and character of duties of beneficence and claims in justice. To go beyond the often confusing and provocative rhetoric of rights and justice, one must first make sense



*T.J. Bole III and W.B. Bondeson (eds.), Rights to Health Care*, vii–ix.

of the translation of unfortunate circumstances into unfair circumstances and show how the claims of those in need can limit the rights of those who possess.

This volume developed out of numerous discussions regarding the standing of rights to health care, duties to provide health care, and claims to the provision of health care grounded in considerations of distributive justice. The plans for this volume began with the 20th trans-disciplinary symposium on philosophy and medicine 'Rights to Health Care', held March 14–16, 1985, at the School of Medicine of the University of Missouri-Columbia.

There are many to whom the editors are in debt for the support of this conference, which initially framed the issues upon which the volume focuses. In particular, the conference received support from the Missouri Committee for the Humanities, the state-based arm of the National Endowment for the Humanities. It was also supported and sponsored by the Program in Health Care and Human Values, the Department of Family and Community Medicine, and the Department of Philosophy, University of Missouri-Columbia. The co-chairmen of the conference were William B. Bondeson and Gerald T. Perkoff. In addition, the conference received the generous voluntary labor and participation of a large number of individuals. The editors wish to express their deep appreciation to all of these institutions and persons.

The papers developed at this conference became the focus of subsequent philosophical and public policy discussions. The current volume draws not only from essays presented at the Missouri conference that have been refashioned over half a decade of dialogue, but also from contributions made by individuals over the intervening years. In particular, the editors wish to express their appreciation to the Liberty Fund, which sponsored a symposium on November 12–13, 1987, in Houston, Texas, hosted by the Center for Ethics, Medicine, and Public Issues, Baylor College of Medicine ('The Profit Motive in Medicine: Contemporary Issues in Historical Perspectives'), from which ancestral versions of the essays by Klaus Hartmann and H. Tristram Engelhardt, Jr., derive. Dr. J. Charles King and the participants at that seminar contributed to the development of many of the issues explored in this volume. Special thanks are due to Mary Ann Gardell Cutter, who worked with the manuscripts over years of discussion among the authors of the essays and the editors of the volume. The editors are similarly in debt to George Khushf, who helped with the final editing. Without their labors and insight, this volume would not have successfully come to completion. Though this volume reflects a half decade of conversations between the contributors and the editors, it is obvious that the debate at both philosophical and public policy levels has not reached closure.

*15 November 1990*

THOMAS J. BOLE, III
WILLIAM B. BONDESON

THOMAS J. BOLE, III

# THE RHETORIC OF RIGHTS AND JUSTICE IN HEALTH CARE

## I. THE NEED TO EXAMINE THE RHETORIC

Discussions of contemporary health care policy are replete with claims about justice or equity in health care allocations and rights to treatment. In 1983 the President's Commission for the Study of Ethical Problems in Medicine and Biomedical and Behavioral Research spoke of society's "moral obligation" to ensure every citizen adequate health care without excessive burden, and concluded that government should meet this obligation ([14], pp. 22, 30–32). In 1990 the American College of Physicians urged that "[a]ll Americans should be able to obtain appropriate health care services", and that there should be a national policy to develop the health insurance coverage to "minimize financial barriers and assure access" to those services ([2], p. 658).

There has generally been a great deal of agreement among policy makers about the obligation to provide some level of health care for all. There is no agreement, however, about the nature and scope of such obligations. Even if people agree that the universal provision of some level of health care ought to be provided, it does not follow that politically organized society ought to provide it. Quite a few societal arrangements would, if realized, obviously save lives. It is not so obvious how to show that resources ought to be transferred and actions regulated in order to realize those arrangements. The funds expended on a meal in a restaurant of distinction would probably have saved the lives of many starving individuals somewhere in the world. If the free-wheeling rhetoric of rights to health care and obligations of justice is to be taken seriously, some justification for their claims is needed. The difficulty at the foundational level is that there is no general moral consensus about how the need by some for food, shelter, and health care creates obligations on the part of others to provide them.

This volume reflects the diversity of claims about rights, duties, and justice in health care, by recognizing the diversity of foundations advanced to secure such claims. Instead of proceeding to a morally



*T.J. Bole III and W.B. Bondeson (eds.), Rights to Health Care*, 1–19.

satisfying conclusion, the volume focuses on the difficulties of grounding the plethora of moral claims made on behalf of those who would benefit were more health care at their disposal. These difficulties provoke foundational issues in ethics. They present the general problem of translating unfortunate outcomes into unfair outcomes so as to ground claims for the services and resources required to help those in need. To translate the unfortunate into the unfair, one must place needs within a moral and/or metaphysical interpretation that construes those needs as generating or grounding such rights. The difficulty is that rights to have one's needs met or one's goods realized, positive rights, usually involve claims on the labors and resources of others. Positive rights circumscribe negative rights, the rights of others to be left alone. In particular, the justification of positive rights involves securing the authority to impose a particular moral vision coercively on those who may not see themselves as owing positive obligations of a particular kind to others, by forcing them to work for others.[1]

This coercive imposition is difficult to justify, because the burden of proof in establishing which sort of rights should be given priority over the other in any particular case seems to lie upon positive, not negative, rights. One's claim to be left alone does not require others to realize one's own goal, whereas one's claim to others' energies or resources assumes that the others either in fact consent to this claim as morally authoritative or ought to do so. A positive right seems to require justification in terms of a moral viewpoint that can claim to be cogent for, and thus to authorize the right-holder's claims against, others, even though they may not share the same premisses as the proponent of that right. No such viewpoint has to be justified by the proponent of a negative right; he can remain sceptical of any premiss advanced, and so of any conclusion derived therefrom. In the face of possible scepticism, the proponent of a positive right has difficulties cogently justifying the particular moral vision that authorizes government to coerce the skeptic's resources and choices in order to realize that positive right.

The generic rationale for such a justification is that positive rights are designed to conduce to the good of society as a whole, the common good, whereas negative rights are not. Indeed, negative rights permit one to use what is one's own not only in morally laudatory ways, but also in morally culpable or morally neutral ones, so long as force is not thereby inflicted upon unconsenting innocent others. The common good, by contrast, is exclusively morally laudatory. Moreover, it is basic to the social nature of morality. Insofar as negative rights ought to be safeguarded by society, they are usually seen as presupposing, and requiring subservience to, the common good. The moral authorization claimed for the state's implementation of positive rights is a special case of that made in behalf of just laws in general: that it either conduces to a normative common good, or is authorized by the consent of the governed. Let us set aside consideration of what consent of the governed means, except to note that if the governed have genuinely consented, then they have ceded their own negative rights, and positive rights need no further authorization.[2] If the common good morally authorizes the state to realize positive rights, the authorization shows why individuals ought to cede their negative rights to a common good to which they do not consent; hence the state would be authorized to sanction their consent.

A morally authoritative common good is not easy to realize, however, because it is difficult to establish a canonical ranking of human goods and goals in terms of secular reason alone. These are terms to which any rational agent (within a particular society) ought to consent, and which do not assume as a premiss some shared belief or commitment that must presume assent in order to be morally authoritative. Any canonical ranking of goods presupposes a standard of goodness. If this standard is determinate enough to permit one canonically to rank likely candidates for the common good, such as equality, equity, social justice, preference utility, and liberty, then the standard itself embodies a determinate notion of goodness that is presumed rather than justified. On the other hand, if the standard does not permit such a ranking, it cannot provide the needed moral authorization. It seems, therefore, that there is no way to make a definitive choice among competing visions of the good or of proper action, except by relying upon a standard of values or a canon of action that itself must be established. In order to choose correctly, one must already have the morally authoritative, higher-level account or vision of the good. One cannot appeal to consequences to resolve the dispute without knowing already how to rank the consequences (for example, consequences for liberty relative to those for security or equality or prosperity). One cannot appeal to preference satisfaction without knowing already how to rank, for example, present versus future consequences, or rationally considered versus emotionally charged consequences. Whether or not the standard permits a canonical ranking, it does not seem to authorize a notion of common good that is morally authoritative upon those who do not assent to it. If no ranking is morally authorized, and, say, the will of the majority or of its elected representatives imposes a particular ranking, it is difficult to make out the moral authority of that will and of its expressions upon those who have not consented to the rules in virtue of which the majority can so work its will.

There is also the specific difficulty in authorizing a common good in terms of the need of some individuals for health care. Why should the common good meet the health care needs of some at the expense of the legitimately acquired possessions of others? It is not as if the pain and disability of the former parties were caused by actions of the latter for which moral redress is due. If it were, there would be reason for moral redress; even in this case, it is not the state but the wrong-doers who are obligated to make restitution, and it is not immediately clear that the state should guarantee that restitution. By and large, however, one's good or ill health is the result of physical and biological processes, and perhaps of social good or ill fortune, but not of actions for which others can be held liable.[3] In what sense, then, can these effects be claimed to be not simply unfortunate but also unfair?

Despite the difficulties in giving convincing answers to this question, practically every representative democracy considers itself to possess moral authority to realize positive rights. In practically all economically developed democracies, health care of some sort is among these positive rights. It is important, therefore, to consider answers to this question. Unless one can show why the unfortunate is unfair, and why it is the state that is authorized to redress the unfairness, there seems to be no moral authorization for the state coercively to reallocate individuals' resources, or for positive rights insofar as their realization depends upon such reallocation. (It does not follow that state allocation of resources would then be *eo ipso* morally illicit; only that one will have to view the moral authorization of resources differently, as is suggested in Section III.) We are forced to scrutinize the rhetoric of rights to health care and of obligations of justice more closely, because our overview of that rhetoric suggests that it is largely baseless.

## II. THE RANGE OF ISSUES

This volume focuses on the perplexities involved in that rhetoric. The volume's essays fall into three categories. The first category sketches the geography of the rhetoric. Its essays display the range of claims and kinds of issues that debates about rights and justice in health care policy involve. The second category attempts to establish various foundations for rights to health care or accounts of justice and health care. Its essays involve the endorsement of a particular vision of the correct distribution of goods, the proper ranking of social desiderata, or the true understanding of social unities such as the state. The third set of essays are critical or skeptical of the second category. They start with the difficulty of establishing a particular canonical view and ask what rights to health care or pattern of just allocation of resources can be secured in the face of fundamental disagreements regarding the status of moral rights and the nature of justice. Unable to discover a canonical web of rights and obligations or just allocations, the essays then turn to consent and agreement to create this web.

The issue looming largest is the relative priority of positive rights and negative rights. Distinguish positive rights by which the state entitles some individuals to the resources and services of others who have not consented or contracted to the state's reallocation, from positive rights in which there is no such problem with the communal resources and services that underwrite them. In the former case the owners of the resources have not authorized the claims to their beneficence; in the latter case they have. Can positive rights in the former sense, which we shall call welfare rights, override negative rights? Or does it even make sense to speak of negative rights apart from a political framework which can, and almost always does, authorize welfare rights?

The answer to these questions turns largely on which side has to bear the burden of proof. On the one hand, to assert that negative rights are inviolable, no matter the consequences for the lives or well-being of the society within which the owners find themselves, is intuitively implausible. If the consequences of not meeting the needs of others in society is, as Beauchamp and Buchanan suggest ([3], [7]), sufficiently grievous for the society as a whole, and if those needs could be met by authorizing welfare rights, the onus seems to be on the proponent of the inviolability of negative rights. On the other hand, welfare rights seem to claim the resources of others who may be unconsenting and innocent of any moral commitment to the claimant, unlike the assertion of control to dispose of one's own (so long as one does not inflict force upon unconsenting, innocent others). To assert one's negative rights is simply to assert the right to be left alone. From this perspective, the proponent of welfare rights bears the burden of proof. He must show that the particular vision of the common good that authorizes some to reallocate and regulate the resources of others is cogent, and that the government is morally authorized to compel that reallocation.

What specific visions of the common good would authorize governments to meet needs of some with the resources of unconsenting, innocent others? The generic answer, we may say with Hartmann, is a sense of social well-being, a sense that permits the disadvantaged not to be so alienated by their unfulfilled basic needs that they are unable to identify, to be in "solidarity", with concerns of the civic community [12]. The moral authorization is the citizenry's duty of "solidarity", to assist those whose life and health would otherwise be lost.

The difficulty is in specifying the answers and establishing specific sorts of duties to provide this assistance. Different specifications have drastically different implications for health care. For instance, to save lives at all costs would require far more coercive reallocation of resources for, e.g., intensive care, than most would tolerate. Even some minimal duty of solidarity would prevent one from claiming that one's own body is solely his property. It would, as Hartmann implies, prevent one from legally taking one's own life, or from helping another to do so. More generally, any specification of the good must justify itself as authorizing the government to coerce the reallocation of resources and regulation of health care.

This volume features several such specific visions of the good of health care:

Veatch's essay [21] offers one such specification, that of egalitarianism. This is the notion that equal moral worth requires that each person have opportunity for equal well-being. Otherwise the more well-do-do would be unconstrained in using their superior resources to exploit the less well-to-do. Since health is a necessary condition for equality of opportunity, egalitarianism implies a health care system in which equal access to health care is provided for all. Indeed, so basic is health supposed to be that one should not even be permitted to barter away one's health care vouchers or credits for other things.

Veatch concedes that the premisses of egalitarianism cannot be proven. They are based upon rational intuition. The difficulty with this concession, Engelhardt counters [8], is to show how the premisses can morally authorize government coercion against innocents who do not share them. The difficulty is acute, because Veatch's premisses contain several features that require cogent argument to override equally intuitive alternatives. One such feature is the assertion that individuals do not in fact own things, not even their own bodies and talents, apart from moral authorization by (the appropriately) politically organized society. If individuals can own property on their own, they can enter into the arrangements whereby inequalities of opportunities result, e.g., between one who receives a good education, or wealth from benefactors, and one who does not. A second such feature is the notion that the moral equality of persons requires equality in the distribution of resources. It may be that different persons are equally deserving of resources, but that some, as the result of gifts, have title to more than others. If we allow that the moral equality of persons requires equal respect of their autonomy, we will allow that, in consequence of the different exercises of autonomy, different individuals will come legitimately to possess different amounts of resources. Third, even if politically organized society were morally authorized to level these differences, there may be good utilitarian reasons not to do so, even within a basically egalitarian scheme. It is difficult to set out how encompassing the egalitarianism should be. Should it, for instance, allow equal access to cosmetic surgery? Or only to serious diseases, and significant extensions of life (in which case the crucial adjectives must be defined)? It might be that one should allow some freedom for individuals to dispose of and to acquire different amounts of resources, and so to allow some to have access to more health care than others, so long as, via taxes and the trickle-down effect of exotic medical technologies, etc., even the least well-off would benefit more than from a strict egalitarianism.

These points are advanced not as objections, but as showing the need to examine the cogency of the foundations for Veatch's, and for any, particular vision of good health care that would authorize the coercive redistribution of individuals' resources. In light of these points we can see the plausibility of the alternative advanced by Daniels [7], and also implicitly defended by Beauchamp [3] and Rhoden [18]. This is the particular vision of the good set forth in John Rawls' *A Theory of Justice*.

Like egalitarianism, it assumes that individuals do not in fact own things. Like egalitarianism, it requires an equal right to the most extensive system of basic liberties for each person that is compatible with similar liberties being available to all others. Unlike egalitarianism, it does not then claim that these liberties can only be legitimately exercised in ways that conduce to an equal distribution of resources. Rather, it claims that a just society can allow differences in the distribution of resources, so long as the result redounds to the benefit of the least advantaged, and there is equal opportunity for access to the resources.

The cogency of the theory depends upon its claim to be the sort of society any rational individual would find just, if that individual had to agree to the structure of a society in ignorance of his social position, natural endowments, and particular historical views about the good. We can also see that the rational contractor does have a vision of the good determinate enough to value liberty more than total equality, or security, or the greatest possible chance of prosperity even if the resulting differences in the distribution of resources might benefit some less than under an egalitarian system. It is a vision of the good that we, like Veatch, might not share. More generally, we might have fundamental theoretical misgivings about why such a theory can morally authorize the restructuring of any established society. The theory abstracts from the culturally and historically conditioned structures of any existing society, and yet claims to be authoritative for refashioning those structures. We may think that the very ahistoricity of the theory is ground, if not to indict it, at least to find it suspect.

In this light we may consider the conditions within which such a theory licenses differences, and the consequences of each of the conditions for rights to health care. If differences in health care are licensed insofar as they benefit the least advantaged, one will have to define the class of the least advantaged. In addition, one will have to decide whether or not the least well off is to be defined simply economically or also with respect to health. Daniels, along with Beauchamp and Rhoden, thinks that differences in health care are licensed insofar as they assure fair equality of opportunity ([7], [3], 18]). This is the range of opportunities that someone with a certain allotment of talents would have had were that person healthy. If a society must redress individuals' disadvantages due to illness and incapacity in order to be fair, it may have to spend a great deal on the very ill and incapacitated before it spends anything on other desiderata, such as education, or income supports, that may allow a greater equality of opportunity for most and at less overall cost. Thus, one may have misgivings with such a theory's ahistoricity not simply at the general level, but also in details. If these misgivings are reasonable, then one may wonder if the theory can morally justify authorizing the government to coerce the reallocation of the resources of some unconsenting innocents in order to provide health care to others.

Such brief animadversions to Rawlsian theory as are made in the previous paragraphs are not, and are not meant to be, telling against the theory as laid out in the essays of Dnaiels and others. They do, however, alert the reader to some of the points to consider in going through the details of these essays. The paragraphs also serve to locate generally the misgivings of several essays about such theories as that of Rawls.

The locus of misgivings is these theories' ahistorical abstractness. Halper, for instance, makes the general point that any right to health care must be qualified by the resources which the body politic can realistically make available [11]. Consequently, Agich observes [1], rights to health care depend upon political will, and this may be more concerned to cut costs than to realize claims to abstract rights. Brody and Sass both lament the fact that talk about rights may occlude attention to measures that effectively help the indigent. Sass points out that to speak of health care rights without stressing responsibilities is in practice to encourage welfare dependence [19]. Brody suggests that the needy be given vouchers, and the choice to use them to meet needs not only for health care, but also for other things that affect personal health, such as nutrition, housing, and clothing [5]. If the voucher is not used responsibly, that is unfortunate. Both Brody and Sass indicate the implausibility of saying that a society must redress individuals' disadvantages due to illness and incapacity in order to be fair. These misgivings reflect the general difficulty of establishing that a theoretical model should be normative for any effective program that meets the concrete health care needs of the indigent.

As an alternative to egalitarian and Rawlsian theories which does not require government reallocation, Friedman offers a preference utilitarianism that he thinks is best met by the free market [10]. Only those who purchase health care or who are objects of charity would receive health care. However, Friedman contends that the free market is so much more efficient than government in meeting needs that it will in the long run better help those disadvantaged by natural and social lot. Such a theory would not satisfy the egalitarian. For him equality in the value placed on each person's life and health is morally preferable to different valuations, in relation to the different abilities to pay, even if the latter were to benefit the least well-off class more than egalitarianism. Friedman's theory could, however, satisfy the Rawlsian, if it in fact worked out. Since the theory relies upon the free market, it would have the added advantage of not having to justify government-coerced reallocation.

Khushf points out [14] that Friedman assumes that politicians are ultimately self-serving, so that the government cannot effectively serve ethical or moral concerns. Buchanan, by contrast, suggests that the government can give organizational expression to moral concerns, and specifically, to the duty of beneficence regarding health care needs [6]. He thinks that the government can also efficiently collect individual contributions, avoiding the free-rider problem, and more fairly allocate resources than uncoordinated charities. Marmor even suggests that the lack of government control has allowed the free market so to drive up health care costs and to reduce access to health care for the needy, that government is morally required to redress what its lack of coordinating policy has caused [15]. However, even if we question Friedman's assumption about politicians, we would still have to show that the government's coercive reallocation of resources is not simply the imposition of some persons' moral views upon unconsenting, innocent others.

In general, theses that propose a theory of justice, or of an equitable society, or of social well-being, must rely upon a particular ranking of values in order to justify the coercive reallocation of resources. That particular standard is the goal that is supposed to be normative for the society; in order to reach it, the political organization of society, the government, is supposed to be authorized to reallocate resources. (Friedman's theory is the exception, because a free-market utilitarian claims that the most effective health care system is one in which there is no or minimal government intervention.) Depending upon how the goal ranks equality, prosperity, security, and liberty with respect to access to health care, the reallocation of resources which the goal authorizes will be dramatically different. The difficulty is that the goal cannot be shown, in terms of secular reason alone, to be normative upon those who are not actually committed to it by contract. The theory is not cogent unless one agrees to the premises, and the premises which declare the canonical ranking of values can always be disputed. Then, as Engelhardt points out [9], the goal cannot justify some to reallocate coercively the resources of unconsenting, innocent others.

## III. AVOIDING THE FANATICISM OF JUSTICE: THE MORAL NECESSITY OF COMPROMISE

At issue in disputes about justice, fairness, and equity in health policy is whether philosophical reflection can discover what we ought to do. A philosophical theory is necessary to show how to translate the needs of some into claims of rights against others who have done them no wrong. The canonical ranking of values contained in a just or fair or equitable health care system is supposed to be authoritative upon those governed by the system whether or not they consent, because it is supposed to be rationally cogent. This cogency is the reason why the theory is supposed to justify the government's coercive reallocation. What may well be the case, however, is that philosophy cannot give argument cogent for those who do not share the premises. Philosophy does not seem to be able to discover the correct theory. More precisely, it provides a plethora of possibly correct theories, depending upon how canonical values are ranked, and intelligent and well informed individuals will disagree about which theory is correct. It is well, therefore, to assess the resulting situation for health care policy and for its moral authorization.

The peculiarity of philosophical theories about the proper health care policy is that they are *a priori*. They constitute goal-based theories of justice or equity that say how individuals ought to allocate their resources in order to attain the good of society-wide health care, no matter whether the individuals have consented to the goal or how the individuals might want to use their resources. Goal-based theories are not constrained by the negative rights of unconsenting and innocent individuals, because the theories claim to say what goods these individuals' resources should promote. It is just this claim about the canonical ranking of goods that provokes objections. Because the moral authorization for redistribution in order to obtain rights to health care depends upon the cogency of the canonical ranking, the objections are difficult to overcome.

These objections are avoided by a distribution of resources that is constrained by negative rights, by a procedure-based rather than a goal-based justice. This sort of distribution is morally authoritative, because it derives not from the rationale behind the ideal pattern, but from the actual consent of the individuals who constitute the community to whom the resources belong. The distribution is just, because it recognizes what is due to the owners of the resources, viz., that they must consent to the distribution of those resources. Such distributions are, moreover, not only effected by private charities; they can be effected wherever legitimately derived communal resources are available for distribution, such as the proceeds from the mineral rights owned by the State of Texas. In such cases the moral authority for the distribution will derive from the consent of the community if made with the informed democratic participation of its members.

The unproblematic nature of this sort of procedure-based justice relative to goal-based systems of justice suggests compelling theoretical and practical reasons for a two-tier system of health care, one tier in which health care is provided on the basis of the individual's ability to pay, and the other in which health care is provided by either private charities or communal resources on the basis of individual need.

Most economically advanced societies are determined to alleviate the health care needs of some of those whose natural lot of health is unfortunate. This determination may be due to charity, as Buchanan contends [3], or, as Engelhardt suggests [9], to provide some insurance for those who suffer misfortunes of health due to natural lot and are unable to pay for the services they need. Given legitimate communal resources available to governments, there is moral authorization to provide this care, so long as the allocation is based upon the informed democratic participation and compromise necessary to determine which needs can be fulfilled. (How such resources were accumulated, and by what processes of consent, are embedded in the historical particularities of each case. Most societies reflect a tacit consensus that there are legitimately derived communal resources available to the government for underwriting positive rights, and one may proceed on the basis of that consensus.) Moreover, since negative rights cannot be morally proscribed altogether, and assuming that there are not enough legitimate communal resources to meet all health care needs or desires, there must remain a second tier of health care, in which some individuals elect to spend from their own resources more on insurance and health care than legitimate communal resources allow for the needy.

The practical reason for the two-tier system of health care is a compromise in the face of theoretical uncertainty about whether there is authority to fully allocate all resources for the publicly funded tier, and, if there is, about what this tier should support. Indeed, for a secular pluralist society to impose one particular health care system is to assume an unjustifiable moral imperialism about the correct way to meet society's health care needs. Such an imperialism might take the (improbable) form of Friedman's free market utilitarianism, or the form of Veatch's egalitarianism, or of a Rawlsian theory that only allows a for-pay tier insofar as it conduces to the benefit of the least advantaged. In this light, the compromise solution not only practically accommodates many of the concerns of contrary proposals, as Khushf points out [14]; it also theoretically justifies a practical accommodation to theoretical uncertainties about the proper scheme of public distribution and the proper accommodation of both public and private ownership.

If the compromise solution is compared with the theories proposed in order to give moral authorization to rights and justice in health care, we see how the compromise takes care of moral authorization. First, it offers an historical solution, rather than an ahistorical, or ideal, solution to the needs of health care. It grants that individuals possess resources of their own and are free to spend them on health care. It assumes that there are legitimate communal resources with which to abet private charity in helping the needy obtain health care. If there are communal resources, there is no need to authorize their acquisition. Their disposition is authorized by the processes of democratic participation and compromise, e.g., by precinct meetings to authorize specific priorities among the health care services to be provided. Second, the compromise solution is morally authorized by the actual persons whose resources these are, not by a rational theory or by insight. Third, the solution is not the discovery of the correct pattern, but the creation of a pattern that seems to those involved to be the best available in the circumstances. Further corrections are made not in order better to attain the abstract ideal, only in order to dispose prudently of the public funds available in light of possible revisions in democratically determined priorities. Fourth, the compromise is fair in virtue of the procedure, which respects the owners in their disposition of resources for health care. There is no need, as in goal-based theories, to see whether the actual outcome measures up to the theoretically ideal of fairness. Each of these four features reveals the moral authorization of the compromising two-tier system to be unproblematic in comparison with goal-based theories.

To speak generically of a two-tier system, however, is not to distinguish the system from any of the advanced industrial countries in Western Europe, the Pacific rim, or North America, save Canada. All of these countries have not only a basic publicly funded social program but also, in addition, an optional, privately funded tier. Only Canada refuses to allow private health insurance to be sold, save for incidentals not covered by its provinces' health insurance plans, and refuses to permit physicians or hospitals to treat both patients who are on such plans and patients who would pay either out of pocket or through private insurance ([13], p. 563f). It is seen as unjust to make the lives and health of those who would pay for additional insurance or services worth more than those who would not or cannot do so. In this sense, Canada instantiates Veatch's plan. Yet, if one is to speak approbationally of a two-tier compromise as morally authoritative, either the above part of this section is utterly irrelevant to practical considerations about morally authoritative health care delivery, or some distinctions must yet be made.

By way of doing the latter, it is worthwhile pinpointing what is problematic about the Canadian system. First, it is questionable whether a government has the moral authority both to proscribe someone from using legitimately acquired resources of his own to obtain additional health care, and to proscribe health care providers from entering into an agreement with that person to supply the care desired. The question would be answered in the affirmative only if Veatch makes a cogent case for egalitarianism. Second, there is the problem, both theoretical and practical, that the recipients are insulated both from the costs and from determining the allocation of the resources for their health care. The practical aspect of this problem is that one cannot continue to supply whatever services are deemed medically required without raising taxes or reducing expenditures for non-health-care government services or, as has been the case in Canada, restraining the use of expensive technological developments and increasing the waiting time to gain access to them (cf. [13], pp. 565–566, 568). The theoretical aspect is that this insulation requires cost shifting, even if it is, as in fact it is, by way of deficit spending, and the parties to which the costs are shifted have not consented. Cost shifting remains a problem unless a cogent case can be made for a goal-based theory which would justify the end to be achieved by the shift of costs. Third, from a practical point of view, a private tier in addition to a publicly funded tier permits those who wish more extensive, or more quickly available, or more amenable medical coverage than the taxpayers deem they can afford, to get it, and to do so in a way which supplies tax revenues for the public sector.

Nonetheless, even if we allow the theoretical superiority of the two-tier system over a one-tier system such as Canada's, we would have to specify those two tiers in some detail, for it is far from obvious that the two-tier system in the United States is better than Canada's in practice.[4] Why not, for example, combine Canada's plan with a private tier? The result would not be entirely egalitarian, but it would be Rawlsian, along the lines advanced by Daniels [7] and defended by Beauchamp [3] and Rhoden [18]. The publicly funded tier would be morally primary, not so much in the sense that it provides some basic minimum health care for all, but in the sense that it attempts to implement what is just or right or equitable or fair. It would have a moral claim upon individuals' resources that is prior to what individuals want to spend on their own for health care. If, however, the arguments in behalf of goal-based justice have been shown to be problematic, two-tier systems which give one tier priority based upon these arguments are equally problematic. What then would a morally unproblematic two-tiers system, one ordained by a procedure-based rather than a goal-based notion of justice, look like? Happily, the two-tier system adopted by Oregon in 1989 [16] serves as an example.

The conceptual core of the Oregon plan is the rejection of the notion that a certain goal of just or right health care is a desideratum to be sought by the public tier of health care financing. The State of Oregon has decided to fund publicly insurance for those whose family income falls below the national government's poverty level. However, what particular services are insured is to be determined procedurally, by a citizen's panel that, after consulting with citizens in town meetings, ranks these services in priority. Since this ranking cannot be changed by the legislature, the legislature cannot allocate funds to procedures or groups favored by politically powerful interests. Whatever revenues are voted go to all eligible recipients and fund those health care services that have sufficiently high priorities to be covered. If there is a revenue shortfall, eliminating certain persons from coverage or reducing reimbursements to providers is prohibited; either more taxes must be levied, or the menu of covered services must be reduced. The legislature's choices are clear, not only in this case but also, e.g., between funding additional procedures and funding other things such as education or housing, which also have an effect upon the health of the citizenry even though they do not count as health care services. One realizes that funding comes out of common resources at the cost of funding other things. Whatever insurance is funded is just, because it is funded by communal funds and the allocation is approved by the citizenry whose funds they are. The end pattern desired is not known before the actual approval of health care funds via the appropriate procedures. The justice of the allocation is procedural rather than goal-based.

The Oregon plan acknowledges in principle the two-tier approach, because it does not in any way hold a health care provider liable for not delivering services that have not been fully funded in accordance with its contractual arrangements with the state. If such services were held to be just or right apart from the procedures which funded them, the provider would be morally culpable, at least; but there is no hint of fault, moral or legal, for not providing what is not paid for. In addition, the restriction of the public tier to those services that are funded for individuals below a certain income level makes clear that the public tier does no more than provide a minimal insurance. This clarity emphasizes the prudence of each person's being in a position to take advantage of the private tier, e.g., through additional insurance coverage.

The Oregon plan sidesteps the problems involved in determining what particular notion of justice or rights should guide public funding of health care. The plan can thus be viewed as a compromise in view of the lack of one obvious normative choice among competing notions. The choice among the competitors is made by the citizens as they, through the panel, authorize the procedures to be funded. The proponents of various notions of what is right or just in health care try to persuade the panel and the citizens in town meetings. In the face of uncertainty about the moral superiority of any one notion, the Oregon plan recognizes the role of limited democracy in setting moral authoritative priorities for support by communal funds, and it recognizes the priority of negative rights, because it holds health care providers and insurers liable only for what is legislatively funded at the levels to which the providers have bound themselves by contract. Given the uncertainty about what the proper scheme of publicly funded insurance should be, Oregon permits its citizens create that scheme, and it legally binds itself to the terms it contracts with private providers.

The Oregon plan differs radically in concept from other two-tier plans, because it gives moral primacy to the private rather than the public tier. That is, it allows the individual citizens to allocate public funding in health care, and it holds private providers liable only to the terms to which they have freely consented. By contrast, in the United States, both at the federal level and in states other than Oregon, it is unclear which tier is morally prior. A hodge-podge of public and private programs such as Medicare encourage cost-shifting from those who cannot, or who have sufficient political power so that they will not, pay for their health care, to payers of private insurance or to the taxpayers at large. The proposal of the American College of Physicians for a "nationwide program to assure access to health care for all Americans" ([2] p. 658) is also unclear about which tier is morally prior. Moreover, in practically all economically advanced countries with a national health care program, e.g., Australia, Austria, Belgium, Canada, Finland, France, Germany, Italy, Japan, Luxemburg, the Netherlands, New Zealand, Norway, Sweden, Switzerland, and the United Kingdom, the fact that an average of over 80% of total health care expenditures were public in 1987, in comparison to 41.4% in the United States ([20], p. 171), suggests, even if it does not show, that the public tier has moral superiority. To the extent, however, that it is unclear that there is a morally authoritative goal of public health care, Oregon's accent upon morally authoritative procedures to articulate that goal, and its respect of negative rights in implementing it, provide an attractive, because unproblematic, alternative. (Oregon is not entirely consistent. It compels private employers to provide employees whose income is above the level covered by public insurance to cover at least those procedures covered by public insurance. But this element of coercion is conceptually detachable from the plan's non-compulsive framework of respect for negative rights and for the procedure of limited democracy.) As the reader sifts through the various proposals and arguments in this volume, he or she will have to determine how solid theoretically they are, and to what extent a two-tier plan of the Oregon type provides an attractive alternative, theoretically as well as practically, to them.[5]

*University of Oklahoma Health Sciences Center*
*Oklahoma City, Oklahoma, U.S.A.*

## NOTES

[1] This authority may be their free and informed consent to the contractual arrangement, in which case the coercion is authorized by their consent to the terms of the contract; but in the case of positive rights created by public health care, it is not clear how those threatened have freely chosen the taxes and regulations which the government imposes, or even the procedures which lead to the impositions.

[2] What the "free consent" of the governed means, in contrast to bowing to the threat of legal sanction, may be difficult to articulate.

[3] It may be the case that some health care needs are caused by social factors, e.g., pollution in Los Angeles, in ways that cannot be explained simply as the health hazards to be undergone as a matter of course in taking advantage of the opportunities of modern society. But in this case, they should be remedied by, e.g., insurance funded by those whose emissions cause the pollution. Similarly, if one is to speak of pregnant mothers whose substance abuse injures the persons that their fetuses will become, those mothers might be morally culpable, but not society as a whole. The general point to note, however, is that these sorts of cases are too insignificant to support health care rights for the needy.

[4] 'Despite the fact that the United States spends more for health care than Canada or Great Britain, Americans are more dissatisfied with their health system, and less

satisfied with their physician care, than citizens of the other two nations.... 61 percent state they favor a Canadian-type national health insurance system' [4], p. 157.

[5] I am grateful to the University of Oklahoma Health Sciences Center for supporting research time used to prepare this introduction. I am also grateful to H.T. Engelhardt, Jr., George P. Khushf, Stuart Spicker, and Richard Wright for comments on earlier versions of this introduction.

## BIBLIOGRAPHY

1. Agich, G.J.: 1991, 'Access to Health Care: Charity and Rights', in this volume, pp. 185–198.
2. American College of Physicians: 1990, 'Access to health care', *Annals of Internal Medicine* **112** (9), 641–661.
3. Beauchamp, T.L.: 1991, 'The Right to Health Care in a Capitalist Democracy', in this volume, pp. 53–81.
4. Blendon, R.J. and Taylor, H.: 1989, 'Views on Health Care: Public Opinion in Three Nations', *Health Affairs* **8**(1), 149–157.
5. Brody, B.A.: 1991, 'Why the Right to Health Care is Not a Useful Concept for Policy Debates', in this volume, pp. 113–131.
6. Buchanan, A.E.: 1991, 'Rights, Obligations, and the Special Importance of Health Care', in this volume, pp. 169–184.
7. Daniels, N.: 1991, 'Equal Opportunity and Health Care Rights for the Elderly', in this volume, pp. 201–212.
8. Engelhardt, H.T. Jr.: 1991a, 'Rights to Health Care: Created, Not Discovered', in this volume, pp. 103–111.
9. Engelhardt, H.T. Jr.: 1991b, 'Virtue for Hire: Some Reflections on Free Choice and the Profit Motive in the Delivery of Health Care', in this volume, pp. 327–353.
10. Friedman, D.: 1991, 'Should Medicine Be a Commodity?', in this volume, pp. 259–305.
11. Halper, T.: 1991, 'Rights, Reforms, and the Health Care Crisis: Problems and Prospects', in this volume, pp. 135–168.
12. Hartmann, K.: 1991, 'The Profit Motive in Kant and Hegel', in this volume, pp. 307–325.
13. Iglehart, J.K.: 1990, 'Canada's Health Care System Faces its Problems', *New England Journal of Medicine* **322**, 562–568.
14. Khushf, G.P.: 1991, 'Rights, Public Policy, and the State', in this volume, pp. 355–374.
15. Marmor, T.R.: 'The Right to Health Care: Reflections on Its History and Politics', in this volume, pp. 23–49.
16. Oregon Legislative Assembly, 65th: 1989, *The Oregon Basic Health Services Act* (*Senate Bill 27* and *Bill 935*).
17. President's Commission for the Study of Ethical Problems in Medicine and Biomedical and Behavioral Research: 1983, *Securing Access to Health Care. A Report on the Ethical Implications of Differences in the Availability of Health Services*, U.S. Government Printing Office, Washington, D.C.

18. Rhoden, N.K.: 1991, 'Free Markets, Consumer Choice, and the Poor: Some Reasons for Caution', in this volume, pp. 213–241.
19. Sass, H.-M.: 1991, 'My Right to Care for My Health – And What About the Needy and the Elderly?', in this volume, pp. 243–255.
20. Scheiber, G.J. and Poullier, J.-P.: 1989, 'International Health Care Expenditure Trends: 1987', *Health Affairs* **8**(3), 169–177.
21. Veatch, R.M.: 1991, 'Justice and the Right to Health Care: An Egalitarian Account', in this volume, pp. 83–102.

SECTION I

# RIGHTS TO HEALTH CARE: THE DEVELOPMENT OF THE CONCEPT

THEODORE R. MARMOR

# THE RIGHT TO HEALTH CARE: REFLECTIONS ON ITS HISTORY AND POLITICS

The right to medical care, in the most egalitarian expression, means equal access to equivalent medical services [17]. In this formulation the distribution of services should vary solely with the degree of sickness. Other socio-economic factors – wealth, race, geographical location, etc. – ought not prevent the same response to similar medical conditions. Treating similarly ill persons similarly does not mean that all illnesses will be treated. Rather, the egalitarian criterion requires that rationing of access – tight or loose – be made by category of ailment, not by the social class of the ill.

There are two features of this formulation that require initial clarification. First, the guarantee addresses medical services, not health. The stipulation that equally ill persons be equally treated does not entail that action be taken such that various socioeconomic groups have equal chances of avoiding illness. Insuring equal access to health status requires far more intervention than guaranteeing that once ill, citizens will be treated equally. Almost all of the twentieth-century debate over the 'right to *health*' in fact has addressed issues concerning not health *per se* but the distribution of access to *medical care* ([4], p. 4). Secondly, note that the right asserted is associated with citizenship, not merit, contribution, or any other indicator of deservingness. At its simplest, the egalitarian argument in medicine asserts the injustice of permitting access to so fundamental a service to depend on anything but the "need" for it.

Like many egalitarian arguments, the right to equal care is open to the criticism of failing to distinguish among individuals on the basis of their *merit*. Some persons deserve better treatment than others, according to this view, though the basis of individual desert varies with meritarian philosophers. Even if the basis of merit were agreed to – a daunting task itself – there are strong arguments against distributing medical care on merit grounds. The equalities of medical need and of merit require, as Outka argues, different responses; those called for by need and those earned by effort ([17], p. 4). Illness – cancer, heart disease, and stroke, to name only the most frightening – strikes some in all groups regardless of merit; the substantial differences in the probabilities of some being sicker



*T.J. Bole III and W.B. Bondeson (eds.), Rights to Health Care*, 23–49.

– the poor for instance – does not invalidate the proposition that persons unequal in merit will find themselves equally in need of care. If merit is largely irrelevant in the origins of illness, it is wrong to apply that criterion in the distribution of the corresponding good. The relevant criteria would be medical condition and the availability of resources for care.

There are two familiar responses to this criticism of the merit principle for distributing medical care. One, recalling the old fable about the industrious ants and the lazy grasshoppers, holds that the prudent, who with foresight and discipline prepare for the possibility of illness, deserve superior treatment. Unlike the fable's inevitable winter, however, the worst crises of health tend to be unpredictable. Second, the capacity to cope with the unpredictable costs has as much to do with wealth – and, in recent years, employment status – as foresight and discipline.

A more profound objection to the egalitarian position arises when the premise of illness' unpredictability is questioned. The assumption that health crises are random, that the patient bears no blame for his condition, is in some cases simply false. Many reckless drivers are maimed in car wrecks, many lungs corroded by nicotine and livers by excess alcohol. If patients are sometimes partially responsible for their medical "needs" ([4], pp. 1–3), then questions of desert are not so irrelevant. Egalitarians concede such patient responsibility. But they argue that employing this knowledge in the allocation of medical care is excruciatingly difficult. Emergency cases bring out how perplexing the issue of access to medical treatment is. They illustrate clearly how people "suffer in varying ratios the effects of their natural and undeserved vulnerabilities, the irresponsibility and brutality of others, and their own desires and weaknesses". Consider the following:

(1) a person with a heart attack who is seriously overweight;
(2) a football hero who has suffered a concussion;
(3) a man with lung cancer who has smoked cigarettes for forty years;
(4) a sixty-year-old man who has always taken excellent care of himself and is suddenly stricken by leukemia;
(5) a three-year-old girl who has swallowed poison left carelessly by her parents;
(6) a fourteen-year-old boy who has been beaten without provocation by a gang and suffers brain damage and recurrent attacks of uncontrollable terror;

(7) a college student who has slashed his wrists (and not for the first time) from a psychological need for attention;
(8) a woman raised in the ghetto who is found unconscious due to an overdose of heroin ([17], pp. 16–17).

Thus even when the merit criterion is relevant for the distribution of medical services, it may for all practical purposes be impossible to apply.

A more precise meritocratic view of medical care can be extracted from the free market tradition in economic theory. If one accepts the view that workers earn their marginal product and that marginal product measures social contribution, then the distribution of goods purchasable by earned income reflects the social contribution of citizens. Many, but not all, free market theorists regard medical care like any other valued item; individuals purchase either care or insurance out of earnings in whatever amounts they choose, given their income levels and "tastes". The free market claim is that government programs to redistribute services like medical care distort individual choice. Different people, it is argued, want varying amounts and types of medical care – beyond some minimum – and treat medical services differently. Some might prefer the most expensive surgery available despite the fact that the more routine procedure is ninety-eight percent as effective; their neighbors might forego the advantage in order to have more disposable income for other purposes. By failing to distinguish care according to the willingness to pay, the argument goes, there is no measure of the marginal benefit of a given treatment to a given person ([6], [20]). Ultimately, free market theorists expect to maximize both freedom of choice and efficiency through the interplay of a medical market left largely to itself; they argue that government should finance no more than some minimum health care to the genuinely (and demonstrably) needy.

There is, of course, some merit in this critique of government intervention in the medical care market. Yet there is an enormous gap between the rhetoric of freedom of choice and the realities of purchasing medical care. The important fact about critical medical care – at birth, near death – is not what one would prefer but what doctors choose and patients can afford. Most analysts do not consider medical care a typical market good. In important cases, consumers (patients) have urgent needs and little information on which to base their choices. Physicians make many of the significant decisions; indeed, their professional discretion is fundamentally based on the knowledge gap between the patient and the provider.

Furthermore, it can be argued that since medical care is so fundamental a need, it is more appropriately viewed as a prerequisite rather than a consequence of societal contribution. What someone can afford – even if it truly measures societal contribution – should be considered irrelevant. This suggestion returns us to something like the egalitarian "right of all citizens to equivalent medical services".

Clearly, the right to medical care is entangled in a number of vexing theoretical problems. The issue of equal treatment reaches to the most controversial questions about the role of the state in responding to social inequality. The argument over whether access to medical care should be a need- or merit-based claim proves particularly vexing when questions are raised about the extent to which illness is randomly distributed as opposed to patient-caused. The right to a minimum amount of health care – a physician's obligation under the Hippocratic oath – raises far less controversial issues than the equal right to equal treatment ([18], [3]). And finally, the structuring of individual options by a government pursuing an ideal of equal access provokes the whole range of free-market arguments against the expansion of the role of the modern state. Some of these theoretical issues were crucial to the American battle over government-financed medical insurance; others remained implicit through most of the debate.

## I. THE RIGHT TO HEALTH CARE: THE POLITICS OF POST-WAR AMERICA

Other contributors to this volume have as their charge the exploration of the philosophical issues surrounding the right to health care. My charge – and special competence – is to discuss the political struggles over this philosophical tenet. In doing so I will not deal except fleetingly with the struggles of the first half of this century. Others, notably Paul Starr in his magisterial work on the *Social Transformation of American Medicine* [22], have charted the "enduring political controversy" among liberals, conservatives, and socialists between 1900 and 1950 about what response, if any, there should be to the profound inequalities in the availability of health care services ([3], p. 304).

My topic here will be the legacy of those struggles in the post-war period. For that purpose, I will address two discrete subjects. The first will be the post-war struggle over the question of whether government should assume responsibility for guaranteeing access to health care to its elderly, the fight that resulted in the passage of Medicare (and Medicaid) in 1965. The second subject will be the past twenty-five years of dispute over the extension of the role of government in health care, questions that initially addressed the advisability of national health insurance and, increasingly since the 1970s, the scope, form, and rationale for limiting the rapid inflation of medical prices and expenditures.

Briefly put, my argument is that the United States legislated substantial extensions of the government's role in financing health care during the mid 1960s, but did so with only the vaguest clarity about its underlying rationale. The result for the 1970s and 1980s – here as in some other areas of social policy – was massive public expenditure on a central item of the modern welfare state but public uncertainty about the ends to be served and the limits to be observed in the government's regulation of the world of medicine. It is no surprise in this context that a Presidential Commission on the ethics of medicine had such a difficult time arriving at a bipartisan consensus about their subject at the very time that the government's role was already massive [3].

## II. THE STRUGGLE OVER MEDICARE

My question here is why government elites chose in the early 1950s to narrow the focus of national health insurance proposals from the general population to the aged, and to restrict benefits to partial hospitalization coverage.

The stalemated Truman national health insurance proposals of the 1948–1952 period provided the immediate backdrop for the choice of the Medicare strategy. The view that the aged were more acceptable to the general public as a deserving group provided the key to the shift to targetted beneficiaries. Likewise, the restriction of Medicare's scope to social security beneficiaries depended upon the judgment, then clearly true, that social insurance programs enjoy considerably greater legitimacy than means-tested public assistance programs. (See Appendix A for competing conceptions of social insurance and the welfare state.)

According to one analytic perspective, the shift from comprehensive national health insurance to the Medicare program represents but one stage in a fluid policy development. The incentives of key actors to promote or oppose this shift are part of the overall portrait of the health politics field of the time. The prominent position within the Federal

Security Agency of such longtime social security experts as Wilbur Cohen and I.S. Falk helps to explain the availability of a social insurance alternative to the Truman plan. It should be remembered that both these officials were advocates of general health insurance, but less sanguine than others about its political feasibility. The access Cohen and Falk had to Oscar Ewing gave a crucial bargaining advantage to those seeking a limited, but more politically appealing, health insurance initiative. The parsimonious explanation of why health insurance for the aged emerged in the 1950s is that it made political sense in a stalemated setting.

Once Medicare arose as a proposal, what shaped its fate? Who were the contestants in that Medicare debate and what was the nature of the contest over time? The debate over Medicare was cast in terms of class conflict, of socialized medicine versus the voluntary "American way", of private enterprise and local control against "the octopus of the federal government". Moreover, though the program most immediately affected the aged, physicians, and hospitals, broad strata of the population were directly or indirectly involved – the families of the aged, all present social security contributors, and the entire health industry.

Not only the themes of the conflict, but the antagonists and their adversary methods illustrate what might be called "class conflict" politics. The dispute over Medicare re-enacted the polarization that characterized earlier fights over national health insurance. The leading adversaries – national business, health, and labor organizations – participated in open, hostile communication and brought into the opposing camps a large number of groups whose interests were not directly affected by the Medicare outcome. In the process, ideological charges and counter-charges dominated public discussion and each side seemed to regard compromise as unacceptable. And both the contest and the contestants over Medicare remained remarkably stable in the period from 1952 to 1964: "two well-defined camps with opposing views, camps with few individuals who were impartial or uncommitted" (26], p. 304).

The static, stereotypical quality of the fight over Medicare is readily understandable when one considers the size and character of the parties to it. Large national associations like the AMA and AFL-CIO have widely dispersed component parts; they function in part as Washington lobbyists for issues affecting the interests of widely disparate members. Hence, they must seek common denominators of sentiment that will satisfy the organization's leading actors without antagonizing large bodies of more passive members. Such large organizations are specialized, with full-time staffs devoted to preparing responses to public policy questions when the occasion arises and in the direction dictated by past organizational attitudes, attitudes that are slow to change and help account for the predictable way in which sides were taken on various Medicare proposals over time. It is not surprising that the debate during this period was stable; mutually incompatible positions on health insurance arose in part from the maintenance needs of large-scale organizations and their leaders.

The enactment of Medicare is appropriately viewed as the result of a complex bargaining game in which none of the relevant executive, legislative, or pressure-group players could fully control the outcome. The key actors all had somewhat different conceptions of the problem at hand. The law emerged from a long, complicated struggle, and its form was not one which any of the major actors intended at the outset.

The bargaining that took place, however, should not be allowed to obscure the vital fact that the election of 1964 had given all the actors less to bargain about [13]. The electoral changes of 1964 reallocated power in such a way that the opponents were overruled. Compromise was involved in the detailed features of the Medicare program, but the enactment itself did not constitute a compromise outcome for the adversaries.

## III. WHAT DID THE PROCESS PRODUCE?

Battles like Medicare are fought in public but settled in private. The national pressure groups made enormous and costly efforts to define the dispute over Medicare in ways acceptable to their members. But within the government bureaucracy, there were continuing efforts to articulate the balance of these rival claims in the legislation proposed to the Congress. The consultation was sometimes explicit and detailed, as when the AFL-CIO and the Blue Cross Association met regularly with HEW officials during the early 1960s. In other cases, consideration of group interest was tacit and intermittent, particularly in the case of the AMA. Overall, the executive proposals sent to Congress reflected a typical pattern: major compromises were built into Medicare bills, anticipating the pressure-group claims that would otherwise have to be balanced in the Congress.

What emerged programmatically was a complex amalgam of health insurance programs.

Part A (*Hospital Insurance:*) The Johnson Administration's Proposal (H.R. I, S. 1)

| | |
|---|---|
| Beneficiaries: | all Social Security eligibles over 65 |
| Benefits: | 60 days of hospital coverage per benefit period, post-hospital skilled nursing home services |
| Financing: | HI taxes on current workers |
| Administration: | federal (SSA), through fiscal intermediaries |

Part B (*Supplementary Medical Insurance: Adaptation of Republican Proposal to Expand Benefits*):

| | |
|---|---|
| Beneficiaries: | elderly who pay monthly premiums |
| Benefits: | physician services (excluding check-ups) |
| Financing: | premiums, general revenues, patient cost-sharing |
| Administration: | federal (SSA), through fiscal intermediaries |

Medicaid (Federal-State grant program for poor families and medically indigent).

The preceding account – adequate for understanding why we have the Medicare program in the form we do – largely ignores the philosophical disputes over the philosophical legitimacy of Medicare itself. The ferocious dispute over Medicare brought out competing conceptions of government's proper obligation to finance health care for the old, but much of the account of propaganda, strategy, and bargaining can proceed in the ordinary terms of pressure group struggles over who should gain and who should lose from the redistribution of financing burdens. This, in itself, is an important commentary on the politics of the right to health care. The advocates of Medicare took pains to mute their philosophical innovation; they appealed both to the commonsensical notion that vulnerable Americans deserved access to health care and to the empirical realities that the old were sicker and less insured than other demographic groups [13]. What they did not advertise was any detailed account of exactly what the right to health care entailed and why the Medicare program appropriately expressed that right. Their opponents, spearheaded by the American Medical Association, waved the flag of socialism in opposing Medicare, but did not dispute that older Americans deserved access to care. Instead, they concentrated on showing how – through individual health insurance, savings, and the charity of medical providers – all the old did not require a massive new program to ensure their access to care [13].

The result was both an impoverished debate about health care rights and a legacy of uncertainty about the grounds for the government's coming to play a major role in the financing of health services across the country.

Had the debate been joined, what might have been revealed? For the advocates of Medicare, there was the twin conviction that medical care was no ordinary market good and that the politics of America required a position markedly short of the egalitarian goal of equal access to medical care. The first position was an assumption without elaboration; the meaning of a right to health care – as against a privilege – was that all should enjoy it. But the scope of the right was left vague. A decent health insurance policy would produce the right result; older Americans could, through Medicare, purchase entry into the mainstream of American medicine. After all, it was the facts of access, not the philosophical ends sought, that dominated the debate. And, what was more, the fusing of social insurance ideas about contributions and the Medicare benefits carried with it the notion that beneficiaries would have "earned" their benefits during their working life [2]. By this device, the advocates were able to portray their remedy as nothing but the social engineering mechanism by which market results – wages – got transformed into an acceptable distribution of access to care in old age. By showing that older Americans were very unequally insured, the advocates could appeal to a widely shared aim – broad insurance for the old – and avoid the details of what right to care Medicare was to enforce.

On the other side, there was a similar absorption in the facts of access and subordination of the philosophical differences. Instead of philosophy, there was red-baiting, the treatment of Medicare as the opening wedge for the feared "socialized medicine". Deeply implicit was the notion that the federal government should help only those for whom private, local, and state remedies had proved inadequate.

For over twenty years we have continued to live with the broad outline of the 1965 legislative result. (Tables 1–4, Appendix B, provide a very summary sketch of the program's beneficiaries, benefits, financing, and administration.) Briefly put, that history is one of rampant medical inflation, the erosion of Medicare coverage despite growing real program costs, and persistent defense of the program against the budget cutting of the 1970s and early 1980s – as against the more vulnerable Medicaid program for the poor.

At the outset of the 1970s, alarm over medical care cost increases stimulated renewed discussion of national health insurance for all. Frustrating patterns of crisis and stalemate remind us of how durable and deep are the forces behind medical inflation. In 1970, Senator Kennedy published a book on American medical care the very title of which was alarmist: *In Critical Condition* [11]. Even the Nixon Administration had its plan – the now forgotten CHIP proposal of 1974 – as an alternative to the Kennedy-Corman and Kennedy-Mills proposals of that period. For a short time in the early 1970s, the Economic Stabilization Program (ESP) clamped down on medical price increases ("the lowest annual percentage increase [in national health spending] in the 1972–82 period, 10%, occurred during the Economic Stabilization Program in 1973" ([7], p. 9). But when ESP's price controls were lifted, the absence of substitute cost restraints resulted in an explosion of medical prices and a rapid rise in public and private health expenditures. So ended the chances for smooth implementation of Democratic or Republican proposals for national health insurance in the mid-1970s.

By the end of the 1970s, the federal health policy debate over federal health policy had largely shifted from national health insurance to cost containment. In 1977, the Carter Administration, alarmed by rising hospital costs, concluded that "clearly, the time has come – indeed it has been here a long time – to bite the bullet on hospital costs" ([16], p. 67). But Congressional stalemate frustrated reform, and the hospital industry's celebrated "voluntary effort" to contain costs proved illusory, as Table 5, Appendix B, makes plain. During the 1970s, medical care reform was like the weather in Mark Twain's formulation: everyone talked about it but no one did much. There was much discussion, many partial programs, but, because of the powerful opposition, there was a marked lack of success in comprehensive reform or in controlling the rapidly rising costs of medical care. Now the problem is worse and all of us paid in the 1980s for the consequences of these failures in the 1970s.

Two decades of relative medical inflation have swelled Medicare's outlays despite, as noted, largely hidden forms of benefit erosion. The costs of health care have increased much faster than have the incomes of the elderly. Between 1970 and 1982, the mean real (inflation-adjusted) income for Americans over the age of sixty-five rose 20%. During the same period, inflation-adjusted co-payments for the Medicare program rose 84%: co-payments and premiums combined increased by over 70% after controlling for the effects of inflation. Thus, as a proportion of their incomes, the elderly's contributions to the Medicare program have increased steadily over time; from about 3% in 1970 to about 5% in the 1980s [9]. (See Table 6, Appendix B.)

The stagflation of the past decade, however, has exacted another, perhaps more crucial, price. As the program matured fiscally, the energy that attended its origins has dissipated. The clear vision of its promoters called for step-by-step expansion once the fundamental legitimacy question of 1965 was settled. Instead, ever since the Vietnam War's escalation, benefit expansion has proven near impossible, and the energies of the Medicare elite have gone into program protection. One result of this is a weakened sense in the diverse publics of American life of what precisely Medicare was for and why it is that Medicare, like Social Security, should not have its benefits reduced at a time of fiscal strain.

The legacy of the 1970s was ferment and frustration. Reformers used to be concerned with planning for expansion in medical research, with hospital construction and modernization, and with increasing health manpower to meet growing needs. In the 1980s, they were increasingly concerned with planning retrenchment. Reformers used to aspire to bring needed medical care to the elderly, the disabled, and the poor. Now they focus on how to cut programs, which public facilities to close, whom to fire, what populations to neglect. Expanding resources for health used to create jobs and better and more accessible services. Now, as prices continue out of control, those who need access to care the most – the sick, the disabled, the elderly, and the poor – are hurt the most. Nowhere is that more vividly illustrated than by the responses to the worsened health inflation of the Reagan era [14, 22].

## IV. PRESENT CIRCUMSTANCES

The Reagan Administration, like everyone else, lamented rising medical costs. They were obsessed with restraining the costs of public programs, not reducing medical inflation. They paid providers less for Medicare and Medicaid patients and, as a result, shifted costs of public programs to workers covered by private health insurance [13], thus threatening the elderly and poor with the possibility of second-rate care. To cut government expenditures they used benefit cutbacks, reducing the number of persons eligible for public programs and shifting costs back onto the sick.

The result has been a noticeable weakening of both Medicare and Medicaid. Because of new federal and state policies, increased numbers of poor people are denied Medicaid eligibility and the care they need. The costs and burden of providing some essential care are thus shifted to local governments and hospitals, with "uncompensated care" a serious issue in some locales. The Reagan Administration invoked the idea of excessive use of medical care by the elderly, the disabled, and the poor as partial justification for budget cutbacks in Medicare and Medicaid. In fact, as much as seventy percent of the cost increases are apparently caused by price increases by hospitals, physicians, and other providers of care, not increased utilization of services ([10], p. 19).

This budget-cutting approach – largely, but not exclusively, Republican – is misdirected. When diagnosing the problem of medical costs in America, the Reagan Administration shared the widespread sense of serious alarm. "Health care costs", they claimed, "are climbing so fast that they may soon threaten the quality of care and access to care which Americans enjoy". They called attention to the fact that "health care funding is one of the fastest rising expenditures in the federal budget". And they shared rhetorically the Democrats' contention that rising costs "are a problem that affects everyone". They cited the elderly who are covered by Medicare but "face the threat of catastrophic illness expense, against which Medicare offers no protection". They bemoaned the fact that "the poor on Medicaid have seen coverage reduced as states have been forced by rising costs to make cutbacks". They referred sympathetically to "workers with employment-based health insurance [who] have received lower cash wages because of the unchecked cost increases for health benefits". And they extended their sympathies to the economic implications for American consumers, who "pay for health-care costs in other hidden forms, including higher costs for the merchandise they buy, since the cost of employee health care benefits must be included in the price of products" [19].

From these shared diagnostic premises, however, the Reagan Administration moved to policies and programs that did not reduce the growing inequalities in health care. The leap from the common diagnosis to their remedy sprang from the misguided faith that "market" incentives are mostly what is needed to reduce costs in the medical world. Benefit cutbacks, in this view, served both as required reductions of federal expenditures and as proper sources of price signals in medical care. The aim was to reduce access in order to lower government expenditures; the proper policy should control costs so as to make access to a decent level of care possible for all Americans.

Since 1981, health policy has been marked by four elements: reduced public medical budgets; reduced benefits and more cost-sharing by recipients of publicly-supported care, a large number of whom are elderly and many of whom are poor; cutbacks in payments to Medicare providers; and an adherence to the belief that excessive insurance drives medical inflation, at the expense of all other issues.

## V. THE POLITICS OF HEALTH CARE

The easiest task in politics, but the most singularly irrelevant, is to figure out what, ideally, a policy should be. In the case of Medicare, the progressive elements in the Democrat and Republican parties might seek the following: comprehensive benefit coverage for the elderly, financing their health insurance with a combination of modest premiums and general tax revenues, and a cost-control system that rewarded medicine for caring, cognitively complex, and efficacious services, while paying less for the uncertain and nothing for the wasteful and harmful. That plan, and a nickel, will get you a five-cent cigar. In the context of the 1990s, the question is what counts as a defensible, feasible cost-control program and how that relates to program changes in Medicare which are both substantively meritorious and politically feasible. This sort of question is not answered by wish lists drawn from the past, nor by handwringing about the complexity of medical care's governance.

The principle that costs require controlling, not shifting, rests on the premise that medical inflation is a serious problem for the country as a whole. Businesses find their products suffering in international competition partly because our medical costs are comparatively high. Labor unions find themselves fighting hard to keep the benefits they have and paying more to stay the same. Young couples face childbirth with tribulations of financial catastrophe; older Americans cannot regard ill health as their sole burden, but must still worry about budgeting for illness despite the large and growing Medicare outlays. And all of us suffer the loss of a precious collective good – the sense that the problem of financing medical care for our old and our poor – let alone ourselves – has not been solved despite the fact that, as a nation, we spend nearly eighty percent more in real outlays than we did in 1970. Ask Canadians in the street – of whatever political persuasion – whether they would exchange their national health insurance for our melting pot of problems; the overwhelming majority consider themselves lucky to have a government program that pays practically all medical care costs and manages to spend per capita some twenty-five percent less than we do.

Opinion polling in the United States reveals not a demand for cost control, but support of it. There is what V.O. Key called a "permissive consensus" on this issue, an orientation toward medical costs that permits but does not require political leaders to fashion support and to mold its shape. Public opinion is, however, both superficially formed and open to rapid transformation. It could support either freezes on doctors' fees within Medicare or broader constraints on physicians. It could be activated for either a New Jersey all-payer system or one restricted to Medicare alone. Talking about what the people will accept as a basis for choice between Republican and Democratic cost controls is misleading. What both parties propose will alienate many in the professions and few in the populace. So, in this case, what is right is possible, even if all forms of serious control of costs will be highly controversial within the professional world of medical care.

The judgment that overall cost control is politically possible does not mean it is easy or inevitable. The current danger is that the preoccupation with deficit reductions will drive out the debate over fundamental medical reform, that the language of competitive markets will divide cost controllers into warring camps, and that the turbulence of the medical care market will distract policy makers from the necessity to frame constraining rules under which that market can work more sensibly. The greatest danger of all, however, is that political stalemate and market adjustments will leave the poor, the unemployed, and the old in great danger of returning to the conditions which prompted Great Society reformers to design Medicare and Medicaid in the first place.

## VI. CONCLUSION

What can one say in conclusion about the right to health care in recent American politics? First, rights language in American health care politics is almost always rhetorical flourish rather than reasoned argument. Both in the struggle over Medicare and the more recent preoccupations with cost control, rationing, and the fate of the poor and uninsured, the working out of what is entailed by a right to health care among all American citizens is very rare.

There are two major exceptions to this generalization. One is the flourishing philosophical specialty of bioethics, complete with journals, think-tanks, and grant applications for spreading more widely the benefits of philosophy. But the benefits at stake here seldom deal with the right to health care in the broad sense, the issue of what level of health care makes sense in a society like that of the United States, and what that has to do with opportunity and cost in education, housing, pensions, and the like. But raise the issue of Baby Jane Doe, and the rights of parents, the child, the state, and others will be analyzed seriously. Raise the question of abortion, and rights language immediately comes to the fore among the lawyers, philosophers, and women's studies specialists who struggle with the conflicts between presumed rights in this area of passionate conviction. But raise the question of co-insurance, Medicare, and the elderly's right to health care, and hardly a philosophical eyebrow rises. Publish a book like *Painful Prescription* – the work by Henry Aaron and William Schwartz [1] that raises the prospect of direct rationing of efficacious care to implement cost control over the next decade or so – and the medical journals, not the journals of philosophy, get excited. It is, as Caroline Whitbeck of MIT wrote in the *Journal of Health Politics, Policy and Law*, a terribly truncated view of philosophy's domain:

> The discussion in contemporary medical ethics is strongly influenced by the patient rights movement that followed on and was strongly influenced by the consumer rights movement, so that it is rights (rather than "utilities", which is a concept equally compatible with Liberal Theory) that are given greatest emphasis. As a result, ethical issues in medicine are construed as turning on the question of the identification of relevant rights and duties of persons occupying various roles, and the correct ordering of these rights and duties and of other values.
>
> • • •
>
> Many ethical issues that arise in medicine find no place within this scheme. Recognition of some of these problems is reflected in the shift in the focus of concern from informed consent to shared decision-making that occurred in the 1982 report of the President's Committee for the Study of Ethical Problems in Medicine and Biomedical and Behavioral Research. Respect for a person's right not to undergo procedures for which she or he has not given consent requires only recognition of the patient's right of veto over any major procedure that the provider *presents* to the patient as an alternative. Shared decision-making, on the other hand, requires full participation of the patient in setting the goals and methods of care and *formulating* the alternatives to be considered for many different aspects of care, and this in turn requires that the patient and at least one key practitioner engage one another in complex communication. The emphasis on *shared* decision-making requires abandonment of the assump-

tion (common to Liberal Theory approaches to medical ethics) that for each decision in medicine there is an answer to the question, "Who should decide?" – i.e., "Who *owns* this decision?" ([25], p. 186)

The price of this is spending more and feeling worse in American medicine. Without understanding what right Medicare was to insure, the American public has been uncertain throughout over whether the program was a success or a failure. Now that restraint in American medicine is everywhere celebrated, the failure to understand those implemented rights means sensible restraint and senseless abandonment of commitments are not clearly distinguished. Indeed, it is a mark of our politics that the thoughtful suggestions of my colleague in this volume, Norman Daniels, never made their way into the suggestions of the President's Commission for the Study of Ethical Problems in Medicine and Biomedical and Behavioral Research. This topic – the right to health care in American politics – is but one expression of how we have developed into a mature welfare state without the benefit of a philosophically acute and reasoned understanding of the road taken and the options foregone or remaining.

*Yale University*
*New Haven, Connecticut, U.S.A.*

BIBLIOGRAPHY

1. Aaron, H. and Schwartz, W.: 1984, *The Painful Prescription*, Brookings Institution, Washington, D.C.
2. Ball, R.: 1985, 'The Original Understanding of Social Security: Implications for the Future', paper delivered at Yale Social Security – Medicare Symposium, April 12–13, 1985.
3. Bayer, R.: 1984, 'Ethics, Politics, and Access to Health Care: A Critical Analysis of the President's Commission for the Study of Ethical Problems in Medicine and Biomedical and Behavioral Research', *Cardozo Law Review* **6**(2), 303–320.
4. Blue Cross Association: 1975, *Conference on Future Directions of Health Care: The Dimensions of Medicine*, sponsored by the Blue Cross Association, The Rockefeller Foundation, and the Health Public Policy Program, University of California School of Medicine.
5. Davidson, S., and Marmor, T.: 1980, *The Cost of Living Longer*, Lexington Books, Lexington.
6. Evans, R.: 1984, *Strained Mercy*, Butterworths, Toronto.
7. Freeland, M.S. and Schendler, C.E.: 1984, 'Health Spending in the 1980s', *Health Care Financing Review* **5**(3), 1–68.
8. Glaser, W.A.: 1985, Syllabus for Advanced Seminar in Health Policy Issues, New School for Social Research, Spring Semester.

9. Harvard University Medicare Project: forthcoming, 'Medicare and the Costs of Health Care for the Elderly', Harvard University Press, Cambridge.
10. *Health Care Financing Review*: Spring 1984.
11. Kennedy, T.: 1972, *In Critical Condition*, Simon and Schuster, New York.
12. Marmor, T.: 1981, 'The North American Welfare State: Social Science and Evaluation', in R.A. Solo and C.W. Anderson (eds.), *Value Judgment and Income Distribution*, Praeger, New York, Ch. 13.
13. Marmor, T.: 1973, *The Politics of Medicare*, Aldine-Atherton, Chicago.
14. Marmor, T. and Dunham, A.: 1984, 'The Politics of Health Policy Reform: Problems, Origins, Alternatives, and a Possible Prescription', *Health Care: How to Improve It and Pay for It*, Center for National Policy, Washington, D.C.
15. Meyer, J.A.: 1983, *Passing the Health Care Buck*, American Enterprise Institute, Washington, D.C.
16. Mondale, W.: 1978, 'Controlling Health Costs, Conference Proceedings', *National Journal*, March.
17. Outka, G.: 1974, 'Social Justice and Equal Access to Health Care', *Journal of Religious Ethics* **2**, 11–32.
18. President's Commission for the Study of Ethical Problems in Medicine and Biomedical and Behavioral Research: 1983, *Volume 2: Securing Access to Health Care*, U.S. Government Printing Office, Washington, D.C.
19. Reagan, R.: 1983, Presidential Message, Health Incentives Reform Program, 129 *Congressional Record*, s. 1717 (H. Doc. No. 98–24).
20. Sade, R.: 1971, 'Medical Care as a Right: Refutation', *The New England Journal of Medicine* **285**, 1288–1292.
21. Senate Finance Committee: 1983, *Background Data on Physician Reimbursement Under Medicare*, staff report, October.
22. Starr, P.: 1983, *The Social Transformation of American Medicine*, Basic Books, New York.
23. Starr, P. and Marmor, T.: 1984, 'The United States: A Social Forecast', in J. de Kervasdoue, J.R. Kimberly, and V.G. Rodwin (eds.), *The End of an Illusion: The Future of Health Policy in Western Industrialized Nations*, University of California Press, Berkeley, California.
24. Waldo, D.R., and Lazenby, H.C.: 1984, 'Demographic characteristics and health care use and expenditures by the aged in the United States: 1977–1984, *Health Care Financing Review* **6**, 1–29.
25. Whitbeck, C.: 1985, 'Review Essay', *Journal of Health Policy, Politics and Law* **10**(1), 181–187.
26. Wildavsky, A.: 1962, *Dixon-Yates: A Study in Power Politics*, Yale University Press, New Haven.

## APPENDIX A: COMPETING CONCEPTIONS OF SOCIAL INSURANCE AND THE WELFARE STATE ([12], Ch. 13)

Policy analysts – and the political actors they address – bring sharply different notions of purpose to the programs that comprise the contemporary welfare state. Whether the concept of the welfare state is itself normatively adequate will not be addressed; for the moment we use the term to describe a group of programs in health, education, and social security that have received very substantial and growing public subsidy since roughly the mid-1930s. The term itself entered our policy vocabulary at about the time of the World War II, and the programs' major period of growth came long after the label became part of the popular lexicon. From the time of their establishment, there have been underlying differences in what such public programs were meant to achieve.

These views – roughly grouped into the residualists, the social democratic, and the Marxist – provide sharply differing standards of appraisal both for the origins of the welfare state and its condition in the 1980s.

### *I. Residualist*

A. One perspective on the program's objectives – called here residualist – characterizes the welfare state as a safety net. The net of social welfare was intended to rescue the victims of capitalism, provide a cushion against certain contingencies (unemployment, sickness, large numbers of children, widowhood), and give subsistence level relief to those unable to provide for their own needs. This view of purpose, originating as it does in the European poor laws, is ubiquitous in modern capitalist nations. Though the popularity of such views differs among those nations, the criteria for judging the efficacy of the welfare state are everywhere affected by these themes. In North America, and in Australia as well, this residualist conception is the staple not only of business and financial elites, but of large proportions of middle- and lower-income populations as well.

B. Most of those who describe the welfare role as residual believe that management of the welfare system should be highly decentralized. In federal regimes diffusion of authority for social welfare to states and provinces has been the rallying slogan of the residualists. In the United States public assistance programs such as the federal-state program of Aid to Dependent Children (later Aid to Families with Dependent Children, AFDC), which was part of the original Social Security Act of 1935, exemplify this ideal. In Canada similar developments took place, reinforced with constitutional requirements that the provinces be responsible for social programs except as provided by constitutional amendment. Advocacy of decentralization presumes that individual families will typically assure the welfare of their numbers. When that fails, institutions close to those families – charitable groups, then local and provincial or state programs – will constitute the safety net protecting against destitution.

C. The metaphor of the safety net suggests the key features of the appropriate welfare policy design. The net is close to the ground and the benefits are accordingly modest – a subsistence that might well vary widely in connection with community standards of adequacy. The clientele are the down and out; the eligibility criteria – whether tests of needs or means – are designed to sort out the truly needy from the rest. There is an implicit notion of potential waste: aid to those who do not need the net to survive. Minimal adequacy, selectivity, localism, and tests of need – these constitute the residualist's controlling notions for evaluating the welfare state.

*II. Social Insurance*

A. So stated, the conception of welfare as residual sharply differs from what I term the social insurance model. The basic purpose of social insurance is to prevent destitution, although a multitude of questions regarding the desirability of redistribution of income and power arise when considering the role of social insurance in the welfare state. Is the welfare state designed to bring about adequacy, equity, or equality? Is it to compensate for past social injustice and misfortune or perhaps to invest in the future? Is it designed to supplement or replace income, to allocate cash, services, or power, and with what balance among them? What links the social insurance advocates, despite their variance in answering these questions, is the rejection of the residualist conception of welfare on the one hand, and hesitancy about how much notions of redistribution of income and power should be the basis of evaluating the welfare state on the other.

B. The central metaphor of social insurance is the insurance card (or book). If the net of welfare is to catch those who have failed, the card of social insurance is to prevent destitution. The aim is simple:

the universalization of the financial security presumed in the higher civil service and among economic elites. The threats to economic security include some obvious ones – involuntary unemployment, widowhood, sickness injury, or retirement – as well as less obvious ones such as a large family. Welfare states have provided for these eventualities at different times, in different orders, and with considerable variation in generosity and terms of administration. Yet, irrespective of form and levels of payment, social insurance programs have rejected as inferior the machinery of selective, means-tested programs: the more universal the entitlement, the closer it is to the model.

C. Contributions during working life, the theory states, "entitle" one to "protection" against the risks of low income. Sometimes the contributions are participation in the larger society's general tax arrangements, as is the case in Canada. Or, as in Britain, contributions are in the form of weekly social insurance payments. These arrangements differ sharply from the percentage of payroll taxes (contributions) of the United States' Old Age and Survivor's Insurance program. Yet overall the idea is to contribute to future protection. Equitable treatment, not the equalizing of incomes, is the model.

D. Clearly, redistribution of income is one consequence of such programs, but it is not the primary aim. And the model of redistribution is not intended to be between socioeconomic classes, but over the life cycle of individuals and their families. The relevant question for the proponent of social insurance pertains to the adequacy of preparation of the citizenry for the predictable risks of modern industrial society. Looked at this way, the welfare state is simply the extension of the principles of private insurance to markets where either the risks are uninsurable or the distribution of income is unlikely to engender widespread insurance purchase.

* * * *

Such a broad characterization is bound to miss the details of one or another nation's mix of programs. Moreover, no nation has only social insurance programs; everywhere both residual and social insurance strategies are mixed in actual social welfare programs. But variation and mixture do not preclude recognition of the evaluative differences in the perspectives. For in the answers to questions about who gets what, when, and where, it makes considerable difference whether the standard is appropriate levels of assistance for the downtrodden or adequate compensation for losses of earning power. And, when the aim is to replace past earnings, the in-kind payments – food, housing, special transport cards – are presumed inferior to cash payments.

So far we have not mentioned redistribution of power in our discussion of welfare states. In the residualist's notion of the welfare state, the powerful take care of the weak, and the state apparatus supplements the limited possibilities of private charity.

Social insurance extends the protection available earlier only to tiny economic elites, but does not transform the structure that created those elites.

* * * *

### *III. Marxist*

From a more radical perspective, aims of help and compensation are themselves questionable, since they constitute adjustments to the harsh realities of industrial society, and not means of transforming society. The standard for the Marxist theorist is equalization, not equity or adequacy of cash payments. Their aim is social change, not evening the distribution of income over the life cycle. And the mechanisms are not ameliorative social programs *for* the people, but income *to* the less privileged. What social insurance advocates count as generous provision is, for the most critical Marxist, illusory, a way to gloss over the contradictions of modern capitalism.

Thus far Marxist theorists have not produced a metaphor that competes with either the safety net or the insurance card. But the image of this revolutionary collective suggests the difference. The aim of programs by the people, not for the people, expresses the compelling notion. And that means a rejection of the charitable societies as well as the harsh administrators of the dole, of the social security office as well as the "helping" professions. Professionalism and paternalism alike evoke the wrath of the Marxist, as well as some social democratic reformers who have grown weary of the routines and restraints of mature welfare states.

## APPENDIX B

TABLE 1
Program structure [8]. Basic facts about Medicare.

The statute is Title SVIII of the Social Security Act (Public Law 89–97, enacted 30 July 1965, as amended subsequently).

Every third-party payment system is simply a flow of money, and so is Medicare:

| Subscriber | Fund | Provider |
|---|---|---|
| *Medicare Part A* | | |
| Payroll taxes | | |
| Employee (1.5% of wages) | Hospital | Hospital |
| Employer (1.5% of wages) | Insurance | (or home health |
| | Trust Fund | agency or |
| | | nursing home) |
| Patient's cost sharing: | | |
| Elderly pay premiums | Medigap | |
| or | | |
| Taxpayers | Medicaid | |
| or | | |
| Elderly patient pays out-of-pocket at time of service | | |
| *Medicare Part B* | | |
| Elderly pay premiums | Supplementary | Doctor's |
| | Medical Insurance | "reasonable |
| Treasury of the U.S. | Trust Fund | charges" |
| Patient's cost sharing: | | |
| Elderly pay premiums | Medigap | |
| or | | |
| Taxpayers | Medicaid | |
| or | | |
| Elderly patient pays out-of-pocket at time of service | | |
| Doctor's extra-billing | | |
| Elderly patient pays out-of-pocket | | Doctor's |
| at time of service | | extra-billing |

TABLE 2 [8]
Beneficiaries.

| | HI | | SMI | |
|---|---|---|---|---|
| | Aged | Disabled | Aged | Disabled |
| Number of persons enrolled, in thousands, 1982 | 26,115 | 2,954 | 25,707 | 2,705 |
| Persons served, in thousands, 1982 | | | | |
| *Part A*: | | | | |
| Inpatient hospital care | 6,338 | 739 | | |
| Skilled nursing facility | 244 | 8 | | |
| Home health agency | 1,074 | 80 | | |
| *Part B*: | | | | |
| Physician care, others' care | | | 16,346 | 671 |
| Outpatient hospital care | | | 7,465 | 982 |
| Home health agency | | | 17 | – |
| Numbers of bills approved for payment, in thousands, 1982 | 14,853 | 1,762 | 166,236 | 21,464 |

TABLE 3

Benefits [8]. Only services from "participating provider" or "approved providers" are paid for.

Inpatient hospital care
- Up to 60 days in each benefit period
- 60 non-renewable "reserve days". If hospitalization exceeds 90 days, these reserve days can be used.
- Cost-sharing by patient for 61st and subsequent days.

Skilled nursing facility care
- Referral from an inpatient hospital immediately or after an interval up to a month at home
- Up to 100 days in each benefit period
- Cost-sharing by patient for 21st and subsequent days

Home health care
- Unlimited number of visits
- No cost-sharing

Hospice care
- Up to two 90-day periods and one 30-day period
- No cost-sharing for most services

Physicians care for covered services
- In hospital, ambulatory settings, home health visits, hospice
- No limits on numbers of services
- Cost-sharing

Medicare does not cover
- Custodial care. Nursing homes other than SNF's.
- Dentures and routine dental care
- Eyeglasses
- Hearing aids
- Meals delivered to the home
- Prescription drugs
- Private duty nurses
- Routine physical checkups

HMO option, beginning in February 1985

If patient enrolls with an approved health maintenance organization, Medicare will pay the HMO in advance a sum equal to 95% of the average cost of all Medicare patients's care in that service area. If the HMO's costs for that patient are lower than its revenue from Medicare, it can either reduce normal out-of-pocket cost-sharing by the patient or give the patient any of the benefits not usually provided by Medicare.

TABLE 4
Financing [8]. Basic facts about Medicare.

Trust funds, in billions of dollars, 1983:

| | HI | SMI |
|---|---|---|
| Total income | 44.5 | 19.8 |
| Total disbursements | 39.9 | 19.0 |
| Balance in fund at end of 1983 | 12.9 | 7.1 |

Sources of income of trust funds:
HI is based entirely on social security payroll taxes.
Half from employers, half from employees.

SMI in 1983:
22.9% from premiums paid by beneficiaries
77.1% from the United States Treasury

In lieu of Medigap and private funds, persons over 65 rely on Medicaid for about:
13% of acute health care costs
42% of nursing home costs

Each year, about 3 1/2 to 4 million persons over 65 receive Medicaid benefits, in addition to Medicare.

27 million people, or 11.7% of the entire population, were 65 years or over in the United States in 1983.

About 4% of persons over 65 have neither Medicare nor private insurance.

New social security rates, effective 1 January 1985
Payroll tax (for OASI, DI, HI): 7.05% on employee
7.05% on employer
Earnings base: $39,600
Maximum possible tax: $2,791.80

Cost-sharing by patients for Medicare Part A (hospital care):

| | Formula | 1985 | 1984 |
|---|---|---|---|
| Deductible | Average day in hospital | $400 | $356 |
| 61st to 90th hospital day | 25% of deductible | $100/day | $89/day |
| lifetime reserve days (i.e., 91st to 150th hospital days) | 50% of deductible | $200/day | $178/day |
| 21st to 100th day in a skilled nursing facility (SNF) | | $50/day | $44.60/day |
| Premium, Medicare Part B | | $15.50/ month | $14.60/ month |

TABLE 5
The outcome of the "voluntary effort" ([14], Table 1).

| Calendar Year | Legislative Action | National Health Expenditures | | | National Hospital Expenditures | |
|---|---|---|---|---|---|---|
| | | (Amount in billions) | % Increase | Percent of GNP | Amount in billions | % Increase |
| 1977 | Carter Hospital Cost Containment Act introduced | $169.2 | 13.1 | 8.8 | $ 67.8 | 13.2 |
| 1978 | A.M.A. and hospital associations argue "voluntary effort" makes Carter legislation unnecessary | 189.3 | 11.9 | 8.8 | 75.7 | 11.7 |
| 1979 | Hospital Cost Containment Act defeated November, 1979 | 215.0 | 13.6 | 8.9 | 86.1 | 13.7 |
| 1980 | "Voluntary Effort" dismantled | 249.0 | 15.8 | 9.5 | 100.4 | 16.6 |
| 1981 | Reagan Administration comes into office. Omnibus Budget Reconciliation Act cuts Medicaid | 286.6 | 15.1 | 9.8 | 118.0 | 17.5 |
| 1982 | TEFRA sets budget cap for hospitals | 322.4 | 12.5 | 10.5 | 135.5 | 14.8 |

TABLE 6
Changes over time in health care costs for elders (millions of Dollars).

| | Year | | |
|---|---|---|---|
| Type of expense | 1968* | 1977** | 1983** |
| *Hospital Expenses* | | | |
| Total Part A Program Costs for aged | $3,767 | $15,207 | $38,624 |
| Part A Co-payments for Aged Beneficiaries | 259 | 973 | 3,318 |
| Hospital Costs for Elders Not Covered by Medicare* | 1,197 | 3,542 | 9,780* |
| *Ambulatory Expenses* | | | |
| Total Part B Program Costs for Aged | 1,490 | 6,013 | 15,771 |
| Premiums for Part B | 677 | 1,987 | 3,834 |
| Part B Co-payments for Aged Beneficiaries (Includes Unassigned Liability | 796 | 2,913 | 7,738 |
| Physician Costs for Elders Not Covered by Medicare** | 116 | 441 | 1,640 |

*Notes*
* 1968 hospital cost data from Stephen Davidson and Theodore Marmor ([5], p. 38). 1968 physician costs data from staff report, Senate Finance Committee [21]. 1968 Part B premium costs from J. Krizay and A. Wilson, *The Patient as Consumer* (Lexington: Lexington Books, 1973), p. 75.
** Data from Waldo and Lazenby ([24], pp. 23–24).
* Uncovered hospital costs for 1983 estimated based on Waldo and Lazenby's [24] reported cost data for 1984.
** Does not include costs for elders not enrolled in Part B. Uncovered costs for physician care for 1977 and 1983 are estimated as residuals remaining when costs of Part B reimbursements, co-payments, unassigned claims and spending by elders not enrolled in Medicare is subtracted from total costs of physician care for those over the age of 65.

SECTION II

# THE RIGHT TO HEALTH CARE: PRESENTATION AND CRITIQUE

TOM L. BEAUCHAMP

# THE RIGHT TO HEALTH CARE IN A CAPITALISTIC DEMOCRACY

Although many societies proclaim the equal worth of all persons and back that proclamation with guarantees of equal justice and rights, large economic disparities persist between individuals and across nations. Twelve percent of the gross national product and over $2,000 per person is spent annually for health care in the United States. Physicians have doubled their incomes since 1973. Yet the poor and uninsured often cannot afford minimally adequate health care. Many of us are therefore perplexed, or at least confused, about the right to health care and about our obligations to insure access to or funding for some level of care for all persons.

However, we may also question the fairness of further shifts in tax burdens in order to assist those deprived of adequate health care. For example, we may doubt whether it is fair that hard-earned corporate and individual incomes should be reduced, inhibited, or taxed in order to redistribute the money through health and welfare programs. Especially troublesome to some is the possibility that a theory of *political* equality might be used as the basis of a theory of *economic* equality of distribution. But their critics have challenged what they view as an expanding "double standard" in free-market societies: an egalitarian political system that dispenses political rights equally, but which is harnessed to a nonegalitarian system of economic distribution.

In many nations – democratic and nondemocratic, capitalist and noncapitalist – there is a firmly established legal right to extensive health care goods and services for all citizens. The prevailing legal situation in the United States is that even if there are solid moral reasons for and no constitutional constraints against enactment of a right to health care, there is no constitutional right to health care. At the same time, there is no constitutional obstacle to the Congress's providing some types of health care to citizens, perhaps modelled on the universal coverage for dialysis [6, 23]. Whether there is or ought to be a *moral* right to more extensive health care remains an open question, and this paper will offer some perspective on why it is as open as it is.



*T.J. Bole III and W.B. Bondeson (eds.), Rights to Health Care*, 53–81.

## I. HISTORICAL CONSIDERATIONS

The idea of a right to health care in the United States – or more generally in capitalist democracies – has uncertain origins, but its history is characterized more by the rhetoric of political proposal than by careful analysis of the meaning and scope of proposed rights. Unlike rights such as the right to life, the right to liberty, and the right to own property, assertions of a right to health care are relatively recent phenomena. The uncertain legal and moral status of this right are shared with other recent proclamations of rights such as the right to food, the right to die, and the right to a minimum wage. There is, however, a long history in the United States and elsewhere of proclamations of a right to *health* – by which I mean a right to be protected against the health-threatening behaviors of others. I shall begin with these preliminary historical considerations and then turn later to the pertinent philosophical problems, distinctions, and arguments that are essential for reflection on the right to health care.

The importance of rights in general and their role in public policy has only recently developed in Western democracies. Until the seventeenth and eighteenth centuries, problems of political philosophy were rarely discussed in terms of rights, perhaps because duties to lord, monarchy, state, church, and God (as well as duties of rulers to subjects) had provided the dominant orientation of political and ethical theory. Pioneering ideas were introduced in this period about protected areas of individual liberty into which the state was not to intrude, expressed as "rights" of the individual (cf. [12]). Proclamations of such rights as those to life, liberty, property, safety, a speedy trial, and the pursuit of happiness subsequently formed the core of major political and legal documents.

Although these developments are commonly associated with John Locke and with certain deontological writers in moral theory, substantively the same conclusions were reached by the great utilitarian writers of the nineteenth and twentieth centuries. They too carved out a strongly protected area for individual liberty into which the state was not to intrude – a philosophy expressed with unmatched eloquence and power in John Stuart Mill's *On Liberty*. All these currents of ethical and political theory conspired to a similar, if not identical, outcome: an emphasis on strong protections for the individual, conceived as the central unit of value. A corresponding *deemphasis* emerged on the state as the central unit of value.

This history was destined to have a deep and abiding impact on the great foundational political documents of the United States. However, the right to health care was not among the proclaimed rights in these documents, and so we need to inquire about the specific history of this right. Carleton Chapman and J.M. Talmadge have documented a history in the United States of discussions about rights to health – and to some extent health care – possibly dating as early as the quarantine laws of 1796 and the 1798 Act for the Relief of Sick and Disabled Seamen ([9]; see also [28]). For example, in 1813, Congress passed laws granting a right to effective cowpox vaccines, which were to be distributed free of charge to all citizens. Subsequently, protective laws controlling adulterated medicines and drugs, hygiene, sanitation, and water supplies were passed. On the one hand, it can be argued that these early public health measures were not provisions of "health care" in the form of, e.g., drugs and treatments. At that time, there were comparatively few effective health care measures, and comparatively minor expense was needed to provide them. Major support systems also existed in charities, families, and religious organizations and their health-connected institutions and families. On the other hand, with the exception of the vaccination laws, these early developments could plausibly be construed as recognitions of rights to be protected from identifiable health hazards caused by the actions of others.

Some discussions of rights to health care occurred in the early 1900s, when attention was focused on European health insurance schemes, in particular Britain's National Health Insurance Act, passed in 1911. These European programs were studied carefully by the American Medical Association, which established its own Committee on Social Insurance in 1916. The AMA interest continued until roughly 1921, when the so-called "liberal" period in AMA history ended, and with it died all realistic hope of securing national health insurance legislation. The shifts in subsequent discussion by this Association are well known, resulting as early as 1969 in an enigmatic statement by its House of Delegates that "It is the basic right of every citizen to have available to him adequate health care" [1].

Serious modern concern about a *universal* and not merely *national* right to health care, along with discussions of other rights at the international level, can be traced to the December 10, 1948, Universal Declaration of Human Rights of the United Nations General Assembly. This document attempts to establish "a common standard of achievement for all peoples". Article 25 specifically mentions a right to medical care and a right to a standard of living adequate to provide for one's health and well-being [29]. This document also illustrates the ambiguous political history of proclamations of rights at the international level, because it does not appear to construe rights as individual *entitlements*. Its preamble suggests that the document is to be read as a blueprint for future actions and declarations of rights, rather than as an assertion of rights that persons in their native countries *now* possess. Thus, the declaration functions as a guide for progressive development toward both rights and material security, rather than as an assertion of prevailing rights or of natural and human rights in the classical sense.

The U.N. document has proved difficult to interpret because of its origins in political rhetoric and compromise. It includes rights taken from classic Western declarations of independence, while also including rights to various goods and services requiring minimum standards of living that were proposed by socialist states. It is a document born of negotiation. Still, it is historically significant for the following reasons: Most of the important formal statements of social and human rights prior to 1948 – especially about health-related matters – were assertions only of rights of independence and noninterference. The American Declaration of Independence (1776) and the French Declaration of the Rights of Man and of Citizens (1789) are examples. Perhaps earlier than any other influential source, the United Nations document broadens the scope of rights to include rights to be provided with goods and services.

The cumulative impact of these and related social developments from the nineteenth to the mid-twentieth centuries – together with the explosion of biomedical technology since World War II – has been to increase our awareness of the importance of health care and of certain alleged inequities in its distribution and payment schemes. Discussions today of "equal access" and "the right to health care" are the direct descendants of this heightened awareness, which is still in search of a consensus position and rationale.

## II. RIGHTS AS JUSTIFIED CLAIMS

I now shift from historical foundations to conceptual foundations. What concept of "rights" is at work when we speak of a right to health care?

Because of their unique historical origins, rights hold a prominent place in moral and political writings and documents in the way they assert claims that demand respect and status. When someone appeals to rights, a response is demanded, and we must either accept the person's claim as valid, discredit the claim by countervailing considerations, or acknowledge the right but show how it can be overridden by competing rights claims. Rights in moral philosophy and political theory are justified claims or entitlements to some good, service, or liberty. As such, rights are analogous to property over which one enjoys discretion. If a person possesses a right, others are validly constrained from interferences or from failures to provide that which one is owed, and the person has discretion whether to exercise a right. Rights are thus to be contrasted with privileges, personal ideals, group ideals, and optional acts of charity, which do not provide a basis for saying what can rightfully be demanded as one's due. A rights bearer is in a position to make demands because of the entitlement.

One reason that the right to health care has a short history can be traced to new social needs for such an entitlement. In many societies and periods prior to mid-twentieth century America, provision of donated or collectively arranged health care was viewed as springing exclusively from the virtues of charity, beneficence, and compassion, rather than from justice-based *rights*. Various social and religious institutions and practices sponsored health care for the needy. These practices were eroded and became inadequate to meet their original goals as technology expanded and costs rose dramatically, while social units such as cities grew larger and more impersonal. The health needs of a significant sector of the population were not met, and in many cases were not even known.

It has been argued that this turn of events, including the new language of the right to health care, "is nothing to cheer about. We would, no doubt, be a better society if we cared enough about each other to preclude strident claims of right and just desert" ([2], p. 24). This position has merit: A society that operates exclusively on natural benevolence and entrenched social practices to meet the health care needs of its members does seem a morally superior society to one in which care must become a right in order to arbitrate the competition for scarce resources. Nonetheless, it is entirely understandable how in a large, technologically advanced society, venerable ways of handling health care needs through benevolence have failed to prove adequate. It is less a problem of there now being too little benevolence than it is a problem of dramatically increased costs of and needs for care that were unimaginable in the past. Nor are these social changes ones that have eventuated only in an unseemly howl for entitlements. In former times little health care planning and resources were needed, even in the way of private health insurance. Blue Cross is still a young organization, by comparison, say, to life insurance companies. Yet today private health insurance – which is a matter of prudence rather than entitlement or justice – is an essential system for meeting most of our health care needs in the United States. In important respects any scheme of *rights* to health care for the general public is nothing more than a *public* system of health insurance, albeit one founded on a welfare principle: rights to Medicare are, as rights, no different from rights to receive an insurance benefit when required premiums have been paid: Anyone eligible under fixed rules and requirements of eligibility may validly demand due services and goods provided by that program.

The philosophical foundation of any right is located in a justifying reason that directly supports the claim of entitlement. One has a right, for example, to a particular seat in a particular section of a stadium if one is holding a legitimate ticket to that location. Only someone with a justified claim has a right to something. The fact that one can argue in clever ways that one ought to receive a good or service is insufficient grounds for a rights claim, because a good argument that *Y* should have *X* does not amount to an entitlement for *Y* to *X*. Even if there are good arguments favoring publicly funded health programs, such as free examinations for breast cancer, it does not follow that anyone has a right to the examinations because of these arguments. This conclusion would not follow even if a pressing *need* for such programs were demonstrated. If there exists no justified claim to *X*, a person claiming entitlement to *X* may be demanding *X* or proposing a right to *X*, but nonetheless has no right to *X*.

Entitlements that do not yet exist seems the most accurate description of some parts of the previously mentioned U.N. statement on universal rights, where we find the notion that rights are guiding ideals rather than actual entitlements. It might be argued that this use of the term "rights" is the prevailing *political* use. In writings on *moral philosophy*, by contrast, the notion of a moral right is usually translated into the language of obligations or into the language of justice. There the language of rights is used to refer to a special area of justified claims that are heavily protected against social and political tradeoffs because they are rooted in moral principles.

However, one must be cautious in formulating this thesis because of the ease with which it may be exaggerated. Ronald Dworkin, for example, has used the model of "trumps" to depict rights, so that true rights are to be overriding and enforced even at a loss to public utility; and John Rawls has argued that "the rights secured by justice are not subject to political bargaining or to the calculus of social interests" ([12], p. xi; [25], p. 4). These ideals express noble protections for the individual, but they are as misleading as they are illuminating. In the instance of the right to health care, these theories would entail that no political bargaining could be involved in setting or restricting the boundaries of the right, and that such a right trumps any competing claim(s) of social utility. As we shall now see, these are implausible ways of framing the right to health care (and, in my view, implausible characterizations of any abstract statement of a moral right).

## III. THE CORRELATIVITY OF RIGHTS AND OBLIGATIONS

How are we to understand the language and basis of rights in moral discourse, and in what respect (if any) is there a relationship between one person's rights and another's obligations? I accept the view that a right entails the imposition of an obligation on others either not to interfere or to provide something, where both the obligation and the right are justified by the same principle or rule. For example, a citizen's right to equal protection or to welfare entails an obligation of government to provide equal protection or welfare, and both alike are justified by a principle of equal protection. This correlativity of rights and obligations appears in law as well as morality: If one person has a legal right and another the corresponding duty, the latter may be held legally responsible, and so liable, for violating the person's right by failure to fulfill the duty. I accept this correlativity thesis for both law and morality, but it needs tightening and qualification or it will suffer from overstatement.

The received interpretation of the correlativity thesis is that obligations establishing that someone is owed or entitled to something necessarily have correlative rights, whereas obligations that fail to establish an entitlement do not have correlative rights. This claim is not controversial, and indeed seems analytically true. However, any stronger formulation of the correlativity thesis – to the effect that duties of other descriptions have correlative rights – has been challenged on grounds that there are various obligations that do not entail rights ([7], pp. 439–440; [20], p. 55). Obligations of charity, love, and conscience, for example, are not owed and do not follow from entitlement; they often function more as services one requires of oneself than as universal moral requirements – much in the way physicians may feel obligated to repay their indebtedness to society for their training but without having any contractual or specific obligation to discharge. Such a weak "duty" or "obligation" – as something self-imposed or self-required – provides the only grounds acceptable to some writers for state-supported health services and goods, because they see the obligation as not owed or required in any stronger sense.

### *A. Two Senses of Obligation*

This weak–strong distinction shows how words like "duty" and "obligation" have come to refer to any required action, whether the requirement derives from another's right (a strong duty) or from some "weak" basis such as conscience, a supererogatory ideal. We therefore should distinguish two senses of being "obligated": (1) obligated by a universalizable or general moral obligation expressing what is owed, and (2) obligated by some self-imposed or contingent stricture such as a rule of conscience or a commitment to charity that is not owed [14, 15]. Of course, much turns here on how we are to establish what is owed, so that one can claim a violation of entitlement if it is denied.

Many aspects of the debate over a right to health care turn on whether sense (1) or sense (2) forms the basis of the obligation to provide health care goods and services. Because sense (2) obligations may derive from charitable (or some similar) ideals, they are only in the weak sense moral obligations, and it does seem odd to say that they are owed to persons. The "strong general obligation" in sense (1) is morally binding on *anyone* in a position to perform an action of the appropriate description. Insofar as an obligation is general – or universalizable on moral grounds – the obligation is strict and the correlativity thesis holds. But only these universalizable or general moral obligations reliably entail rights; special obligations do not always entail rights, and self-imposed requirements are optional and never entail rights.

If this general correlativity thesis is correct, little is distinctive about rights as a moral category. As with obligations, the moral basis for their assertion is found in underlying moral principles (or rules). Although it remains controversial whether rights are based on duties, duties are based on rights, or (as I believe) both are justified equally by the same underlying principles, I shall circumvent this controversy here by limiting my claim to the view that the principles in a general moral system *both* impose obligations *and* confer correlative rights.

*B. The President's Commission*

This whole line of argument may seem unduly complicated and unnecessary for present purposes, but in fact it is vitally important for analysis of the right to health care. If society's obligations are strong obligations (sense 1), then rights to health care are entailed; if the obligations are weak (sense 2), then rights to health care are not entailed. In the most comprehensive statement of position on policy, the President's Commission for the Study of Ethical Problems in Medicine and Biomedical and Behavioral Research has taken the understandable, although I believe mistaken, view that only the weak sense of "obligation" is at stake in discussions of health care distribution and equity, and that there is therefore no right to health care:

> The government's responsibility for seeing that the obligation to achieve equity is met is independent of the existence of a corresponding moral right to health care. There are many forms of government involvement, such as enforcement of traffic rules, ... or to promote biomedical research, that do not presuppose corresponding moral rights but that are nonetheless legitimate.... In a democracy, at least, the people may assign to government the responsibility for seeing that important collective obligations are met ([23], p. 34).

The Chairman of the President's Commission, Morris Abrams, has said that it was both prudent and right that the Commission opt for the thesis that there is no right to, only an obligation for, health care. However, this view seems to me either confused or incompletely developed. The President's Commission offers no supporting argument other than question-begging comments such as that "a person may have a moral obligation to help those in need, even though the needy cannot, strictly speaking, demand that person's aid as something they are due" ([23], p. 34). The Commission seems to treat the "obligation" to provide health care much as the role and justification of charity hospitals have traditionally been conceived. The Commission sees the obligation as merely "self-imposed" (sense 2 above), and therefore seeks to avoid a commitment to the language of rightful claims.

The Commission's *motivation* is clear: it discovered that approximately ten percent of the American population is unprotected by any form of health insurance, and took the view that a decent society is one in which such a situation is not allowed to persist. American values traditionally have not, in theory, allowed financial restraints to lead to harms from inadequate health care. The belief has prevailed that a society condoning such restraints on the unfortunate, while permitting health care advantages for the wealthy, is callous and uncaring. The Commission argues, in short, from premises rooted in a conception of a good or decent society, rather than from a conception of what justice demands or which rights citizens possess.

The Commission viewed the "social obligation" to provide health care as falling on many devoted social groups, from the family to institutions of religion and charity, but the Commission also envisioned a major role for the federal government as the institution of ultimate resort and final responsibility. The Commission proposed a standard of "equitable access to health care", which "requires that all citizens be able to secure an adequate level of care without excessive burdens" ([23], pp. 4–5, 22). This standard appears to propose – quite confusingly – an egalitarian-inspired *strong* moral *obligation* of equal access for all citizens (*cf.* [31], p. 114). While the meaning and level of commitment involved in "equitable access" would have to be filled in by reference to allocational possibilities and priorities within the government and health-care system, this does not detract from the obligation of equal access. In any general system, there will always be some level of uncertainty about the quality and quantity of the goods and services to be made available.

Although the general idea behind the Commission's statements is clearly to design a system that eliminates financial barriers to care, it would have to be specified whether "equal access" means (1) equal entry accessibility (ability to enter into the system, but not reimbursement or funding), (2) a decent-minimum threshold of care or funding, (3) equal care or funding, except for optional items, or (4) an entirely egalitarian system of both accessibility and funding. A demanding, but also pragmatic, position is that we are obligated to strive for the ideal specified in (4), while admitting that something more like (2) or (3) will be the actual outcome in most circumstances, especially if the affected health problem is minor or some particular health care good or service is expensive or serves a tiny minority of the population.

Naturally, such determinations about the level of commitment to care are precarious without vast empirical information about particular societies and their situations. A way out of this difficulty of policy is to rely on just *procedures*, such as congressional funding, cost-benefit analysis, and the like. Perhaps it is for this reason that the President's Commission looks to "fair, democratic procedures" as the proper way to make the choices among specific options regarding the level entailed by "adequate health care" ([23], p. 42). But, in the end, procedural solutions cannot substitute entirely for substantive ones, and inevitably we must return to the central question, whether the obligation to provide health care is an optional goal resting on our social values – like biomedical research – or rests on some stronger basis of obligation.

I agree with the Commission that biomedical research is a self-imposed social "obligation", but is the obligation to provide health care to the seriously ill and injured in a different category, one based on strong moral, rather than weak or self-imposed obligation? Has the Commission – as I think – confused self-imposed "obligations" to provide health care with the strict obligation to provide health care? When the Commission summarizes its views on the subject, the obligations appear to be strong rather than weak ([24], p. 29). The Commission seems to want it both ways: The obligation is both strong and weak – weak if rights are at issue, and strong if the question is the obligation to ensure equal access to health care.

There is, I suspect, an underlying reason why the President's Commission and many who have contributed to literature on the problem of rights and duties seem hesitant, ambivalent, and confused in their pronouncements. There is no *sharp* distinction between the class of strong-obligations-therefore-rights and the class of weak-obligations-therefore-no-rights. The obligation to assist unfortunate and vulnerable persons, for example, needs careful specification in a context before it can be determined whether the obligation is strong or weak. Similarly, the obligation to provide health care will need careful refinement and specification of scope before its status as strong or weak can be assessed.

This theoretical problem of drawing the boundaries of strong and weak is too tangential to warrant further discussion here. It should suffice to register a dose of scepticism about the possibility of drawing any sharp line in the moral life not only between strong and weak obligations, but also between *obligation* and that which is meritorious but *beyond obligation*. I do not say that we should smudge these distinctions into one sweeping category of "moral good", because that maneuver would err in the other direction, reintroducing us into the same confusions we find in the Commission. But the idea of a continuum running from strict obligation (required acts) on one end, to heroic and saintly supererogatory acts (non-required acts) on the other end – with many stops in between – is an idea whose merit is yet to be tested in contemporary moral theory.

However this problem should ultimately be handled, I shall simply register a modest disagreement with the President's Commission through the following formulation: Any moral premises that successfully support the claim that we have rights to health care correlatively support the conclusion that society (or the appropriate agent) has the strong general obligation to provide the health care; and whatever considerations successfully support the claim that we have strong general obligations to provide health care to individuals correlatively support the conclusion that individuals have rights to health care. It therefore does not matter whether these issues are framed in terms of rights or in terms of general obligations, because the outcome will be identical. I therefore disagree with the Commission that "a right is not a logical corollary of an ethical obligation of the type the Commission has enunciated" if that obligation is understood as a *general* obligation (*all* of which I take to be strong), and not merely one that is self-assumed ([23], p. 32).

## IV. ARGUMENTS SUPPORTING THE RIGHT TO HEALTH CARE

The expression "*social* obligation" means that society as a collective agent has a duty to perform acts and may be validly criticized should it fail to perform them. We need now to examine whether there is a *strong* social obligation to provide health care in the United States, and therefore a right to it? I shall explore this question by developing what I believe to be the two best arguments for an affirmative answer. Later some weaknesses in and limitations of these arguments will be addressed. One argument is based on *collective protection and social investment* and the other on *fair opportunity.*

### *A. The Argument from Collective Protection, Investment, and Return*

The first argument is premised at the outset on analogy and a principle of consistency. It starts from the position that the general range of protection rightly afforded by government – and already afforded in most nations – naturally extends to health care. The government is constituted to protect citizens from risk to the environment, risk from external invasion, risk to the public health, risk from crime, risk from fire, the risk of highway accidents, and the like. We do not ask people to have their own firefighters, crimefighters, jet pilots, etc. It seems a natural extension that government would protect against risks to health and thereby would meet health care needs, especially given the significance of this benefit to individuals and to society (see [19], pp. 278–287; and also [30], pp. 86–94; [10], pp. 52–53; [18]). Moreover, the advantage of securing vital collective goods through modest but enforced duties to contribute is itself a secure principle of modern political life. (We shall return to this problem in the section below on libertarianism.)

It does not, however, follow from this argument that there is an entitlement or right to health care. Most enforced collective functions are governmental and pertain to what are generally classified as *social* goods, whereas health care presumably is a *private* good. On the one hand, I find the public/private distinction particularly woolly when it comes to distinguishing public health from private health. The distinction between public health and medicine has never been sharp, and I think it cannot be made so. On the other hand, I will concede the premise that *some* private/public distinction is viable and, moreover, that some private goods, rather than public goods, would follow from the right to health care. Under this premise, the argument from collective protection still needs supplementation if it is to convince those who have reservations about using collective goods to support any form of private need.

The required supplementary argument is to be found in a conception of social investment that proceeds from the following premises: The collective social system that funds the education of physicians and medical research – and not merely the education and research of professionals in public health – is heavily subsidized by public funding. Vital parts of that system could not exist without such funding. Like any system of protective insurance, investment, and return in which the public is involved, a decent return on one's contribution is anticipated, and that return should come in this case in the form of protection of health or alleviation of illness and injury. In short, public support of the system of training, research, and technological application in health care is through the taxation of individuals, and not simply by private donation or by personal funding. This support is premised on a system of returning the investment equitably to those who fund the system, at least in the form of collective protections.

This argument should be sufficient to overcome the earlier objection that protection of individual health is relevantly dissimilar to public health measures. However, I admit that the argument is limited and unfocussed, and that any obligation or right that it establishes to health care and health protection would require more extensive analysis and qualification than can be attempted here. In particular, one would have to examine carefully the underlying analogy to financial insurance and investment schemes, as well as the nature of the contribution that is made and the type and extent of return that may reasonably and rightfully be expected. For example, consider our social sponsorship of the drug research that is funded through NIH and regulated by FDA. We certainly expect FDA and NIH to protect our health in various ways, but we do not expect the federal government to provide us with drugs – not even expensive anti-cancer drugs. We can scarcely claim such a right to drugs based on our contribution toward drug regulation and research. One might take a similar view about, for example, cancer research more generally. The investment might be for the purpose of finding cures or treatments rather than dispensing medicines or providing treatments once efficacious therapies have been found. In the case of physician education, it might be argued that the investment is in training, and even protection of the public health, but not in services or distributive systems.

There is also the issue of *which parties* are obligated to provide the care. Perhaps the reciprocal obligation falls more heavily on health care deliverers than on the government. Deliverers accept public funds, and thus seem to acquire an obligation of debt, gratitude, and fair play to supply health care that is directly correlative to a right to health care (*cf.* [8, 10, 27]). This proposal has merit, but only if the parties understand the terms of the arrangement. For example, students at the U.S. Military Medical School (USUHS) understand that acceptance of their funding creates an obligation to serve the military and its patients. Apart from such a contractual arrangement of consent, it is hard to imagine how a workable public policy could be developed that required health care deliverers to discharge the obligation to serve the public. Nor does the deliverer's obligation to society now entail that the obligation be discharged in this way.

There is also the deeper problem of whether there should be a public subsidization of health-care activities at all – e.g., whether there should be any form of subsidization of medical education. This issue presents the further problem for the social protection and investment argument whether we would conclude that there is no right to health care *if* there were no public subsidization of these activities. This inference is unwarranted, because there may be other premises (in justice, say) on the basis of which a right to health or health care can be legitimately founded. So, I reiterate that this first argument is underexplored in my presentation of it, and that it establishes but a programmatic basis for the right to health care. Nonetheless, human society as we know it is a cooperative enterprise controlled by the contractual terms of cooperation. The justice of the terms of cooperation is a perennial concern of philosophy, and justice could not be achieved if collective protection, reciprocity, investment, and the like were dismissed as irrelevant considerations.

Some philosophers would contend that I am fallaciously appealing to justice and to rights in this argument. They would hold that I have a hidden and unexamined premise to the effect that we *ought* to use collective protection measures and *ought* to invest for health returns, a premise that has no more of a foundational status in our moral viewpoint than, for example, does the premise that we ought to use collective protection measures and investments for retirement. Collective social investments in, say, nuclear power plants and personal investment of retirement funds are matters of *decision*; they pertain to the kind of society members want or choose to assume, and do not turn on matters of rights, justice, or moral obligations. Or, to put another slant on the same point, what is owed as a matter of justice in health care is *owed* only *after* the decision has already been made to devise a social distribution scheme for health care, but whether the scheme itself ought to be provided is a question external to the moral claim that health care is due as a matter of justice or right.

This is an objection to, or (as I prefer to say) a refinement of, the social investment argument, with which I have considerable sympathy. I shall return later to this problem of the role of decision in social allocation schemes, after I examine the second argument.

### *B. The Argument from Fair Opportunity and Need*

Central to a number of theories of justice is a principle of fair opportunity, according to which (in appropriately general and abbreviated form) no person should be denied social benefits to primary goods because of disadvantaging properties for which the person is not responsible. The justice of social institutions is to be gauged by their tendency to counteract lack of capacity and opportunity caused purely by luck of birth (family and class origins), natural endowment and health status, and historical circumstances (accidents at some point or over the course of a lifetime). To use distributive principles that fail to correct the results of fortuitous advantages and disadvantages would be to treat people in accordance with differences they do not merit or deserve and for which they bear no responsibility. Disadvantages produced by the lottery of life are from the moral point of view arbitrary, and insofar as disease or injury induces these disadvantages, resources should be used to counter the effects of injury and disease ([25]; [11], pp. 158ff).

Health care needs are randomly distributed and not under the individual's control in many instances. Moreover, the need for health care affects the seriously sick and injured far more significantly than other members of the population, reducing opportunities still further. Unlike most other needs – for food, shelter, recreation, and the like – the costs can be staggering in the health care sector, and increasingly uncontrollable as the person's health status worsens. In general, lack of opportunity and the impact of handicap are magnified in significance as injury and disease increase and go unchecked. If a person were able to care adequately for his or her health needs by private initiative, private resources, or private insurance schemes, then health needs, like recreational needs, might be viewed as purely individual responsibilities. But, for a variety of reasons, health care needs will never be adequately covered through private resources and insurance schemes, and *cannot* be procured by a significant percentage of citizens.

The heart of the argument from fair opportunity is that a denial of access to those desperately in need is unjust because needs for health care – unlike, for example, government subsidized recreation in the national parks – can have a direct and profound effect on opportunity, quality of life, and functional capacity. Even were there no effect on opportunity in the sense of career opportunity or opportunity for personal advancement, the effects of an absence of health care on pain, suffering, premature death, and the like would be serious moral concerns, giving health care a special status that most needs cannot claim. The fact that they are undeserved afflictions, and to some extent satisfiable needs, in a context in which perhaps thirty million Americans cannot secure health care, suggests a special place in our scheme of values for health care.

However, this argument, like the first, is more suggestive than complete. To fill it out, one would have to be more specific about the principles of opportunity at work, as well as about their application. Principles of fair opportunity have always had their most convincing application in contexts of the provision of education and access to political participation, where they have achieved near unanimous acceptance in our culture. It must be candidly admitted that as the importance of opportunity is progressively broadened to regions such as health care that are beyond the zones of political participation and education, the argument from fair opportunity may become progressively less convincing; indeed, the very principles at work may not be identical. Health care, like food, may be essential to survival and thus may be a fundamental need, but what is the *opportunity* envisioned by the satisfaction of these needs? The opportunities that education and political participation provide are not obviously the same as the opportunities that food and health care provide. Moreover, if the goods to be provided are scarce, as opposed to nonscarce, this condition too may make a difference to the argument. These matters would have to be resolved through patient philosophical analyses before one could be convinced that the argument from fair opportunity is entirely satisfactory.

### *C. Further Limitations of These Arguments*

There are ways other than those thus far mentioned in which one should be cautious about drawing sweeping conclusions from the above two arguments (from social protection and investment and from fair opportunity). First, the fundamental premises and conclusions in these arguments do not entail that the responsibility for public funding falls on any particular social agency. Whether it is a local, state, or national responsibility is not a question into which I shall venture, and therefore I shall have nothing to say about national health legislation (*cf.* [8] and ([24], pp. 25ff.) for some speculation on these matters).

Second, these conclusions also say little about the precise *scope* of our obligations, or about the rights correlative to them. This is the more important and pressing matter of social concern and health policy. Any full account of our obligations – worked out in a systematic program – would have to restrict, by careful argument, the scope of claims about a right to health care. Consider, for example, proposals advanced by Charles Fried and others that we ought to provide to all a decent minimum of goods and services, whereas we are not obligated to provide more than a decent minimum ([17], pp. 29–34). This proposal is not to be understood as a right to the best possible care, because the best available treatments are often prohibitively expensive and sometimes are luxuries rather than necessities. While the arguments I have displayed march to the conclusion that we are collectively obligated to provide at least a decent miminum of health care if persons are desperately in need, the scope of this decent minimum is scarcely even addressed in my presentation of the arguments. To resolve this matter, we would need to fashion a carefully delineated framework of principles for macroallocation (see [4]). Unacceptable social costs might follow from a program with an extensive set of financial demands, and considerations of justice might conflict with those of efficiency and utility. Increased taxation could frustrate or abuse the legitimate framework of expectations for personal use of earned income. Here allocation problems inevitably become central to claims on behalf of a positive right to health care. Not *whether* there is such a right, but the *limits* to the right, is the primary problem, and it can only be treated by some principled or procedural way of identifying which or what kinds of health care services we should devise to discharge our obligations ([5], pp. 127–129).

## V. LIBERTARIAN OPPOSITION TO THE RIGHT TO HEALTH CARE

I shall now consider a position resolutely opposed to social protection and investment as well as fair opportunity arguments, and that, indeed, impugns all defenses of the right to health care. The argument derives from premises in the so-called libertarian theory of justice, especially premises in libertarian arguments that deny positive rights, allowing only negative rights.

### *A. The Libertarian Theory*

The libertarian theory has its roots in writers such as John Locke and Adam Smith, who have supplied the locus classicus of the justification of the free-market economic system. Although the classical writers never broached the question, the scheme of funding for health care that we have traditionally embraced in the United States was founded on these assumptions. We have not employed this theory to the entire exclusion of all others, but it is the underlying rationale for most parts of the system as we have known it. What makes the theory *libertarian* is its advocacy of distinctive processes, procedures, or mechanisms for ensuring that liberty rights are recognized in economic practice – typically the rules and procedures governing economic acquisition and exchange in capitalist or free market systems.

As Adam Smith classically described capitalist economic systems, people acting in an individually self-interested fashion exhibit behavior patterns that collectively further the interests of everyone in the larger society. Such a system presumes a model of economic behavior that attributes a substantial degree of freedom to individual agents. People are envisioned as freely entering and withdrawing from economic arrangements in accordance with a controlling perception of their own interest. People freely *choose* to contribute to economic arrangements, and because contributions are freely chosen, they can be considered morally relevant bases on which to discriminate among individuals in distributing economic burdens and benefits. This vision underlies Robert Nozick's characterization of the fundamental material principle informing libertarian theories of justice: "From each as they choose, to each as they are chosen" ([22], p. 163). Unlike Rawls and the supporters of the fair opportunity rule, Nozick's principle would not attempt to counteract life's disadvantaging properties.

In seeing free choice as central to an account of justice in economic distribution, libertarian writers often accept an individualist conception of economic production and value. They maintain that people should receive economic benefits in proportion as they freely contribute to their production, a view that assumes the possibility of recognizing meaningful distinctions between individual contributions to production. The industrious and imaginative hospital administrator, for instance, would from this perspective be contributing far more to a for-profit corporation's success than the similarly exemplary cafeteria bus-boy or secretary, and the administrator therefore deserves – on grounds of justice – the proportionately greater share of the profits.

Libertarian theorists reject the conclusion that utilitarian and egalitarian standards provide valid normative requirements for distribution of goods and services such as health care. People may be equal in a host of morally significant respects (e.g., entitled to equal treatment under the law and equally valued as ends in themselves), but the libertarian believes that all other theories inevitably entail practices that coercively extract financial resources of individuals through systems of taxation. Individuals are illegitimately used to larger welfare goals of the state, and their property is expropriated for these purposes. Equality or utility should not be advanced from this perspective at the expense of individual rights, an ordering that inverts proper moral priorities: rights to goods and services (positive rights) are wrongly given priority over both individual property rights and liberty rights (negative rights). In effect, basic rights are sacrificed to the larger public interest by using one set of individuals to benefit another set of individuals; but no moral grounds would justify the sacrifice ([22], p. 32f).

The libertarian thus views utilitarianism as an entirely perverted theory of morals. Rights function for the libertarian as constraints against allowing the maximizing of social utility to proliferate at a cost to individuals; rights are not the end result of some calculation of social utility. It is, then, a basic violation of justice to act on the principle that people deserve equal or utilitarian economic returns. In particular, libertarians believe that the fundamental right to own and dispense with the products of one's labor as one chooses must be respected even if its unrestricted exercise leads to ever expanding inequalities of wealth and resources.

The libertarian theory is meant as a challenge to many of the assumptions underlying political realities in contemporary industrial societies, through a rejection of *all* distributional patterns that might be imposed by material principles of justice. In this view, justice consists not in any particular distributional outcome, but in the unhindered operation of just procedures. There must be no pre-planned "pattern" of justice independent of the demands of the procedures in a free market economy. Any outcome is just, as long as it results from the consistent operation of the specified procedures. (For Nozick there is no pattern of just distribution independent of the procedures of acquisition, transfer, and rectification; justice is served whenever individual rights are respected in the protected operation of these procedures.)

This claim has been at the center of philosophical controversy over libertarianism. Many of the most influential theories of justice, including the egalitarian and utilitarian theories, react vigorously to the libertarian commitment to pure procedural justice as the sole important criterion. This perspective continues, I believe, to play an influential role in arguments about the right to health care in a capitalist economy. Accordingly, let us now examine one form in which such a libertarian framework has been applied to the right to health care.

### *B. Negative and Positive Rights*

Libertarian writers on the subject of a right to health care often invoke a distinction between positive and negative rights. This distinction is based on the difference between the right to be free from something (a liberty right) and the right to be provided with a particular good or service (a benefit right). The motivation behind the invocation of this distinction by libertarians is to defeat theories that would coerce one group of persons to pay for goods such as the health care needs of another group. Thus, even minimal taxation is impugned as a way of supporting health care distribution, which is the equivalent of denying all possibility of a legitimate right to health care.

However, negative or liberty rights sometimes require state interventions using general revenues, a complexity that tends to obscure any neat positive/negative distinction. For example, consider various rights to have one's health "protected". This right might or might not include health *care*, because the health protection/health care distinction does not reflect a firm difference between exclusive classes. But to assert a right to protection *is* generally to claim more than mere freedom from interferences that negatively affect health; it is to assert that the state is obligated to enforce the rights of citizens by using collective financing and state agencies actively to protect them against dangerous chemicals, emissions, polluted waterways, the spread of disease, and the like. A claim that some rights are rights both to freedom from interference *and* to active protection is perhaps best understood, then, as a dual claim to a negative right to freedom and a positive right to goods or services. These rights are analyzable as complex, containing both negative rights and positive rights within their broad scope. Typically libertarians do not shy away from positive rights of *this* description, if there is a significant *liberty* deserving protection.

Nonetheless, the distinction between positive and negative rights has been employed by several writers in order to develop a programmatic libertarian position on the issue of rights to health or health care. For example, it has been argued that:

> A "right" [in classical historical writings] defines a freedom of action ... The greatest perversion of the concept of rights occurred during the Presidency of Franklin Delano Roosevelt, when "right" was surreptitiously transferred from the *freedom* to pursue a value to the *value itself*: now all Americans had the right to a job, the right to a house, [etc.]... Modern politicians have [subsequently] reduced [the idea of rights] to utter absurdity. The Democrats, in their 1972 platform, promised to secure the right of the American people to health ([26], p. 11).

This form of justification is used to defend carefully circumscribed social health programs asserting a right to be protected against controllable health hazards that are socially caused. Public health programs such as toxic substance and environmental pollution control, occupational safety regulations, and sanitation are among obvious examples of justified protective measures provided by the state. But when these writers discuss a government-funded *health* program, they maintain that there is no state obligation to provide *health care* or *health insurance*. That is, at most there exists a right to *health*, in the sense of a right to be protected by the state from risks or hazards to health that are the result of the individual or collective actions of others.

There are several problems and confusions in this line of argument. First, the historical claims on which it rests are dubious, as we have seen. The earlier discussion of the history of rights claims indicated that laws implicitly granting *positive* rights were initiated long before the Roosevelt period. Second, even if it were true that in early Western philosophical writings and political constitutions the emphasis was exclusively on negative rights, the *concept* of rights employed in such documents is not confined to negative rights. It is no perversion of the general *concept* of rights that positive claims be included, and there is no powerful theory anywhere in contemporary ethics to the conclusion that rights are exclusively negative (cf. [21]). The available arguments show only that an *extensive* network of positive rights would unduly constrict negative rights.

Third, these programmatic proposals seem to suppose that the class of socially caused injuries and diseases is plainly distinguishable, which is far from the case. The inherent complexity of the causal links is now well established, and emphasis is increasingly being placed not merely on biological factors in the etiology of disease, but on the *social* and environmental bases of disease and ill health as well (see [3], p. 1976; [27]). If ill health is broadly rooted in socially induced causes such as environmental pollutants and infant feeding practices, then an equally broad socially funded health care or health insurance program based on a *negative right* to health would follow. The program would provide preventive and curative goods and services for numerous controllable and treatable diseases – possibly including, for example, alcoholism – on grounds that society is responsible for what happens to its members. The class of diseases protected by a negative right presumably expands as evidence concerning the causal role of social factors expands. The proponent of negative rights thus would saddle us with the uncomfortable, and in the end impossible, task of determining the extent to which social rather than biological or individually-induced factors contribute to the etiology of disease and injury.

A program of national health based exclusively on such a negative-rights basis would also *exclude* the obligation to control, prevent, or treat various conditions for which we have generally felt most strongly obligated to provide state benefits. These would include conditions that are significant *public* health problems. The bulk of these conditions might be genetic – for example, cystic fibrosis, sickle cell, hemophilia, and PKU – but this class would also include the effects of natural disasters and diseases associated with the aging process. Because the entire issue of whether a condition qualified for protection and services would turn on causal agency and responsibility, in cases of unknown etiology there would always be an impossible difficulty in determining which way to turn the presumption of causal responsibility. Thus, an exclusively negative conception of the right to health invites chaos in health policy and is unsatisfactory both on theoretical and practical grounds (see [5], pp. 125–126).

### *C. A Final Assessment of Libertarian Theories*

Although the underlying assumptions of various libertarian reservations about free-market restrictions have long been indigenous features of Anglo-American economic arrangements, I would challenge their capacity to prevail over the two arguments from collective protection and fair opportunity developed earlier in this paper. In concluding this section, I shall offer several reasons why the libertarian theory is, in the final analysis, unconvincing.

An initial problem has to do with the general role of principles of nonmaleficence and beneficence in morality. The libertarian believes that protecting autonomy always takes priority over protecting against harm and providing benefits. I do not deny that moral principles protecting autonomy ought to have a treasured and protected status, but so should those designed to prevent harm. The libertarian willingly allows that harm produced *by the actions of others* provides good and sufficient grounds for restriction of the autonomy of those others and for compensation to the individual(s) harmed. But there remains the issue of the weight of moral principles that protect against harms not caused by others and that compensate for those harms once they have occurred. For example, municipal and state-supported hospitals are typical examples of our prevailing use of collective funds both to protect against harms and to provide benefits. The libertarian has the burden of proof to show that this use of collective funds is contrary to, rather than supported by, morality.

Morality is concerned with the *harmfulness* of harms, not merely with the *causal origins* of the harms. Any strong moral obligations of nonmaleficence or beneficence that protect against harm would, of course, entail rights to the correlative protections. A major problem for libertarians is that their doctrine compels them to gloss over the following important thesis: It does not matter from the perspective of one who might suffer harm (and, in some cases be compensated for the harm) whether the origin of the harm is (1) *intentional*, as in an act of malice such as attempted murder, (2) *accidental*, as in an automobile accident, or (3) *natural*, as in an earthquake. Suppose in each case there is no person to be held accountable, and therefore no right to assistance or compensation can be enforced. Are we to say, as libertarians propose, that in case (1) the occurrence is unfair, in case (3) merely unfortunate, and in case (2) the fairness or misfortune is dependent on the circumstances (e.g., the absence or presence of negligence)?

To take this view leaves a burden of proof on libertarians to show that although the harm may be identical (loss of a limb, say) and the needed health care identical in cases like 1 through 3, in one case there is a right to be protected and in another case no such right. The libertarian might, of course, deny that there is any obligation or right in any of these cases (1–3), but then we are returned to the fundamental problem of the libertarians' lack of a proof that principles of nonmaleficence and beneficence cannot override rights of autonomy.

If it is once conceded that we can and should, as a moral matter, protect against (some types and level of) harm (and compensate for the harm) independent of the causal origins of the harm, it is a short step to the conclusion that there is also a right to positive benefits to such goods as health care. The conclusion is shorter still because of the conceptual uncertainty that surrounds the distinction between the obligation to avoid harm and the obligation to benefit. For example, giving persons wheelchairs and medications is a benefit, but one aimed also at protecting against harms that would otherwise afflict the affected persons.

The heart of the libertarian's argument is an appeal to the violation of autonomy created by *coercive* (as contrasted with voluntary) collective schemes. Such schemes are the inevitable consequence of strong principles of nonmaleficence and beneficence of the sort I accept. In order to show that there is something so special about principles of liberty or autonomy that principles of beneficence and nonmaleficence can never override them, a libertarian theory of rights would have to be expressed so that it *always* protects autonomy at the expense of welfare, no matter (1) the *level* of the coercive constraint placed on autonomy (0.0001% of one's salary through taxation, e.g., would be excessive for the libertarian because coercive) or (2) the significance of the benefit conferred upon others (life-saving technology, say). I believe that no reasonable theory of morals would hold that liberty is an absolute value, above life and above all forms of welfare; and yet this seems precisely the final libertarian solution. The libertarian is right to insist that respect for autonomy is a necessary condition of morality, but this does not make autonomy an absolute principle. Our shared general scheme of moral values cannot possibly accommodate such a claim of absoluteness.

Presumably the libertarian finds the importance of liberty (or autonomy) in its status as the condition of a meaningful human life. But if this be the justification of its protected status, it seems an equally good justification of a similarly protected status for health care, because health is also the condition of a meaningful exercise of liberty (cf. [2], p. 41). The libertarian thus needs some justification of a free-market economy for health care other than the classical justification for liberty itself.

It will not do for the libertarian to reply that a right to health care would prove severely intrusive in its impact on the property rights of citizens. No doubt *some* egalitarian schemes – e.g., some radical egalitarian conceptions of equal access – could prove to be severely constraining, but supplying a decent minimum of health resources is not likely to have a severe impact on liberty in the American economy. These problems of predicting impact are partially empirical questions, but again the burden of proof is on the libertarian to show that the consequences of a carefully controlled system of macroallocation governing a right to health care would eventuate in unacceptable harms to property and consequently to liberty.

The rights not to be coerced and to private property are rights of great importance, but not so important or precise in scope as to be absolute. Nothing about either right suggests more than prima facie status. Accordingly, any moral right – such as the right not to be harmed – that is weightier in the circumstances can override the right not to be coerced or the right to hold property. If the libertarian retreats at this point to the superficially more appealing position that only a *certain level or range* of coercion through taxation is prohibited by the right not to be coerced (while a smaller measure is permitted), much has been conceded merely by the acknowledgement that some degree of coercion is permitted. More importantly, the concession shows how moral considerations such as assistance and protection against harm *can* have weight *even if* they require coercion in the form of taxation of property.

This conclusion must not be taken, of course, to suggest that coercive schemes of taxed contributions to prevent harm and to assist *always* defeat the right against coercion. To mention but one among several reasons why taxation for such purposes might be unjustified, the risks and costs of coercion might well outweigh the possible benefits and averted risks of the coercion: From my perspective, this slant on the issues returns us to the fundamental problem of the allocation of resources, which I believe is a far deeper moral and social problem than that of prevailing rights (see [5]).

### *D. Weaknesses in Alternative Theories*

The line of argument that I have taken in criticizing the libertarian theory prompts some related questions about the moral acceptability of *all* the leading, rival theories of justice in contemporary philosophy. One possible view is that the theories are irreconcilably opposed, springing as they do from rival starting premises and eventuating in intractable and interminable disagreements. Another view is that we may be able to take the best of each theory, dispensing with unpalatable parts. Perhaps within our political framework an economic system should be fashioned as libertarian and a social system as egalitarian. Perhaps. But such a neat bifurcation seems to me politically and socially naive. I would maintain that we have no available theory of justice to bring such diverse accounts into unity. We shall have to limp along as best we can with these competing conceptions, using bits of each as we can see our way to doing so.

General theories of justice, we should admit, are not sharp instruments for handling detailed problems of public policy, such as the scope of our obligations to provide health care. Such theories introduce principles intended to order our diverse judgments about right and wrong as consistently and harmoniously as possible. The theorist ideally starts with the broadest possible set of our considered moral judgments about justice, and then erects a theory that reflects those judgments. Then, by a process of supporting example and opposed counterexample, the theory and its principles are reflectively adjusted until as congruent as the theorist can make them.

This procedure, which effectively involves testing our basic moral judgments for coherence through the construction of a moral theory, is at the methodological heart of moral theory, at least as currently practiced by many of its leading proponents. However, because the method leaves considerable latitude for choice of initial judgments and supporting examples, we should not be surprised when competing theories emerge. These opposed theories of justice illustrate how far apart we still are, as philosophers and as a community seeking consensus over the right to health care. We shall therefore have to reflect with increasing subtlety in the near future, if we are to reduce the gaps between our most general theories of economic justice and to put them to work so that they are aids rather than hindrances in the attempt to evaluate the merits of the idea of a right to health care.

## CONCLUSION

I accept the two arguments developed in this essay to the conclusion that there is a right to health care, but I also believe that details in working out the scope of this right will not be forthcoming independently of a detailed allocational scheme that establishes patterns of access and forms of eligibility. The justifiability of specific social expenditures – not some natural, inalienable or preexisting right – will determine the scope of this right. The idea of rights as trumps is a myth of modern philosophy. Rights to goods and services are not absolute, and they weaken as they come into conflict with needs for other resources that individuals can claim by virtue of some other, presumably equally justified, allocational commitments. Only negotiation within the culture, not some philosophically sound argument, will ultimately resolve these issues. At the present time we have neither a philosophical nor a negotiated solution adequate for addressing these allocational questions with the needed depth.

Moral principles and judgments often do not establish a firm basis for public policies. Usually this occurs not because moral considerations are unimportant, but because there are conflicting moral demands and no single moral perspective is determinative. In such cases a moral *decision* concerning the weight of competing, well-defended moral claims is required, and this decision in turn fixes the acceptable policies. It would be convenient if moral principles could always fix outcomes in the way rules of sport determine winning and losing outcomes, but often they cannot be settled so explicitly and must be resolved by moral deliberation, negotiation, and decision. In such cases it is neither unreasonable nor unfair if wisdom and prudence are found in two or more competing positions. I believe this to be the current situation in the debate over "national health legislation" and "the right to health care", and I expect the situation to be perpetuated well into the future. This lack of finality should not be a reason for despair or skepticism, however, because reasoned arguments are still possible. We should all be better off for having accepted the challenge to debate these matters through essays such as those found in this volume.[1]

*Georgetown University*
*Washington, D.C., U.S.A.*

NOTE

[1] I owe several important observations and criticisms in this paper to Ruth Faden, Dan Brock, James Childress, and Allen Buchanan.

BIBLIOGRAPHY

1. American Medical Association, House of Delegates, Chicago, 1969.
2. Arras, J.D.: 1984, 'Utility, Natural Rights, and the Right to Health Care', in J.M. Humber and R.F. Almeder (eds.), *Biomedical Ethics Reviews 1984*, Humana Press, Clifton, New Jersey, pp. 23–46.
3. Beauchamp, D.: 1976, 'Public Health and Social Justice', *Inquiry* **13**, 3–14.
4. Beauchamp, T.L.: 1982, 'Morality and the Social Control of Biomedical Technology', in W.B. Bondeson, H.T. Engelhardt, Jr., S.F. Spicker, J.M. White, Jr. (eds.), *New Knowledge in the Biomedical Sciences*, D. Reidel Publishing Co., Dordrecht, Holland, pp. 55–76.
5. Beauchamp, T.L. and Faden, R.: 1979, 'The Right to Health and the Right to Health Care', *The Journal of Medicine and Philosophy* **4**, 118–131.
6. Blackstone, W.T.: 1976, 'On Health Care as a Legal Right: An Exploration of Legal and Moral Grounds', *Georgia Law Review* **10**, 391–418.
7. Brandt, R.B.: 1959, *Ethical Theory*, Prentice-Hall, Inc., Englewood Cliffs, New Jersey.
8. Camenisch, P.: 1979, 'The Right to Health Care: A Contractual Approach', *Soundings* **62**, 291–310.
9. Chapman, C.B. and Talmadge, J.M.: 1971, 'The Evolution of the Right to Health Concept in the United States', *Pharos* **34** (1), 30–51.
10. Childress, J.F.: 1984, 'Rights to Health Care in a Democratic Society', in J.M. Humber and R.F. Almeder (eds.), *Biomedical Ethics Reviews 1984*, Humana Press, Clifton, New Jersey.
11. Daniels, N.: 1981, 'Health Care Needs and Distributive Justice', *Philosophy and*

*Public Affairs* **10**, 146–179.
12. Dworkin, R.: 1977, *Taking Rights Seriously*, Harvard University Press, Cambridge, Massachusetts.
13. Feinberg, J.: 1973, *Social Philosophy*, Prentice-Hall, Inc., Englewood Cliffs, New Jersey.
14. Feinberg, J.: 1970, 'The Nature and Value of Rights', *Journal of Value Inquiry* **4** (4), 263–267.
15. Feinberg, J.: 1981, *Rights, Justice, and the Bounds of Liberty*, Princeton University Press, Princeton, New Jersey.
16. Frankena, W.K.: 1976, 'Some Beliefs about Justice', in K.E. Goodpaster (ed.), *Perspectives on Morality: Essays of William K. Frankena*, University of Notre Dame Press, Notre Dame, Indiana.
17. Fried, C.: 1976, 'Equality and Rights in Medical Care', *Hastings Center Report* **6**, 29–34.
18. Hume, D., 'On Suicide', as reprinted in S. Gorovitz *et al.* (eds.): 1976, *Moral Problems in Medicine*, Prentice-Hall, Inc., Englewood Cliffs, New Jersey.
19. Jones, G.E.: 1983. 'The Right to Health Care and the State', *Philosophical Quarterly* **33**, 279–287.
20. Lyons, D.: 1970, 'The Correlativity of Rights and Duties', *Nous* **4**, 45–57.
21. Nagel, T.: 1975, 'Libertarianism without Foundations', *Yale Law Journal* **85**.
22. Nozick, R.: 1974, *Anarchy, State, and Utopia*, Basic Books, New York.
23. President's Commission for the Study of Ethical Problems in Medicine and Biomedical and Behavioral Research: 1983, *Securing Access to Health Care*, Vol. 1, U.S. Government Printing Office, Washington, D.C.
24. President's Commission for the Study of Ethical Problems in Medicine and Biomedical and Behavioral Research: 1983, *Summing Up*, U.S. Government Printing Office, Washington, D.C.
25. Rawls, J.: 1971, *A Theory of Justice*, Harvard University Press, Cambridge, Massachusetts.
26. Sade, R.: 1974, 'Is Health Care a Right?', *Image* **7**, 11–19.
27. Sidel, V.: 1978, 'The Right to Health Care: An International Perspective', in E.L. Bandman and B. Bandman (eds.), *Bioethics and Human Rights*, Little, Brown, and Co., Boston, Massachusetts.
28. Stevens, R.A. and Stevens, R.: 1974, *Welfare Medicine in America: A Case Study of Medicaid*, The Free Press, New York.
29. United Nations: 1973, 'Universal Declaration of Human Rights', in *Human Rights: A Compilation of International Instruments of the United Nations*, United Nations, New York.
30. Walzer, M.: 1983, *Spheres of Justice: A Defense of Pluralism and Equality*, Basic Books, New York.
31. Wikler, D.: 1983, 'Philosophical Perspective on Access to Health Care: An Introduction', in President's Commission, *Securing Access to Health Care* [23], Vol. 2, Appendix F, pp. 104–151.

ROBERT M. VEATCH

# JUSTICE AND THE RIGHT TO HEALTH CARE: AN EGALITARIAN ACCOUNT

## I. INTRODUCTION

This essay is meant to be an account of an egalitarian theory of justice applied to health care. It will in part be a descriptive account of how those who favor some version of a qualified "right to equal health" or a principle of distribution of health care based primarily on need have supported their views. In addition, it will in part be a normative account of the kinds of arguments and assumptions that might best support the egalitarian position and what qualifications might be necessary.

It is not a knock-down, drag-out logical argument proving the egalitarian interpretation of the principle of justice, i.e., the principle that benefits (and burdens) must be distributed fairly. In fact, one central thesis of this essay is that no definitive, rational defense of this *or any other* interpretation of the principle of justice is possible. It will be an account of the kinds of arguments and assumptions needed to support the egalitarian interpretation for determining what is ethically required in distributing health care as well as what is just in general.

By "an egalitarian account of the principle of justice applied to health care", I refer to any of a number of closely related positions that among other things include the moral rule that justice requires that persons be given an opportunity to have equal net welfare insofar as possible and that, applied to health care, justice requires that persons be given an opportunity to have equal health status insofar as possible. This paper attempts to explain how persons might reach that conclusion and the extent to which they can defend their conclusions against those who reach alternative conclusions. For technical reasons having to do with the argument over whether rights or the correlative obligations are prior [13], I prefer to carry on the discussion in terms of a small set of *prima facie* duties or principles. However, insofar as one can say that the principle of justice conveys an obligation to distribute health care to certain people, one can also say derivatively that those people possess an entitlement right to the health care for which justice calls.



*T.J. Bole III and W.B. Bondeson (eds.), Rights to Health Care*, 83–102.

*Prima facie* duties or principles are right-making characteristics of actions or practices. These are qualities of actions or practices that tend to make them morally right. Since more than one *prima facie* principle may be relevant to a particular action or practice and they may conflict, it will also be necessary to determine, by relating various *prima facie* principles, what is right overall or what is one's duty proper [20], p. 19). (In the absence of conflicting *prima facie* duties, a *prima facie* duty is, in fact, a duty proper.)

The debate over what constitutes a just health care allocation is made more complicated by the fact that while sometimes justice is taken, as is used in this essay, as one *prima facie* right-making characteristic of an action or practice, in other cases the term is used more loosely. It is used in a broader sense, what Aristotle refers to as "complete" justice [18]. Justice is, in this broader sense, the ethically right distribution on balance, taking into account the full range of *prima facie* principles. It appears that when John Rawls, for example, speaks of justice, he has in mind not a single *prima facie* principle, but something akin to "the right distribution on balance". This is true regardless of the fact that at places he explicitly identifies justice as one among many ethical principles ([19], p. 17). His concept of justice seems to incorporate a complex mixing of concern for liberty, equality, and welfare. As such, he and others who use the term "justice" in this broader sense do not identify clearly what the specific right-making characteristic is that pulls in the direction of more equal distribution, an equality that can be offset for certain counterbalancing reasons. A true egalitarian account of justice will identify a specific *prima facie* principle of justice, which then will have to be related to other, potentially competing principles. In this essay I hope to provide such an account and demonstrate that even when the balancing of principles is complete, the egalitarian principle of justice leaves a major mark on health resource allocation.

## II. THE PREMISES OF EGALITARIANISM

Any ethics of distribution will stand upon a set of fundamental premises. While some people claim that they are offering arguments for their fundamental premises, this will ultimately prove impossible. It is far better for the defenders of egalitarian and anti-egalitarian positions to state their premises boldly so that the points of potential disagreement can be identified. Egalitarianism rests on three basic premises, three assumptions that drive its ethics of distribution:

(1) Human beings are of equal moral worth in the sense that no human deserves a claim to more than or less than an equal share of available resources.
(2) The natural resources of the world should be seen as always having had moral strings attached to their use. They have never been "unowned" and available for appropriation and use without conditions attached.
(3) Human beings have a *prima facie* responsibility as moral agents to use the resources of the world to move society toward a distribution of resources that is more equal.

These three basic beliefs about the moral nature of the world lead to a conclusion that there is a *prima facie* duty to arrange resources so that people have opportunities for equal welfare. These are starting premises. I know of no definitive arguments for any one of them. They are, however, deeply rooted in many religious and secular philosophical world views.

They are, most conspicuously, among the core beliefs of the Judeo-Christian world view. The first premise reflects the view within that tradition that there is an infinite center of value in comparison to which all finites are equal in their finitude. The second premise reflects the view that the earth was created by a Creator God and the resources therein belong to humans only contingently to use according to certain preexisting requirements. The third premise conveys what theologians would call the doctrine of stewardship, that humans are obliged to take positive steps to insure maintenance and restoration of more equal distribution of resources. It would be foolish to deny the important Judeo-Christian historical roots of these premises.

It is equally important to realize, however, that some secular philosophical commentators hold essentially these same starting assumptions. Many philosophers hold that humans are equal in their moral worthiness of an equal share of available resources and that resource distribution should reflect that equality ([3], pp. 171–174; [7], p. 272; [16], p. 34; [24], pp. 43, 48; [25]). A secular thinker can hold that there is an infinite value in comparison to which humans are equal in their finitude. Many secular thinkers, including those committed to ecology as well as egalitarianism, hold that there are prior claims on natural resources that contravene human appropriation without strings attached [4]. Many secular thinkers hold that humans have *prima facie* obligations to take positive steps to provide for the welfare of their fellow humans ([11], Ch. 4). Thus there is nothing peculiarly theological about the starting premises of egalitarianism.

Of course, none of this leads to any definitive *argument* in favor of egalitarianism. There is no reason to presume that one who was not committed to these assumptions would adopt them. If the burden of proof is on egalitarians to establish in any definitive scientific or logical way these three starting premises, they must fail. Before conceding victory to the anti-egalitarians, however, we should see if they are in any stronger position. Any world view, egalitarian, anti-egalitarian, or somewhere in between, requires certain starting assumptions. It is as hard to prove that power or talent justifies distribution as that resources should be distributed equally.

Anti-egalitarians must accept at least one of the premises counter to those cited above. They must accept – on faith – that humans are not of equal moral worth in that at least some persons are entitled to more of the world's resources than others, or they must accept – on faith – that there are no preexisting claims of the community on the natural resources of the world, or they must accept – on faith – that humans have no obligation to use the resources of the world to establish greater equality. Without at least one of these assumptions, the anti-egalitarian position collapses as surely as the egalitarian position does without its assumptions.

If each side needs its assumptions and in principle the assumptions are not subject to logical or empirical proofs, then it will make an enormous difference who bears the burden of proof. It might be argued that whoever would make a positive assumption bears the burden of proof. After all, those who assumed before the space era that the far side of the moon was not made of cream cheese did not feel obliged to prove their position, while they plausibly would hold that defenders of a cream cheese theory would bear the burden of proof.

That probably does not work, however, as a way of settling the burden of proof issue pertaining to the assumptions of the egalitarian and anti-egalitarian camps. One problem is that either side's assumptions can be stated negatively, thus avoiding a burden of proof. The egalitarians might claim there is no evidence for differences in moral status among humans while the anti-egalitarians might claim there is no evidence for equality of moral status.

Moreover, in many cases the positive assumptions that undergird a worldview are so convincing that their deniers bear a burden of proof. Those who believe in the existence of an external reality do not have to prove it; the solipsist, though offering negative assumptions (the denial of external reality), bears the burden.

If the problem were one that lent itself to a pluralist solution, we would simply let each group adopt its own pet assumptions relevant to distribution ethics and be done with it. We would cluster on Engelhardtian islands ([8], p. 132), each of which would function with its own starting premises for a theory of distribution. In fact, however, we cannot retreat to a pluralism of fictional islands. Some health care policy must be adopted for the United States and, eventually, for the world. Decisions must be made upon which to run the Medicare system. DRG weights must be assigned. If I am correct, all those who are inclined toward the Judeo-Christian tradition, and a large group of those who are not but who share its starting assumptions, are willing to accept the view that one of the right-making characteristics of actions is providing people opportunities for equality. In this limited, carefully stated way, egalitarian justice is one among the *prima facie* principles of a normative ethical theory seen as plausible by the majority of the population, and is reasonably included in an ethic of health care allocation when the society must make a forced choice between including it and excluding it. An ethic of allocation that included only the principle of autonomy or only the principle of beneficence would simply be implausible.

For some, justice is a principle that is, in effect, one among several principles of conduct that are derived from an overarching utilitarian ethical theory. They are rule-utilitarians with a principle of justice in their system ([2], p. 214). For others, in a more straight-forward deontological way, justice, as well as promise-keeping, truth-telling, autonomy, and perhaps other principles, are right-making characteristics of actions independent of whether they also happen to be principles that tend to maximize the good ([20], p. 21; [23], pp. 250–287).

What is important to this egalitarian interpretation of the principle of justice is that justice cannot be reduced to some other right-making characteristic. It cannot be, as the act utilitarians would have us believe, that striving for greater equality of net welfare is simply a device for maximizing the aggregate amount of net good. Striving for greater equality of net outcome may often maximize the aggregate good, but it need not. Even when it does not, according to the egalitarian, an action is justice-promoting insofar as it equalizes net welfare. (I add this qualifier because of the various possible interpretations of the principle of justice, i.e., of what it means to distribute *fairly*. Remember Ross's view that justice is the distribution of happiness and misery to the virtuous and wicked respectively.

It cannot be, as the libertarians would have us believe, that the ethics of distribution can be reduced to the principle of autonomy combined with a gargantuan theory of private property such that the right distribution of health care is the one that results when people exercise their autonomy to make deals based on the accidents of the distribution of power and resources in the world [8, 17]. It may be that rational people seeking to create a peaceable community would agree to such liberty as a side-constraint of actions. It is not clear, however, that they would agree to only liberty as a side constraint. Nozickeans are quick to tell us that liberty is *a* side constraint ([8], pp. 123–124; [17], pp. 33–35). They are much slower to tell us what the other side constraints are. Rational people seeking to create a peaceable community might also agree to a principle of equality of net welfare – that is, to a principle of egalitarian justice. They might do so, first, because it might be the only way that rational persons could assure themselves of a peaceable community. Otherwise those who are losers in the natural lottery may well band together against those who were the winners. More fundamentally, even if it could be established that rational self-interested persons would impose only freedom as a side-constraint in order to create a peaceable community, this would not necessarily say anything about what is ethical. The critical question here is whether we are willing to accept the position that whatever rational, self-interested people agree to constitutes an arrangement we would call "ethical". Critics of the Nozickean contract theory in which side constraints are generated by actual contractors reflecting their own special unique positions, interests, power, and status are not willing to concede that mere agreement among actual contractors necessarily results in an ethical system. Using the methods of G.E. Moore's open question test for analyzing the meaning of the term "ethical" [15], it seems meaningful to say "The contractors have agreed on (thus-and-so), but is this contract ethical?", demonstrating that "ethical" does not mean "what the contractors decide" – if it did, the above question would be answered affirmatively by definition. Insofar as contractors want their community to be an ethical one, they would reasonably add constraints to their method such as a veil of ignorance or some other device for incorporating an equal regard for all persons. The result would be the addition of a *prima facie* principle of egalitarian justice in addition to a principle of liberty.

For Nozickean libertarians to establish that an ethics of health care distribution should be based solely on the principle of autonomy (i.e., liberty), they have to establish, first, that liberty is the only side-constraint rational people would choose and, second, that even if it is, such a single-principled society is an ethical one. The alternative system – one that includes an independent principle of justice that requires *prima facie* that persons have an opportunity to equality of net welfare – is much more plausible. It is a principle that is included within the entire history of the Judeo-Christian world view and among a large group of secular thinkers as well. Some notion of the uniqueness and inviolability of the individual and a commitment to equality is also included in every key document of the American heritage: the Declaration of Independence (with its acknowledgement that all are created equal), the Constitution (with its purpose of establishing justice as well as securing the blessings of liberty), the Gettysburg Address (with its celebration of a nation dedicated to the proposition that all are created equal as well as conceived in liberty), even the pledge of allegiance (with its liberty and justice for all). It is not that justice as an independent principle takes precedence over liberty. That is not the issue for the moment. Rather it is that liberty cannot stand alone as the only constraint on human conduct as the libertarians would have us believe. If some set of premises must be selected for establishing a society's health care policy, the overwhelmingly more plausible set of premises is the one that includes the independent egalitarian principle of justice.

## III. THE EGALITARIAN FORMULA

What we have done thus far is introduce the egalitarian premises on the agenda. One among the right-making characteristics of action for an ethically plausible society is an independent principle of justice that requires that we strive toward greater equality. This is especially important for the United States, where it is affirmed explicitly and is therefore important for actual practice in this country. That still leaves us a long way, however, from an egalitarian health care policy.

### *A. The Concept of Equality of Net Welfare*

The first task is to understand more clearly what the concept of equality entails and whether it is viable. The starting point of a concept of equality is equality of moral worth. Egalitarians, however, insist on pushing much further. They see some type of equality of outcome as *prima facie* required. Often that is expressed in terms of equality of net welfare or its equivalent ([1], p. 71; [6]; [7]; [23], p. 268).

It is often pointed out by anti-egalitarians that, if welfare is measured by some subjective consideration such as happiness, then it could be a very difficult task to produce equality of welfare [3, 12]. Moreover, it is not clear why it would be ethically appropriate to do so even if we could. Someone who had very expensive tastes and who was very hard to please would command enormous resources just because he or she was so finicky. It seems unreasonable that justice demands enormous expenditures just to make the finicky and the easily pleased equally happy.

If equality of welfare is the goal of the egalitarian, it surely must be on a more objective basis. Yet it is very controversial whether there are objective standards for determining what contributes to personal welfare. One solution, proposed by Ronald Dworkin among others, is to strive for equal resource consumption rather than equal welfare [6, 7]. As long as we are dealing with resources that people need about equally – resources such as food, clothing, shelter, leisure, recreation, and so forth – and as long as it is reasonable to assume that marginal resources increase peoples' welfare about equally, then equal distribution of resources would at least roughly produce equality of welfare.

One charge raised against egalitarians is that they would produce a world in which everyone was the same ([12], p. 131). These anti-egalitarian critics seem to be making an aesthetic argument here, thinking that sameness would make life boring. If, of course, exactly the same amounts of the same food, clothing, etc. were distributed to each person, sameness would result. Now it is by no means clear that we are obliged to make the world aesthetically pleasing (in this case, interesting), but even if it were, there is no reason why egalitarians would insist on the same for everyone. All that even a radical egalitarian reasonably would strive for is an equivalent amount of welfare for each person. This might be measured by equivalent amounts of resources for each distributed as some generalized medium, such as money, or some more restricted medium such as food stamps or housing vouchers. An egalitarian should not even object to trades from one kind of good to another, assuming that the marginal increase in welfare for each party of the trade is about the same. The result is a reasonable approximation of equal net welfare, but not sameness.

### *B. Equal Outcome Versus Opportunity for Equal Outcome*

There is one additional qualification that egalitarians should be willing to accept. Assume egalitarians are committed to the position that justice is one *prima facie* principle of ethics and that justice requires striving for greater equality of net welfare among individuals. It makes no sense to strive for equality of net welfare (even objective net welfare, if such could be determined) in the special case where an individual is given his fair share and squanders it ([10], p. 127). What, for example, should happen if a person is given a fair allotment of food and chooses to consume it unwisely or waste it? Of course, if the unwise choice were the result of true lack of capacity, of mental inability, such that the person were not substantially autonomous, adjustments would have to be made. What, however, of the person who has the capacity to choose, but chooses foolishly leaving himself or herself with less welfare on balance?

One response might be that the welfare he gains by the unorthodox use of the food compensates for the agony he suffers when he has to go hungry. Even if that is false, however, it seems that all egalitarian justice requires is an opportunity for as much food as others. If that opportunity is squandered and the now starving individual demands more food, surely even egalitarians would recognize no claim (assuming the person was not mentally deranged when he wasted his food). All egalitarians are really committed to is an opportunity for equality of net welfare, not actual accomplishment of equality [22]. Of course, thinking of the distribution problem as one of equal distribution of resources leads to the same conclusion. All that persons are entitled to is an equal share of the resources, even if they do not use their resources in a way that produces equal outcomes.

### *C. Equal Welfare and Equal Health Status*

Thus far the egalitarian position has been sketched to the point where it appears to be committed to the position that one right-making characteristic of a social action or practice is that it provide an opportunity for equality of objective net outcome (probably most easily measured by equality of resources). That, of course, still does not lead us to equality of health status or equality of health resources or a right to the health care needed to be as healthy as others.

It turns out that adding health care to the egalitarian position complicates matters considerably. Especially if we accept the position that persons need have only equivalent welfare, not identical lives, is there any reason why persons should not be permitted to trade some of their health care for other elements of welfare they value more highly? Is there any reason why egalitarians would oppose swaps involving health care any more than they would oppose swaps in other areas?

At least in an ideal, fair world, I do not see any theoretical reason to oppose such swaps. At the practical level, especially in a less than perfectly just world, there may be very good reasons. If the goal is equality of outcome, giving persons equal amounts of food stamps, housing, clothing, etc., and then letting them swap would probably lead to an end state where people are better off but nevertheless still about equal. If health care is considered, however, problems are much more complex. People start out with health statuses that are very unequal. The only way for the egalitarian to take this into account is to distribute health care on the basis of need rather than equality.

If, however, care (or a certificate entitling one to care) is distributed on the basis of need and people are then permitted to trade for other goods until Pareto optimality (i.e., the position where further trades no longer improve the welfare of each party) is reached, we will be left in a world where people will have very unequal health statuses. This kind of inequality will, of course, be just, since the less healthy persons will have had their opportunities for health, but purposely traded them away. The problem this creates is that it will quickly become extremely difficult to distinguish between the unhealthy persons who are justly unhealthy (because they have received a certificate for health care and traded it away) and others who are unhealthy and have not yet been compensated. An elaborate bookkeeping system would be required, tracking persons throughout their lives, to determine whether they have received the just amount of care. Since health care interventions are normally probabilistic and justice would require giving a second round of care to persons who have poor health status after a first round of care, the person who had traded away his entitlement might even argue that he still has some care coming since he might have been a loser in the lottery in the first round of care.

A straightforward solution to this problem is simply to differentiate health care from other social goods. While most goods – such as food, clothing, and shelter – could be handled by a monetary allotment and trades could be permitted, it would be far more practical to provide health care out of a separate resource pool with persons getting the care (or insurance for the care) they need to have an opportunity to be as healthy as others [5].

In such an egalitarian world, everyone would get an equal entitlement. It might come in the form of an insurance policy with a specified range of coverage (say a $2000 per person per year package with prohibitions on or compensations for any special illness-based policy limits). Certain differences in coverage might even be tolerated, as is suggested by Enthoven and others advocating "consumer choice plans" [9]. As long as everyone had benefits priced at $2000 and received coverage specified in the plan, somewhat different treatment plans would exist, but everyone would have an equal opportunity to be healthy.

Conceived of in terms of equality of resource distribution, a slight adjustment would have to be made. We might think of all persons receiving social resources (money, services, etc.) as well as biological resources (physical and mental abilities). The principle of justice would then *prima facie* press society toward a policy in which everyone would then receive an equal share of the social resources to cover their needs and desires outside the medical sphere plus an insurance policy to cover their health care needs. Those who were winners in the natural lottery and thereby received a large share of the biological resources would, by that very fact, consume less of the resources in the pool for health insurance. Considering social and natural resources, everyone would come out equal.

Thus, for strictly pragmatic administrative reasons we might want to differentiate the equal treatment in the health sphere from equal treatment in other spheres. An analysis of resources for education would reveal that problems exist there that are quite similar to those in health care. Different people receive vastly different amounts of natural resources (mental abilities). Those who receive abundantly will need far fewer social resources to achieve opportunities for equal outcome than those who are underendowed. For similar reasons, therefore, people ought to receive the education they need to give them an opportunity of function as well in society as others, just as they receive health care needed to give them an opportunity to be as healthy as others.

These pragmatic arguments for differentiating health care and education would seem to apply even in an ideal world where resources are distributed fairly. In the real world, where people have vastly different amounts of resources for historical reasons, there is another reason why some might insist that health care be distributed so as to give people equal opportunity to be as healthy as others. It is often pointed out that in such a world, if least well off, starving, homeless, persons were given a chance to trade their health care certificates for other goods, they would be pressed (some say coerced) to sell, thus giving up health care they desperately need ([14], p. 74). This is often given as an argument against permitting trades.

I agree that trades should not be permitted in these circumstances, but the argument is much more complex than it appears. The defenders of trades in an unjust world acknowledge that the poverty-stricken would be forced to sell their health care. They defend this result, however, on the grounds that if the poverty-stricken are very poorly off, they should at least be able to improve their lot by trading something they desire less (health care certificates) for something they desire more (food or shelter) ([14], pp. 73–75). Even if they trade for what appears to be less important goods (cars or recreation), they must at least be better off given their priorities (or else why would they have traded?), assuming they are not mentally deficient or psychologically unstable. Anyone who values autonomy or justice or both, so the argument goes, should look favorably upon a social arrangement whereby the least well off can improve their position, even if it means getting little or no health care.

The problem is a real one. In some worlds, perhaps in our own, it may be possible to gain something approaching a fair allocation of health care for the poor, even though nothing approaching welfare support in other areas such as food and housing is possible. Since commonly marginal utility decreases, it stands to reason that many of the poor could stand to improve their overall welfare if they traded some of their relatively good health care coverage for some food or housing that they need desperately. A principle of autonomy would surely support such a policy of permitting trades. Moreover, if justice demands improving the lot of the least well off or making them more equal to others in net welfare, then trades would seem to be required by justice as well.

I find this a powerful argument, one that even egalitarians should consider seriously – assuming that the underlying injustice of the world is taken as a given. That is the problem, though. It requires conceding defeat on the broader social problem of injustice in society. It requires the creation of social institutions that acknowledge that injustice is part of the world in which we live.

People differ fundamentally on the extent to which they believe social practices should be adopted that require such concessions. Some people say that since moral evil is inevitable, social institutions should reflect that fact. They develop ethics of a "just war", ideal prisons, and policies that tolerate unequal health care. Others are more idealistic. They refuse to participate in institutions that require conceding defeat on their ideals. They become pacifists, refuse to carry guns, etc. They insist on living the proper ethic for the ideal at least in those areas where they can. Ernst Troeltsch distinguishes between what he calls absolute and relative natural law to describe holders of the two positions [21]. Holders of the relative position attempt to create moral laws acknowledging the evil of finite humans. Holders of the absolute position insist on putting into practice where they can moral laws that would exist in an ideal world. These people would be reluctant to permit people to trade off their health care entitlement if, in an ideal world, they would not do so. The assumption here is that, since the poor would be getting something approaching the share of health care they would get in a just world, we should not create an institutionalized practice of trades that would leave them with less health care. Rather, energies should be spent on getting more acceptable levels of support in the other spheres. They would insist that justice requires that people get the just amount of health care even if it means they are somewhat less well off on balance. This would leave the world glaring at the unfairness in the other areas of life and perhaps hasten the day when the unfairness in these other areas will be corrected, whereas if persons were permitted to trade their health care for food, some of the pressure might be taken off the campaign for justice in the allocation of food resources.

While some idealists adopt this position because they hope that their example will drive society to make the improvements needed for social practices to be more fully ethical, others adopt this stance simply as a matter of conscience. They might concede that their example is unlikely to change the practice – that their example of pacifism or vegetarianism will not reform the world. They might say that even without the change they would desire, their consciences force them to favor the ethical behavior that would accord with the ideal, at least in the spheres over which they have control. In Kantian style, they would be acting such that they could will that their "maxim" (rule of conduct) be adopted universally.

Where one comes out on this matter will depend on how much he or she believes the real world should be shaped by the ideal. If one is an idealist, trades would not be permitted in the real world, for this reason as well as for the more practical reasons addressed above. For both kinds of reasons, health care would be differentiated from other social goods and would be justly distributed when persons are given an opportunity to be as healthy as others.

## IV. *PRIMA FACIE* EQUALITY AND ACTUAL EQUALITY

For an egalitarian, this is what justice requires: opportunities for equal net welfare (best measured by equal resource consumption). For practical and perhaps theoretical reasons as well, health care would be differentiated out so that justice would require giving persons an equal opportunity to be as healthy as others.

That is what justice requires for the egalitarian, but is that what is ultimately the morally right course? I have consistently maintained that egalitarianism is the position that there is an independent *prima facie* principle of justice that requires opportunities for equality of objective net welfare. Any egalitarian I know also maintains, however, that there are other *prima facie* principles as well. What one's actual duty or duty proper is will depend upon how one relates the demands of the principle of justice with the other ethical principles. Among standard lists of *prima facie* principles are the principles of autonomy, truth-telling, promise-keeping, and avoiding killing, as well as the principles oriented to consequences: beneficence and non-maleficence ([2]; [20], p. 21; [23]). What should happen, for example, if someone holds that there is a duty to keep promises and health care services have been promised to some persons beyond those that would be needed to give them an opportunity to be as healthy as other persons? The government, for example, may have promised persons that as part of Medicare they are entitled to end-stage renal disease treatment, even though some such persons are not as poorly off medically as others suffering from other diseases for which there is no Medicare coverage. Justice would appear to require diverting some care to others in greater need while promise-keeping would require opposing the shift. What one ought to do in the final analysis will depend upon how promise-keeping and justice are related.

Efforts have been made to rank-order the principles. That project, most would agree, has not been successful. There is no absolute priority for justice or promise-keeping or autonomy, in spite of certain efforts to identify ethical side-constraints such as liberty. While liberty is one side-constraint, it is not the only one. If there are other side constraints as well and they sometimes come into conflict, what can one do except balance the competing claims and side with the principle that appears to be more weighty at the moment ([20], p. 41)? That actually concedes less to the egalitarian's critics than it might appear, as we shall see momentarily.

While there seems to be no possibility of giving absolute priority (lexical ordering) to one deontological principle (or side constraint) over others, it may be possible to establish a priority in relating these deontological principles to the principles that focus only on maximizing aggregate consequences: the principles of beneficence and non-maleficence. If beneficence and non-maleficence can be balanced off against autonomy or truth-telling or justice, then it is always possible, at least in theory, that enough good consequences will justify sacrificing the principles of autonomy, truth-telling, etc. This may happen for both act- and rule-utilitarians. In principle, slavery would be justified if only enough good would come of a particular instance of it, not only to outweigh the bad consequences to the enslaved (and to others), but also to outweigh the violation of the principle of autonomy. A rule against slavery would have to give way to a rule that permitted certain classes of utility-maximizing exceptions. The research of the Nazis would have been justified if only they had designed experiments really capable of producing enough good for enough people, even though a smaller group were done immeasurable harm and their autonomy was violated.

In order to avoid this seemingly unacceptable conclusion, it is necessary to give the deontological principles – autonomy, promise-keeping, truth-telling, avoiding killing, justice, etc. – joint priority over beneficence and non-maleficence while they are balanced among themselves (cf. [19], pp. 40–45; [23], pp. 291–305). Beneficence and non-maleficence would then come into play only when the other principles are fully satisfied or when they conflict in such a way that there is a tie. On this basis it can be said that rights (derived from the deontological principles) trump mere consideration of consequences.

Still, it might be argued that this absolute priority is too strong –

clearly, we should not honor a promise of small importance to the promisee when only breaking it could, for example, save the lives of many. (Assume in this case the promisee cannot be contacted in order for him to release the promise). But while it seems we should break the promise, this does not argue for a less stringent priority of deontological principles over beneficence and non-maleficence. *Justice* demands that we break the promise, because, if we do not, those who would die as a result would be made much worse-off and, therefore, would no longer have opportunities for equal net welfare. Similar objections may be met in this fashion.

## V. THE INFINITE DEMAND PROBLEM

This leaves two serious problems for the egalitarian. First, if autonomy is reintroduced at least as a co-equal principle, does not the egalitarian give away so much that the resulting health policy would not really give opportunities for equal health status at all? Second, if justice trumps beneficence, would not the entire system collapse at the point where we come up against the least well off group of persons who have an incurable disease, a disease which would create an infinite demand on the system ([10], p. 128)? The two questions can be addressed simultaneously.

The problems may not be as severe for the egalitarian as it at first appears. There are at least three, perhaps four, ways out of the infinite demand problem. First, although it appears that justice would require giving all of society's resources to the worst off group that has an incurable disease, that is really not the case. If all of society's resources were diverted to this group, others' welfare and health status would decline dragging them down toward the least well off group. As that group gains resources and others lose resources, the lines could actually cross, so that some other group becomes least well off. For example, it is not clear that we would have to divert polio immunizations away from the presently healthy in order to do research to "cure" blindness. Surely the blind are not as healthy as others and egalitarian justice would require diverting some resources to try to help them, but if enough resources were diverted so that it cut into the healthy persons' supply of polio vaccine, the healthy would be medically worse off. In fact, if enough doses were diverted, polio could become rampant and all would be at serious risk. At some point the previously healthy would be at a high risk and might actually be worse off than the group of blind persons. Justice would then require diverting resources from the better off blind persons in order to benefit the now least well off persons at high risk for polio. Justice itself sets its own limits to the infinite demand problem.

Second, other deontological principles also set limits, although not severe ones. Autonomy, for example, would permit persons to use their own private resources as they see fit. This ethical claim would have to be set off against those who want to create a stiff tax to benefit the unhealthy. Since the two principles can be traded off against each other, some compromise would result.

The reason that that does not concede too much to the libertarians is that, first, autonomy is balanced against justice. More importantly, however, the principle of liberty only permits persons to dispose of goods they possess justly. Even Nozick concedes this much ([17], p. 151). If, however, one of the core assumptions of our society is and always has been that there never has been any "unowned resources" out there to be possessed as a starting point of purely private property, then conceding autonomy concedes much less than the libertarians have believed [4]. It might give persons the right to choose their own life styles, prolonging medical treatment – but it does not give people the right to dispose of as they see fit the resources for which they are society's temporary custodians. Autonomy or liberty, it turns out, is not much of a check against egalitarian interpretations of the principle of justice.

There is one way in which the principle of autonomy does set an important limit on the infinite demand problem. Rawls's second principle supports practices that permit inequalities provided those inequalities redound to the benefit of the least well off groups ([19], p. 302). While Rawls is often thought of as offering an egalitarian critique of utilitarianism, his position is actually much less egalitarian than that of a full-fledged egalitarian. After all, he explicitly claims that justice requires inequalities in a special set of circumstances.

In contrast with Rawls, real egalitarians hold that justice is a *prima facie* principle that pushes practices in the direction of greater equality – exactly the opposite of what Rawls's principle of justice requires in the special case where inequalities improve the lot of the least well off. These real egalitarians would say that *justice* requires moving toward equality of net welfare rather than moving away from it, even when the least well off would benefit by inequality.

If egalitarians also recognize a *prima facie* principle of autonomy, however, they would permit those who have a claim of justice to waive that claim when they wanted to. One plausible instance would be a circumstance when the net welfare of the least well off would be increased by waiving their claim based on justice. If, in cases when the least well off would significantly improve their lot, they waive their justice-based right to health care, then the result would look like a Rawlsian distribution. The egalitarian's ethical account of the situation, however, would be very different from the Rawlsian's. While the Rawlsian says that justice requires the inequality, the egalitarian would say that this is a situation where justice is reasonably sacrificed in order to act on some other principle – the principle of autonomy.

As a test to see if one is an egalitarian or a Rawlsian, one should ask whether it makes any moral difference whether it is the least well off or the elite who are making the argument that equality should be waived in order to benefit the least well off. Under Rawls's analysis it makes no difference (since justice itself requires inequality regardless of who favors the inequality). Under the true egalitarian account of justice it makes a critical difference. Since justice generates rights claims for the least well off, it is only the least well off who have the autonomy-based right to waive their claims. The elites arguing for high salaries or special health care for the elite are, in principle, given no weight. The Reagan administration's argument for trickle-down theory as a way of benefitting the poor by rewarding the rich is ruled out in principle as well as on possible empirical grounds. For a real Rawlsian, the trickle-down position must be defeated on the empirical ground that rewards for the elite do not trickle down or that other practices would yield greater benefits for the least well off. For the egalitarian, the least well off are the ones who should be permitted to consent to inequalities. In such cases justice is sacrificed for the principle of autonomy upon which the consent of the least well off is based.

Promise-keeping is another deontological principle that could pose a serious limit to the infinite demand problem. If others had been promised health care, then promise-keeping would provide a check against the infinite demand. If, however, promises have not been made foolishly, then even promise-keeping will not provide a serious problem for the egalitarian. Thus justice itself as well as the other deontological principles (autonomy and promise-keeping) provide checks against the infinite demand problem, but not checks that are so serious that the implications of egalitarian interpretations of the principle of justice are completely defeated.

If these deontological principles do not provide enough control on the infinite demand problem (without completely overwhelming the demands of justice), then one may have to reconsider the lexical ordering of the deontological principles over the principles of beneficence and non-maleficence. We could then easily provide checks against the infinite demand of egalitarianism, but in doing so we would concede a great deal. We would surrender not only the absolute priority of justice over utility, but the priority of autonomy and the other deontological principles as well. Utilitarians would apparently find this acceptable, but it is a compromise neither the egalitarians nor the libertarians nor anyone standing in the Judeo-Christian or secular liberal tradition should be willing to make.

*Kennedy Institute of Ethics*
*Georgetown University*
*Washington, D.C., U.S.A.*

## BIBLIOGRAPHY

1. Ake, C.: 1975, 'Justice as Equality', *Philosophy and Public Affairs* **5** (1), 69–89.
2. Beauchamp, T.L. and Childress, J.F.: 1983, *Principles of Biomedical Ethics*, Second edition, Oxford University Press, New York.
3. Bedau, H.A.: 1971, 'Radical Egalitarian', in H.A. Bedau (ed.), *Justice and Equality*, Englewood Cliffs, New Jersey, pp. 168–180.
4. Brody, B.A.: 1981, 'Health Care for the Haves and Have-Nots: Toward a Just Basis of Distribution', in E.E. Shelp (ed.), *Justice and Health Care*, D. Reidel, Dordrecht, Holland, pp. 151–159.
5. Daniels, N.: 1979, 'Rights to Health Care and Distributive Justice: Programmatic Worries', *Journal of Medicine and Philosophy* **4**, 174–191.
6. Dworkin, R.: 1981, 'What is Equality? Part 1: Equality of Welfare', *Philosophy and Public Affairs* **10**, 185–246.
7. Dworkin, R.: 1981, 'What is Equality? Part 2: Equality of Resources', *Philosophy and Public Affairs* **10**, 283–345.
8. Engelhardt, H.T. Jr.: 1981, 'Health Care Allocations: Responses to the Unjust, the Unfortunate, and the Undesirable', in E.E. Shelp (ed.), *Justice and Health Care*, D. Reidel, Dordrecht, Holland, pp. 121–137.
9. Enthoven, A.C.: 1978, 'Inflation and Inequity in Health Care Today: Alternatives for Cost Control and an Analysis of Proposals for National Health Insurance', *New England Journal of Medicine* **298** (12), 650–658.
10. Fried, C.: 1978, *Right and Wrong*, Harvard University Press, Cambridge, Massachusetts.
11. Gewirth, A.: 1978, *Reason and Morality*, University of Chicago Press, Chicago.

12. Lucas, J.R.: 1971, 'Against Equality', in H.A. Bedau (ed.), *Justice and Equality*, Prentice Hall, Inc., Englewood Cliffs, New Jersey, pp. 138–151.
13. Macklin, R.: 1976, 'Moral Concerns and Appeals to Rights and Duties', *Hastings Center Report* **6** (5), 31–38.
14. Menzel, P.: 1983, *Medical Costs, Moral Choices: A Philosophy of Health Care Economics in America*, Yale University Press, New Haven, Connecticut.
15. Moore, G.E.: 1903, *Principia Ethica*, Cambridge University Press, Cambridge, England.
16. Nagel, T.: 1973, 'Equal Treatment and Compensatory Discrimination', *Philosophy and Public Affairs* **2**, 348–363.
17. Nozick, R.: 1974, *Anarchy, State, and Utopia*, Basic Books, New York.
18. Ostwald, M. (trans.): 1962, *Aristotle: Nicomachean Ethics*. Bobbs-Merrill Company, Inc., Indianapolis.
19. Rawls, J.: 1971, *A Theory of Justice*, Harvard University Press, Cambridge, Massachusetts.
20. Ross, W.D.: 1939, *The Right and the Good*, Oxford University Press, Oxford.
21. Troeltsch, E.:1925, 'Das stoische-christliche Naturrecht und die moderne profane Naturrecht', in *Gesammelte Schriften* Band IV, Verlag J.C.B. Mohr (Paul Siebeck), Tuebingen, pp. 166–191.
22. Veatch, R.M.: 1980, 'Voluntary Risks to Health: The Ethical Issues', *Journal of American Medical Association* **243**, 50–55.
23. Veatch, R.M.: 1981, *A Theory of Medical Ethics*, Basic Books, New York.
24. Vlastos, G.: 1962, 'Justice and Equality', in R.B. Brandt (ed.), *Social Justice*, Prentice Hall, Englewood Cliffs, New Jersey, pp. 31–72.
25. Williams, B.A.O.: 1971, 'The Idea of Equality', in H.A. Bedau (ed.), *Justice and Equality*, Prentice Hall, Englewood Cliffs, New Jersey, pp. 116–137.

H. TRISTRAM ENGELHARDT, JR.

# RIGHTS TO HEALTH CARE: CREATED, NOT DISCOVERED

As Tom Beauchamp acknowledges, in many philosophical disputes arguments can be suggestive even if they are not conclusive [1]. The result is that one must often rely on fair negotiations to shape the character of public policy rather than on some method for discovering what public policy should be. Still, suggestive arguments can guide negotiations even where they cannot determine conclusively what the results should be. The more this is so, the more rights to health care are created rather than discovered. In this fashion, Beauchamp introduces one of the major questions regarding justice in health care: is justice to be understood through a process for fashioning rights to health care or through a body of defended doctrines which establish a pattern of proper health care distributions?

Robert Veatch aids us by offering an account of how to discover an answer. As such, it is an instructive example of moral reasoning – even where it may fail. To begin with, Robert Veatch has provided a very forthright and very helpful exposition of an egalitarian approach to resource allocations for health care. The very clarity of the exposition gives ground for embracing a non-egalitarian public policy regarding health care. As Veatch himself notes, there is "no definitive, rational defense of this *or any other* interpretation of the principle of justice" ([6], p. 83). However, in order with moral justification to employ coercive force to impose a particular view of justice, one will need both (1) to establish the moral correctness of a particular view of justice and (2) to establish moral authority for the use of force in its imposition. Veatch incorrectly assumes that the question is simply a choice between two or more views of justice, one of which will be endorsed, and that therefore one is authorized to accept the most likely candidate, presuming one can give some general account that will distinguish more likely from less likely candidates as the basis for a coercive state policy of allocating health care resources.

Veatch also suggests that a 'libertarian' approach is simply one among various competing positive accounts: one valuing freedom, the others equality, fair opportunity, utility, etc. Here, however, is where the



*T.J. Bole III and W.B. Bondeson (eds.), Rights to Health Care*, 103–111.

distinction between freedom as a value versus freedom as a side constraint does some of its best service ([5], pp. 30–34). One is moved to a libertarian position not so much because of a successful argument on behalf of valuing liberty or private property, but because, where answers are unclear, individuals are free to choose on their own. Which is to say, (1) the more one despairs of the establishment of a particular view of justice and/or (2) the more one doubts the authority to use force to impose one's particular view of the allocation of resources upon unconsenting innocents, the less plausible it becomes that one has the moral warrant to achieve by force such an encompassing endeavor as an egalitarian health care system, and the more one must acquiesce in the free choices of physicians, patients, for-profit hospitals, not-for-profit hospitals, etc. Note that all the non-egalitarians need to establish their position are (1) a minimal area of personal freedom and (2) a minimal amount of private property and services. People will be free to exchange their property for the services of others, leading to non-egalitarian resource allocations. Another way to put this is that the more one despairs of discovering the correct pattern of justice, the more one must create it through peaceable negotiation, where the source of authority will be the consent of those involved.

Insofar as there is skepticism regarding the ability to discover the morally canonical character of such allocations and/or the ability to derive moral warrants to impose such allocations by force, one must derive authority for common actions from negotiation and consent. If one is interested in resolving issues in a non-arbitrary manner that does not rest on force, such negotiation must be actual negotiation, not simply an appeal to a mythical original contract or covenant. In such circumstances, the right not to be used without one's consent becomes the cardinal negative right, one which does not depend on a particular concrete view of justice or the good life, nor on some concrete view of positive rights. If one cannot discover what one ought to do through conclusive rational arguments, the only alternative for resolving disputes when moral strangers meet (i.e., individuals who do not share the same view of moral life or justice) other than simply on the basis of force will be mutual consent and agreement. Respect for the wishes of others (i.e., gaining their consent) becomes *the* side constraint, because it becomes *the* source of moral authority, if moral authority cannot be derived conclusively from reason or from a general provision of the grace of God. Requiring the consent of others before they are used will not ensure the establishment of a peaceable community, but it will provide the one source of authority for a society that spans individuals with different moral viewpoints and who may belong to different moral communities. This is not the place to discuss the extent to which general consent can be presumed [2], except to note that none such can be presumed with regard to any general, all-inclusive, egalitarian health care system. However, it must be noted that limited democracies have been endorsed, not so much out of endorsing liberty as a value, but out of recognizing the consent of the governed as the sole source of political authority, given the weakness of reason and the silence of God.

These points can be rephrased. The minimal notion of the moral community as one that is based on mutual respect not force requires the consent of others before coercing their services unless and until there is a successful moral argument authorizing such force. The 'libertarian' position is the fall-back position when one cannot clearly establish an end-state theory of justice and/or the authority to impose it by force. Because I agree with Beauchamp and Veatch that there are no clearly convincing arguments to establish a particular end-state account of justice (including one that pursues liberty as a value), I conclude that both Beauchamp and Veatch's arguments must fail. The non-egalitarian position is secured on none of the so-called "faith" claims to which Veatch refers, but rather on the basis of a minimal skepticism regarding the capacities of reason to establish a morally canonical pattern for the distribution of resources or the authority to impose such a pattern. Does the 'libertarian' bear the burden of proof, when the 'libertarian' account is not justified in terms of some overriding value given to liberty, freedom, or autonomy? The burden of proof must be shouldered by the individuals or communities using coercive force on behalf of a particular view of justice or the good life. Those who endorse a 'libertarian' view based on a skepticism regarding the capacity of reason to establish a proper way of distributing resources and/or the authority to realize a particular pattern of distributions have no such burden to bear. They can turn instead to mutual consent as the source of joint authority for common actions, despairing of general rational arguments that purport to establish a particular moral viewpoint and the moral authority coercively to establish it. The force user, the one who imposes coercively a particular moral viewpoint, is the one who has the burden of proof.

There are other difficulties with Veatch's arguments. First, Veatch confuses deserving or meriting something with having title to it. One may very well own things which are not deserved. For example, I may develop a computer program and share it only with close friends. Or I may give it to a total stranger. Those individuals may not have merited it. But once I have given it to them, they have title to it and this possession may advantage them. Indeed, just because someone owns more than others, and others may have greater merit, it does not follow that one does not have just entitlement to one's own possessions.

Second, Veatch does not make out his first premise. He does not show that equal moral worth requires equal possession of wealth or resources. Equal moral worth requires, rather, that persons should be equally the object of respect through forebearance from unconsented interferences. More than that will require justifying a particular canonical concrete account of justice and the moral life, and no one has accomplished that.

Third, Veatch suggests that non-egalitarians must presume that all property is privately owned. However, they need only show that some private property exists. A non-egalitarian view is established if one compromises in uncertainty and holds that not all property is either private or communal possession. Democratic capitalism may very well be an example of an attractive compromise between two extreme positions: those who hold that all property is private and those who hold that all property is communal.

This leads to the fourth difficulty with Veatch's argument. He presupposes that if property is not privately owned, it is then owned by particular societies or particular governments. But particular societies will have as much problem establishing their title to goods as will particular individuals. Skepticism regarding claims to private property or governmental property claims may lead to charging a worldwide rent on the use of natural resources on the premise that land and material resources belong to all. However, absent a clearly established theory of justice and a clearly established warrant for coercive imposition of that theory of justice by force, the proceeds of that rent would need to be paid equally to all in as fungible a form as possible ([2], pp. 127–135). This, of course, is not what Robert Veatch envisages.

A part of the difficulty with egalitarianism lies in the assumption that one should or may impose a tidy order on the affairs of men and women within societies organized as states. The more one is skeptical regarding conclusive arguments in these areas, the more one must allow individuals to contract freely for their health care. One moves from discovering the proper pattern for health care distributions to the problem of creating patterns for health care distributions. Moreover, since the desires of different individuals are likely to be different, one is brought to accepting numerous forms of insurance against losses at the natural and social lotteries, with different scopes of payment, unhindered by encompassing price controls such as would exist within all-payer systems. The price of recognizing that one does not have *the* moral arguments to establish *the* theory of justice, and/or *sufficient* moral authority to impose one's own views of justice by force, is acquiescence in, or toleration of, a diversity of individual choices.

A final set of points regarding Veatch's appeal to the Judaeo-Christian and American traditions is in order. First, the monotheistic metaphor of the Judaeo-Christian tradition may entrap one in seeking a single solution for societies that encompass people with divergent visions of the good life. A polytheistic metaphor may be more apt in suggesting that we should tolerate the worship of many gods and goddesses – i.e., tolerate many approaches to deploying resources for health care. A single uniform encompassing health care system suggests an outcome that would be both tyrannical and stultifying, where all would be compelled to embrace a single view of justice. In any event, islands for pluralism are possible insofar as individuals are free to fashion and choose different health insurance systems.

These considerations lead to another observation regarding religious images and traditions: originally the Christians avoided the use of force in coercing others to comply with their Christian viewpoint. Surely this was forgotten with the persecution of heretics and pagans in the fourth century. But the commitment to force and violence in the imposition of a world view is not clearly part of the roots of Christianity. The distribution of one's resources to the poor or the holding of goods in common was not imposed by force on unconsenting non-Christians, at least not by the early Christians (*Acts* 2:45).

Also, regarding American notions of equality and justice, neither the American Declaration of Independence nor the compact styled the Constitution of the United States of America contains a statement that equality in rights to forbearance entails equality in the possession of goods and/or welfare. Rather, there is in the American tradition a commitment to equal protection against the unconsented intrusions of others, including the intrusions of the government.[1] The American experience was formed in terms of the limits of reason and the limits of social authority. There was a clearly stated concern to be protected from the tyranny of majorities[2] and to insure that private property would not be taken without compensation.[3] The American Constitution with the Bill of Rights was created precisely to protect material inequalities peaceably achieved by giving equal rights to the forebearance of others.

Beauchamp draws from the uncertainty of argument in these areas the conclusion that a diversity of structures may be created by negotiation, albeit guided by suggestive philosophical arguments. The fact that the arguments are usually suggestive, not conclusive, leads to a basis for tolerating numerous conflicting outcomes and the recognition that answers will be as much created by fair negotiation as discovered. My differences with Beauchamp's views concern (1) the limits on the powers of the sovereign, (2) the character of just taxation, and therefore, (3) the likelihood that governments could with moral right be *the* prime contenders to effect encompassing schemes for the allocation of health care resources. Further, even if there is a strong basis for the social causation of diseases, it does not follow that taxation is the way to effect compensation. Instead, one must determine which actual individuals or groups are responsible for which actual damages. But even when one can determine which actual individuals have caused which actual damages, it does not follow that the state is obliged to be forthcoming with resources in order to make whole those who have been injured. That obligation falls to those who are responsible for the injury. The obligation that corresponds with a forbearance right falls on those who fail to forbear. The violation of negative rights by third parties does not create positive rights to either police protection or aid in achieving goals of retributive or distributive justice. It is for reasons such as these that Robert Nozick does not attempt to ground the state's provision of police protection in claims to a basic right to equal, adequate, or minimally decent levels of police protection.

A clarification is in order here insofar as Beauchamp presumes that, since Nozick's libertarian theory allows a tax for police protection, it will have difficulty in disallowing a tax for welfare, since both require identifying a hierarchy of costs and values, and the authority to impose it. One should note that in *Anarchy, State and Utopia* Nozick develops a justification for redistribution payments for police protection as a compensation owed to people who are forbidden to take justice into their own hands, because they would be unreliable enforcers of justice ([5], pp. 101–119). The redistribution is not in virtue of a right to beneficence, but is in virtue of a compensation due to unreliable enforcers of justice because they are not able to exact justice for harms done to them. On the one hand, they do not have a right to be unreliable enforcers of justice, yet on the other hand, they have a right to the justice and protection they could secure themselves, were they left to their own devices. Nozick thus seeks to secure a redistributive scheme for police protection that is not based on redistributive goals. Put somewhat Procrusteanly, since individual enforcement of justice (one would presume outside of individual actions in immediate defense of life, health, integrity, and property) is less reliable than what is provided via a societal police system, Nozick believes he has a justification for the state's monopoly on force and the consequent need for the provision of police protection in a redistributive fashion.

Whether these arguments of Nozick succeed or not, one can imagine other schemes for the financing of police protection. For instance, wealthy individuals and large corporations will have a major interest in law and order and therefore in funding both police and the military. Even if individuals are not provided full police protection when they fail to pay their monthly police utility bill in the libertarian utopia, all, even they, will benefit from the general interest of most in securing police services. Those who are interested in pursuing this public good will indirectly aid all by the level of law and order they secure. Moreover, insofar as states and communities have services to sell, they can accumulate resources to use in defending themselves and providing police protection for the indigent. In the end, defense on as large a scale as is presently undertaken may still be quite feasible within libertarian constraints. As David Friedman has pointed out in his study of Iceland from 930 to the 13th century, a community can function quite efficiently in the absence of the public enforcement of justice [3].

The extent to which a society may be able to provide publicly supported rights to police services or health care will depend on the extent to which it owns resources (at least in the absence of special arguments such as Nozick's). As Beauchamp acknowledges, the social creation of wealth does not in and of itself create societally owned wealth, absent either (1) strong special arguments or (2) actual contracts among actual groups of individuals. Individuals may, because of the advantages of common labor, work together in various cooperative endeavors leading to their mutual enrichment without any moral claims arising to each other's property. But, actual claims on actual individuals require actual bases for showing that a true debt exists in more than a metaphorical sense, or in a sense that requires more than a polite thanks. Further, the plausibility of claims to a basic right to health care goes aground on a skepticism regarding a canonical ranking of basic social desiderata. If one does not know how one must canonically rank liberty, equality, prosperity, and security, one will not be able to give an account of basic rights to beneficence, or to goods and welfare. One will be left with the task of creating, not discovering, rights to both police and health care. Such politically fashioned rights will not rest on basic human rights, but on decisions of particular groups regarding the employment of their communal resources.

My response to the papers by Veatch and Beauchamp is more a *via negativa*, appealing to the limits of reason, than an attempt to justify a particular program for the allocation of scarce resources. It is as important to acknowledge the things we do not know, as to underscore the things we do. One might recall that this is how Socrates understood the statement of the god at Delphi that he was the wisest man in Greece. Socrates did not think he knew what in fact he did not know (*Apology* 21d). There is much to be learned from recognizing the limits of our capacities to establish particular views of the good life and to impose them with moral authority on unconsenting individuals [5]. These limits lead to conclusions regarding the kinds of societies we have the moral right to organize through coercive force. They also lead to a special distinction between positive duties, which are founded on a particular ranking of social desiderata, on a particular moral sense, and negative duties as the duty not to use others without their permission, because persons through their consent are the one legitimate source of moral authority when arguments to establish particular theories of justice or views of the good life fail. Here negative duties are not just positive duties restated.

Even where their papers may not succeed in establishing their positions, Beauchamp and Veatch by their clear and insightful articulation of their views help us to see better the limited nature of the human condition, where the limits are not just those of finite resources, but limits to the authority to use force in the pursuit of even the most important and alluring goals.

*Baylor College of Medicine*
*Houston, Texas*

## NOTES

[1] One might think of the often-quoted dissent by Justice Brandeis, which has been taken as one of the eloquent arguments for individual rights to privacy. "The makers of our Constitution... ought to protect Americans in their beliefs, their thoughts, their emotions and their sensations. They conferred, as against the Government, the right to be let alone – the most comprehensive of rights and the right most valued by civilized men." *Olmstead v. United States*, 277 U.S. 438, 478 (1928) (Brandeis, J., dissenting).
[2] There are many reflections going back to the very early history of the United States focusing on concerns about the tyranny of the majority. One might consider the remarks by James Madison in *The Federalist*, No. 51, where he argued that a republic must be so structured that a society is broken into many parts so there will be "little danger from interested combinations of the majority."
[3] The American concern to protect property rights is substantial and pervasive and enshrined, *inter alia*, in the Fifth Amendment of the Bill of Rights.

## BIBLIOGRAPHY

1. Beauchamp, T.L.: 1991, 'The Right to Health Care in a Capitalistic Democracy', in this volume, pp. 53–81.
2. Engelhardt, H.T.: 1986, *The Foundations of Bioethics*, Oxford University Press, New York.
3. Friedman, D.: 1979, 'Private Creation and Enforcement of Law: A Historical Case', *The Journal of Legal Studies* **8**, 399–415.
4. Friedman, D.: 1984, 'Efficient Institutions for the Private Enforcement of Law', *The Journal of Legal Studies* **13**, 379–397.
5. Nozick, R.: 1974, *Anarchy, State, and Utopia*, Basic Books, New York.
6. Veatch, R.M.: 1991, 'Justice and the Right to Health Care: An Egalitarian Account', in this volume, pp. 83–102.

BARUCH A. BRODY

# WHY THE RIGHT TO HEALTH CARE IS NOT A USEFUL CONCEPT FOR POLICY DEBATES

## INTRODUCTION

In recent years, one of the crucial concepts often invoked in policy debates about the provision of health care is the concept of the right to health care. Some have argued that health care is a privilege and not a right. Others, probably the more prevalent group, insist that there is a right to health care and that it has significant implications for the ways in which health care is provided and for the level of governmental funding of health care for the indigent. Regardless of the position being advocated, there are few serious discussions of the organization and financing of health care that fail to invoke, either explicitly or implicitly, the right to health care.

The main purpose of this paper is to argue that discussions of health policy, including the discussion of the organization and financing of health care for the indigent, would be more fruitful if references to the right to health care were deleted. I shall argue for this claim by arguing for four subsidiary claims. They are: (a) the right to health care seems important in policy discussions because it seems to be central to dealing with the question of society's obligation to fund health care for those who cannot afford to pay for that care; (b) it is actually unhelpful in dealing with that question, because reference to the right to health care fails to give us any basis for dealing with any of the real problems connected with that obligation; (c) moreover, discussions about the right to health care imply, and sometimes explicitly affirm, that health care is something special, something to which people have a special right, but discussions about these health policy questions go better if that assertion is rejected; (d) finally, it is likely that there are many additional moral considerations that should play a major role in shaping health policies. In particular, there are other moral reasons than the right to health care for insuring that at least some health care is provided to those who need it but who cannot pay for it independently.

One point about the logic of the argument needs to be kept in mind



*T.J. Bole III and W.B. Bondeson (eds.), Rights to Health Care*, 113–131.

from the very beginning. I am not arguing in this paper either that there is a right to health care or that there is not a right to health care. This paper is neutral on that topic. What I am claiming is that the question of whether there is this right to health care should not be viewed as a central question in the moral examination of the important issues in health policy in which it is usually invoked.

## PART I

The first of my claims is that the right to health care seems important primarily in those policy discussions concerned with the question of the provision of health care to those who cannot afford to pay for it. In order to motivate and substantiate this claim, I need to begin by reviewing certain important facts about the history of American health care. Much of this will be familiar [1, 10], but it nevertheless needs reemphasizing.

The story that we need to consider begins with the tremendous medical and surgical advances of the latter part of the nineteenth and first quarter of the twentieth century. During that period of time, the question of access to health care began to become important for two reasons. To begin with, health care became more valuable as the newly developing surgical and medical techniques enabled physicians to do more by way of curing (or at least alleviating) the problems of their patients rather than simply diagnosing them. At the same time, however, the cost of obtaining these services began to rise, and the problem of how to pay for them became a great problem for many Americans.

The first national response to this problem was the rise of health insurance. We will not here review the history of the rise of Blue-Cross and Blue-Shield and other private insurance schemes during the 1930s, 1940s and 1950s. The crucial point that we want to make is that America's initial response to the demands for this more effective but more expensive health care was to insure that most Americans could have access to health care because they were covered by privately funded health insurance schemes, schemes that were usually developed for groups of employees. This meant that those who were no longer employed because they had retired or who were only marginally employed and therefore indigent were not covered by this initial response. This meant that there was a real gap in the initial American response to this problem. By the early 1960s, while about seventy percent of America was guaranteed access to the ever more expensive but more effective health care which they desired through these private insurance schemes, some thirty percent of Americans were left with a major access problem.

Naturally, those who advocated assuring access to health care for those in need of it were dissatisfied by this result. Initially, in the post-World War II period, this led to the advocacy of a national health insurance scheme. When this approach failed, attention turned to developing schemes to cover the elderly and the indigent. A great landmark in this development was the passage of the Medicare and Medicaid programs in 1965. In the debate surrounding the passage of that legislation, those who argued for it often appealed to the right of all to receive the health care they needed. On the other hand, those who opposed the Medicare and Medicaid programs often did so with the argument that health care was a privilege rather than a right.

We now understand even more clearly than in the past that the passage of Medicare and Medicaid did not close all of the gaps in coverage for Americans. There remain a group of Americans who are under sixty-five, who are ineligible for Medicaid, but who are too poor to pay for their own health care, who do not have adequate access to health care. The debate about what to do with them has begun to assume a familiar structure. Those who advocate new programs to cover these Americans often put their claim in the form of a right to health care, arguing that it is members of this group (some 12–13% of all Americans) who have a right to health care, but whose right to that care is not being realized because of their lack of any public or private health insurance coverage. Those who oppose the development of new programs often insist that these residual groups are not being deprived of some health care to which they have a right.

Obviously, this is a very short and simplistic account of what is a much more complicated history. Nevertheless, I think it illustrates the crucial point that I wish to make. The rise of effective but expensive health care has been a real problem for those of limited means. America has responded first with the development of private insurance schemes primarily for the employed, and then with the development of Medicare and Medicaid. It is still confronting the question of what to do with the remaining people of limited means not covered. In each of these stages of discussion, those who have advocated more extensive social schemes to insure the delivery of health care to the needy have invoked the right to health care. Those who have opposed the development of such more extensive schemes have often tried to challenge that claim of the right to health care. The right to health care has therefore been viewed as important to policy issues because it has been viewed as central to the question of how much health care we are going to make sure that everyone has regardless of their ability to pay.

There is one striking counter example to this historical thesis that needs some comment. This is the work of the President's Commission in its volume entitled *Securing Access to Health Care* [9]. One major thrust of that volume is the clear claim that when private forces fail to insure equitable access to health care, "the ultimate responsibility for insuring that society's obligation is met, through a combination of public and private sector arrangements, rests with the federal government" ([9], p. 5). In short, the President's Commission is certainly to be listed among those who would expand, if necessary, the role of the government in assuring adequate access to health care for all those who need it. Nevertheless, in an important discussion on pages 32 through 35 of that volume, the President's Commission chose not to rest its case upon any appeal to the right to health care. It said:

> The Commission has chosen not to develop the case for achieving equitable access through the assertion of a right to health care. Instead it has sought to frame the issues in terms of the special nature of health care and of society's moral obligation to achieve equity, without taking a position on whether the term obligation should be read as entailing a moral right ([9], p. 32).

How, then, does the thesis I would advocate handle this apparent counter example?

The first thing we need to note is that the Commission's discussion on pages 32 through 35 is very obscure. It begins by noting that a right to health care is neither legally nor constitutionally recognized at the present time. It points out that a great many federal and state statutes that fund or regulate health care have been interpreted to entail that there are statutory rights of various groups to health care benefits. But the point that it makes is that neither the Supreme Court nor any appellate court have found a constitutional right to health care. Grant that this is true. It is equally true that neither the Supreme Court nor any appellate court have found a constitutional obligation of society to provide health care. At the same time, there is no question that there are many federal and state statutes that have been interpreted to provide statutory obligations on the state or on the federal government to provide certain beneficiaries with health care. So examination of the current legal and constitutional situation provides no basis for distinguishing between the right to health care and the social obligation to provide health care.

The second argument that the Commission used was the claim that it is not a logical consequence of an ethical obligation of the type it postulates that there is a right to health care. It said:

> In a broad sense, to say that society has a moral obligation to do something is to say that it ought morally to do that thing and that failure to do it makes society liable to serious moral criticism. This does not, however, mean that there is a corresponding right. For example, a person may have a moral obligation to help those in need, even though the needy cannot, strictly speaking, demand that person's aid as something they are due ([9], p. 34).

The Commission is here trying to explain how there can be a social obligation to provide health care without the indigent having a right to that care. But the explanation seems particularly inappropriate. When the individual person has a moral obligation to help those in need, there is no particular needy person that has a right to that help, because the individual can fulfill his obligation by helping other needy people. Nothing analogous to that is present in the case of the social obligation the Commission postulates. If, as the Commission believes, society has an obligation to provide some level of health care to all of the indigent, then it would seem that, as a correlative to that obligation, the group of the indigent have a right, which they hold against society as a whole, that that health care be provided. In short, it seems, contrary to the opinion of the Commission, that if the Commission is right about the ethical obligation to provide health care, then it is wrong about the existence of the correlative right to the health care in question.

This last point undercuts the Commission's third argument, the argument that "it [the right to health care] is not necessary as a foundation for appropriate government actions to secure adequate health care for all" ([9], pp. 32–33). The Commission's point is that such programs can be justified on the grounds of the obligation without appealing to the right. If we are correct in our previous suggestion that the two are correlative, then the Commission has provided no independent basis.

Two hypotheses are possible. One is that the Commission just reasoned badly on this whole topic. It thought that it could make out the case for the necessary government programs without appealing to the right to health care, but it was wrong. Another hypothesis is that it understood very well that the obligation it postulates gives rise to a correlative right to health care, but it found it politically expedient to try to make its case without invoking the language of the right to health care. Knowing the philosophical acumen of the staff and advisors to the Commission, I prefer the second hypothesis to the first. Whichever is true, however, I think that the Commission's discussion needs to be looked at as a special event, one that does not count against a more general claim that the advocacy of the right to health care has been intrinsically tied up with the advocacy of extensive governmental programs to insure that all those who need health care receive that care regardless of their ability to pay. Moreover, I know of no other major policy debate in which the right to health care figures prominently. So I conclude that my thesis (a) is correct, and that the right to health care seems important in policy discussions because it seems to be central to dealing with the social obligation to fund health care for the medically indigent.

## PART II

I turn now to the second of my major claims. This is the claim that the concept of a right to health care is unhelpful in dealing with the crucial policy question connected with the social obligation to fund health care for those who cannot afford to pay for that care, the question of how much health care should be provided to those who are in need of that care but lack the ability to pay for it on their own.

One quick answer to that question would be that we are obliged to provide all of the health care from which they would benefit. This answer has an immediate attractiveness. After all, if someone could benefit from some health care, that presumably means that they are suffering from some health problem and need the health care to restore their health. Is that not a sufficient reason why they should get that health care?

The standard difficulty with that answer is by now familiar. It is that such a policy is likely to result in a tremendous increase in America's health care budget, and such an increase is unacceptable. In short, it is just too expensive to accept that particular answer. American health care, by the end of 1990, consumed over twelve percent of our Gross National Product. This is a tremendous increase from earlier levels of four to six percent, and this type of increase simply cannot continue. A major component of that increase was a social commitment to provide all of the health care from which people could benefit, regardless of their ability to pay for it privately. We cannot continue that social commitment.

This way of reasoning is much too simplistic. It assumes, without any argumentation, both that the increased expenditures in recent years on health care is bad and that it would be silly for us to continue to support that rate of increase in future years. I have argued elsewhere [6] that these claims might well be mistaken. I have argued that much of the increase in real per capita expenditures on health care represents nothing more than more Americans getting better health care regardless of their ability to pay for it and that this may be a good thing. I have also argued that, challenging the second of those assumptions, further increases might simply represent the continued need to provide new and better but expensive medical technologies for an increasingly aging population with greater needs for health care. So I think that the standard response to the claim that everyone should receive all the health care from which they would benefit, the response that simply talks about the increase in health care costs, is too simplistic.

There is, nevertheless, a good reason to reject that answer. The claim that everyone should receive all of the health care from which they will benefit is simply insensitive to the fact that someone else must pay for the health care, and pays no attention to the question of whether those others really have the obligation to fund that much health care. There are, after all, a great many things from which people could benefit and which they are not in a position to find on their own. Others may be under an obligation to provide some portion of all these other things, but they are surely not under an obligation to provide everything from which anyone could benefit. The same thing seems to be true of health care. The increase in health care costs shows us how much we can be called upon to provide if we provide the health care from which everyone can benefit, and reminds us that the argument for that major social obligation is unclear. Moreover, it reminds us that we have other social obligations and other social and individual goals beside providing health care. Those who insist that we have an obligation to provide all of the health care from which people could benefit regardless of their ability to pay for it simply fail to take into account how the obligation to provide health care (and the correlative right to the health care) fits into a network of other rights, obligations, and values.

The second position that is often put forward is the egalitarian claim that everyone is entitled to any form of health care from which they could benefit providing that anyone is receiving it. The suggestion is that there might be some forms of health care which are too expensive to provide to everyone who could benefit from them, but that it would be wrong for anyone to receive that health care. This egalitarian position is much easier to dismiss than the first position. After all, it would insist that individuals could not pay out of their own pocket for forms of health care that society deems too expensive to provide to all who could benefit from it. It would, in fact, limit the liberty of the affluent with no corresponding benefit to the less affluent. Moreover, it would result in the absurd conclusion that the affluent could use their extra funds on frivolous pursuits, but could not use them to secure extra health care.

Considerations of the type that we have discussed have led most commentators to say that the right to health care and the corresponding social obligation to provide it to those who cannot pay for it for themselves is a limited right. Considerations of this type have led most commentators to introduce the notion of an adequate minimum level of care, with the suggestion that the social obligation to provide health care and the correlative right to that care is the obligation to provide the health care that falls within this adequate minimum level of care. The remarks of the President's Commission are very helpful in this context. They make the following claim:

> Whether the issue is framed in terms of individual rights or social obligations, it is important to recall that society's moral imperative to achieve equitable access is not the unlimited commitment to provide whatever care, regardless of cost, individuals need or that would be of some benefit to them. Instead, society's obligation is to provide adequate care for everyone. Consequently, if there is a moral right that corresponds to this obligation, it is limited, not open-ended ([9], p. 35).

This view leads me then to the central point that I want to make in this section. It is clear from the above brief summary that introducing the concept of the right to health care (and the corresponding social obligation to provide that health care) does not by itself take us far in dealing with the important policy questions. To be sure, those concepts do lead us to the conclusion that society is mandated to provide some health care to those who cannot pay for it for themselves. That, however, leaves all the important policy questions open. What forms of health care are we obligated to provide? To what health care do the indigent have a right? What is this adequate minimum level of care?

I believe that most people who have thought about these questions have had the wrong model of how they arise as important policy questions. I think that they have primarily looked at certain discrete forms of health care and have raised the question as to whether it falls under the adequate level of care to which everyone is entitled. Thus, there has been considerable discussion as to whether expensive neonatal intensive care is something to which every neonate is entitled if they are in need of it, regardless of the ability of their parents to pay. Questions have been raised about whether society is under an obligation to provide kidney dialysis for all those who could benefit from it, or whether that obligation should be limited by the patient's underlying medical condition or by their age. Recently, a favorite example for discussions of this sort has been the question of organ transplantation. Should organ transplantation be part of that minimum level of care to which all are entitled regardless of their ability to pay [1]?

I do not want to deny the importance of those discussions. But I think that we get a misleading picture of the important policy issues that we need to face if we focus on these examples. Focusing on these examples makes it sound as though the important policy issue is whether new and very expensive forms of high technology medicine need necessarily be provided to all those who could benefit from them, regardless of their ability to pay. It enables us to miss the more ongoing everyday set of questions that totally permeate the provision of health care. Let me explain my point.

Suppose that we distinguished two forms of health care that are often provided on a routine basis and whose provision might be challenged. One type of health care is purely wasteful health care. This would involve forms of health care that are of no benefit to the patient. It is easy to be opposed to the provision of wasteful forms of health care even to those who are willing, perhaps out of a lack of understanding, to pay for them, and even easier to oppose their provision to those who cannot afford to pay for them. A second, and very different, type of case is raised by those forms of health care that provide some limited benefits at some considerable cost. We might call these forms of health care potentially cost ineffective forms of health care. Should these be provided to all who might obtain the limited benefits from them, whether or not they are able to pay for them? This is a much harder question, and a question that permeates the everyday practice of medicine. My point is that we get no help in thinking about this question from the perspective of the right to health care.

Perhaps two recent examples may help bring out this point more fully. The first of these involves a study [8] of the routine use of chest X-rays upon the admission of patients to a hospital. The authors did not want to suggest that spending money on these routine chest X-rays was a wasteful health care expenditure. They recognized that there was a percentage of cases, remarkably small (4%), in which routine chest X-rays picked up a problem not known about before, one that was treatable upon its discovery. So each patient who received the routine chest X-ray did receive the statistical benefit of improving their chances of getting health care for all their problems, including ones unknown upon admission. However, the cost for all the X-rays required to find these treatable problems was very high. Is that a reasonable health care expenditure? And, more relevant for our purposes, does the benefit in question justify that expenditure in the case of patients who are not paying for the health care themselves? Is that part of the minimum level of care to which they are entitled?

A second example brings out the same point. Women with a history of genital herpes who become pregnant require screening to ascertain whether they have an active infection close to the time of the birth of their child. If they do, they need to have a cesarean section in order to avoid transmitting genital herpes to their child. There are two ways in which they can be screened. One, which is much less expensive, is simply undergoing a physical examination. Unfortunately, this will not pick up cases of subclinical infection. The second is undergoing weekly viral cultures in the last weeks before delivery. This more expensive approach will pick up additional cases of recurring genital herpes and prevent some cases of neonatal herpes. How much does this cost? A recent study [2] suggests that the cost of each case of neonatal herpes prevented by such screening is $1,840,000. Is this a reasonable health care expenditure? More crucially for our purposes, is this one of those forms of health care that ought to be provided as part of that minimum level of care to which we are all entitled?

The nature of the question posed by these sort of cases is relatively straightforward. There are a very large number of diagnostic and therapeutic interventions that can be performed which share the following characteristics: they will all produce some benefit in a modest percentage of cases. In most cases, however, they produce no benefit. We do not know in advance which cases are which, so we can only gain the benefit by using them in every case. The actual cost of producing the benefit for each case in which it is produced is quite high. Nevertheless, in each case in which the benefit is produced, it is a considerable health care benefit. Each of these examples forces us to consider whether medical practice should include these procedures. One possible way of dealing with this question would be to ask the patient. If the patient is paying for that health care, then this is at least a possible solution. Whether it is a good solution is something we can put aside for now. The crucial point is that this suggestion makes no sense if the patient is not paying for the care. At no cost, the patient has every reason to ask for the extra procedure. Are we obligated to provide those extra procedures if the patient cannot afford to pay for it? Does the patient have a right to these extra forms of health care? I do not begin to see how we can answer these questions. But it is questions like these which occur throughout every aspect of the provision of health care and which are central to shaping the provision of health care for the indigent. All of this leads me to the conclusion that when we discuss the real ongoing policy questions about the provision of health care to those who cannot afford to pay for it, we receive no help from thinking about the social obligation to provide health care and the correlative right of the indigent to receive that health care.

My claim in this section is therefore clear. America, as a society, has introduced schemes for insuring that most Americans receive health care whether or not they have the ability to pay for it. The rise in health care costs has indicated to us that our policy discussions need to focus on the question of how much health care we are going to provide to those who cannot afford to pay for it. That question is not just a question about a few esoteric forms of expensive high technology medicine. It is, instead, a question about the many diagnostic and therapeutic inventions that might be used with some benefit but at a considerable cost. These day by day decisions are far more important to shaping the structure of American health care than some much discussed esoteric decisions. I know of no discussion of the right to health care and the correlative social obligation that provides us with a framework for resolving these questions. This enables me to conclude that we are not helped in the important ongoing discussions of the real policy problems by references to the right to health care.

## PART III

I turn now to an entirely different set of considerations that lead me to doubt the importance of the right to health care in discussions of health care policy. These issues center around the question as to whether health care, as it has traditionally been understood, should for policy purposes be identified as something special to which people have a right that must be met by developing special policies for the financing and provision of health care.

We need to begin by distinguishing health care as it has traditionally been understood from a broader notion of health care. Health care as it has been traditionally understood is the sort of thing that physicians, nurses, and other allied health professionals provide. It consists primarily of diagnostic procedures, medical and surgical interventions, and a variety of rehabilitative efforts. What all of these things have in common is that (hopefully) they contribute to longer and healthier lives.

These are not, however, the only things that contribute to longer and healthier lives. They may not, in fact, be the things that most contribute to longer and healthier lives. Other things which play a central role, perhaps even a more central role, include adequate food, safe housing, sufficient clothing, the elimination of environmental and work place hazards that promote illness and death, education that opens up a variety of opportunities for more healthful living, etc. The indigent have as much difficulty obtaining these goods without social programs designed to help them as they have obtaining health care as understood in the narrow sense.

There are a variety of reasons that people give why the provision of health care in the narrow sense should be viewed as special. They point out that the provision of these traditional forms of health care preserves life, provides one with the bodily capacities essential for pursuing opportunities and plans, helps one avoid pain and suffering, etc. All of these are no doubt reasons why health care as traditionally understood is very important. They are equally good reasons why these many other goods are also important. In fact, if we understand health care as those measures that preserve bodily health and increase life-expectancy, then it is difficult to see any basis for distinguishing the provision of health care in the narrow sense from the provision of these other goods.

Reflections on this point suggest that the development of programs to aid the indigent by providing them with the means to prolong their life and improve their bodily health is going to have to involve the development of programs that address these many different needs, and properly balance them in a setting of budgetary constraints. Such a development is not going to be well served by the development of programs that focus in on health care in the narrow sense while neglecting these many other needs essential for good health. But the right to health care addresses itself only to the narrow conception of health care. It is therefore in no position to help us design properly balanced programs within our budgetary constraints to meet the needs of the indigent.

This point has so far been made in a very abstract fashion. Let me now try to give a concrete example to illustrate the point. Consider the provision of neonatal intensive care to the low birth-weight newborn children of indigent parents. Should this be part of those goods and services which society provides to the indigent regardless of their ability to pay? We need to keep in mind that studies have demonstrated again and again the tremendous effectiveness of neonatal intensive care in decreasing mortality and morbidity for low birth weight neonates, although this is certainly obtained only by heavy expenditures [3]. I am struck by the fact that the discussion of this question from the perspective of the right to health care focuses either on the lives saved, arguing that this should clearly be part of that health care to which all are entitled regardless of their ability to pay, or else focuses on the tremendous costs, arguing that these costs are sufficient reason to not provide this form of health care even to those who would benefit from it. What is left out in these discussions is the crucial point that the indigent people in question often fail to receive other goods more central to both their health and the health of their children. The clearest example of this is the failure to provide adequate programs of prenatal care. Even putting that example aside, one might well wonder whether the class of the indigent are better off obtaining the benefit of access to neonatal intensive care or obtaining access to, for example, better housing. I know one neonatal intensive care unit located in the midst of some of the worst shacks in the United States.

I do not have any easy answer to the policy question of what mix of goods society should be providing to the indigent. But I am convinced that we need to be looking at that question, and not just at the question of what forms of health care, understood in the traditional sense, we should be providing. If our focus turns to that broader question, then we are not going to be helped at all in these policy decisions by references to the right to health care.

This point that I have just made is really a generalization of the point that I made in the previous section of this paper. I argued there that the appeal to the right to health care was of little help in policy discussions because it provided us with no guidance for deciding which forms of health care were to be covered by the right to health care, for deciding which forms we are obliged to provide to all who would benefit from them regardless of their ability to pay. The discussion in that section focused on choices among traditionally recognized forms of health care. The point I have just made is that the choices that we need to make are even more general. We need, within some budgetary constraint, to develop policies for providing to those who cannot afford to pay for them some, but unfortunately not all, of those things that would improve their health and increase their life expectancy. These choices are even more difficult than the choices that I discussed in the previous section, because they are choices among more disparate goods. But they are the choices that we need to make. The appeal to the right to health care, which really only involves one of the goods in question, is in no position to help us think about these issues.

It is these sorts of considerations that lead me to the conclusion that the important policy questions that we need to address about the provision of goods to the indigent would be better resolved if we stopped thinking of health care, as traditionally understood, as something special. There are a variety of goods which would be beneficial to the indigent because, like traditionally understood health care, they increase life expectancy and improve health, and we need to develop policies for providing some, but not all, of them. The policy discussions about how to do this do not benefit from identifying one in particular as something to which people have a special right.

There is another way of making this point. The second way is somewhat more controversial, because it involves adopting a particular approach to the provision of goods to the indigent. It is an approach that I have argued for elsewhere [4, 5], and I do not propose to set out the arguments for it again here. I do want, however, to sketch this approach, and then to explain how it too leads to a policy analysis which benefits because it does not treat traditional health care as something special.

The basic idea behind this new scheme is as follows: society's entire redistributive activities would be funded by a proportional tax on wealth. There are special moral considerations that would help to determine what an appropriate tax rate of that sort would be. The funds in question would be distributed to the indigent, in proportion to their degree of indigency. The notion of indigency is also a notion which could be defined in terms of the same special moral consideration. Those receiving the funds would then make their own choices among the various goods open to them which are valuable for preserving their life, for promoting their health, and for advancing the goals that they have. Society's main restriction on the expenses would be to insure that the interests of minors and other incompetents are not neglected in the decision-making for them by their parents and their guardians.

If such a scheme were adopted, then we would no longer have to decide what forms of health care, understood in the traditional fashion, would be provided to the indigent, and how this would fit in with the social provision of other valuable goods to the indigent. We would no longer have such a decision to make because society would no longer be making that decision. It is those indigent citizens, whose funds have been supplemented by a social redistribution of money, who would be making the decision for themselves.

Therefore, if such a policy approach were adopted, the right to health care would no longer be an important concept for policy discussion. After all, we saw in section one of this paper that its main role is to help as a basis for the social obligation to provide health care, understood traditionally, to the indigent. If the scheme I have proposed were adopted, then society would no longer be in the business of doing that, since its main redistributive program would simply be the redistribution of cash. So the right to health care would be of no use in policy discussions.

There is an obvious connection between the two main lines of arguments which I have employed in this section. One of the main reasons I have offered for moving redistributive policy in the direction in which I have suggested is that I fail to see any principled basis by which our society could decide between the various important goods that indigents need so as to fashion an appropriate program for helping them in a way that is compatible with just budgetary constraints. Readers sympathetic to the policy scheme that I have advocated will therefore see why the right to health care is of little importance to policy discussions. Readers not sympathetic to my particular policy conclusions, but aware of the arguments offered at the beginning of this section, will at least see that once we broaden our policy discussion so as to recognize the need for a trade off between health care, traditionally understood, and other goods that the indigent need, the right to health care will become increasingly less important in policy discussions.

## PART IV

In this final section of the paper, I want to offer some additional reasons for being skeptical about the centrality of the right to health care in health policy discussions. What I shall be arguing for is the claim that there are many other moral reasons besides the right to health care that need to be invoked when the right to health care is usually invoked.

In order to see the point of this final argument, it may be helpful to begin by imagining a critic raising the following objection to my argument until now: what you have really been arguing for until now is the claim that the right to health care does not help in the examination of all those questions surrounding the decision of how much resources should be devoted to providing what sorts of health care for those who cannot afford to pay for it. Still, the right to health care needs to be invoked regularly in discussion because it is *the* foundation of the idea that society has any obligation to provide health care. Unless we begin with the notion of the right to health care, we have no reason to worry about all the policy issues we have been discussing, for without the right to health care, there is no moral basis for the claim of the indigent to health care. What I shall be arguing for in this final section is that this claim of the critic is mistaken. In particular, I shall be suggesting that there are many other important reasons why the indigent should receive health care. These reasons include the utility of providing them with that care, reverence for the value of life that requires that the care in question be provided, and the fact that the provision of such care is a compassionate response to the needs of the indigent.

One good reason for adopting any social policy is that the adoption of the policy promotes important benefits for members of society, benefits that outweigh the costs of adopting the policy. This claim is the truth that remains from the utilitarian literature after utiliarianism is rejected as an inadequate general moral theory. It asserts that the benefits, even when they outweigh the costs, are not necessarily sufficient to justify the policy but merely provide a moral reason for adopting the policy. It asserts that there may be other moral reasons. These are the ways in which this claim differs from the standard utilitarian claim. The crucial thing to know for our purposes is that this weaker claim is sufficient to help us understand one of the reasons why we ought to adopt a policy of providing health care to the indigent. The indigent benefit greatly from the health care that is provided to them, for their life is extended and their health improved by the provision of that care. To be sure, there are costs incurred by the rest of society, the cost of providing that care. Nevertheless, to the extent that we are convinced that the benefits to the indigent outweigh the costs to the rest of us, that provides us with a reason for providing the care. So the right to health care, even if it exists, is not the only basis for the provision of health care to the indigent.

One of the fundamental values of our society is respect for human life and health. Some refer to this value as the value of the sanctity of life. The language of sanctity carries with it, often intentionally, the suggestion that human life is of the highest value, may never be disregarded, and must be preserved at any cost. One need not hold this very strong view about the value of human life or health to be committed to that value. Many, while not prepared to commit themselves to the sanctity of life, are nevertheless prepared to commit themselves to the value of the respect for human life, a value that calls upon us to preserve human life and health at considerable (but not necessarily any) cost in most (but not necessarily all) cases. Those committed to this value may then see it as still a second reason for the provision of health care to the indigent regardless of their ability to pay. The life and health of the poor person is, they would argue, a matter deserving of great respect, and it is this respect that drives us as a society to fund health care for the indigent. We could, no doubt, try to say that this should be the special obligation of hospitals and physicians, but there is no reason why they have a special obligation to bear this burden. Socially funded programs for reimbursing providers for providing that care is the way in which we as a society show our respect for the life and health of the indigent. So we have then a second reason independent of the right to health care for providing health care to the indigent regardless of their ability to pay.

One of the widely accepted virtues is the virtue of compassion. What is this virtue? We can say that compassionate behavior consists of behavior which attempts to alleviate the suffering of others. Now illness and the approach of death certainly represents the basis for much human suffering. The attempt to alleviate that suffering by providing appropriate health care constitutes compassionate behavior. There is, therefore, no reason why we cannot view social programs to fund health care for the indigent as a way of our society's displaying appropriate compassionate behavior. To be sure, a particular suffering person is not directly before us as we pay our taxes, but there is no reason why compassion must only be shown in acts of alleviating the suffering of particular individuals. There is no reason why social policies to aid whole classes of suffering people cannot be viewed as justified precisely on the grounds that they are compassionate policies. So we have then still a third moral reason for funding health care for the indigent who cannot pay for their health care themselves.

There are those who would object to all three of these lines of argument. They would remind us that social funding of the health care of the indigent is not a question of voluntary charity. Social funding of the health care of the indigent involves requiring people to support that health care by requiring them to pay taxes. There is an element of coercion in the social funding of health care for the indigent. These critics would argue that it is only right to health care that could justify the use of state coercive mechanisms to fund health care for the indigent. All of these other moral reasons, even if legitimate, will only be reasons why the behavior of individuals who help provide that care is morally commendatory. They cannot provide any reason, however, for justifying a government funded program of health care for the indigent. Only a right to health care could justify the use of coercion in such a program.

The issues raised by this criticism are obviously far too extensive for discussion at the end of a paper on a different topic. Let me, however, just say the following by way of response. The presupposition of this objection is that only a right on the part of the recipient would justify a government action to compel citizens to fund a program the benefits of which would be received by others. This is much too strong a claim. It grows out of the libertarian tradition and represents, at least to my mind, a helpful response to the excesses of the welfare state's view that any general good can justify government coercion. One need not, however, accept its very stringent limitations in order to avoid the excesses of the welfare state. The search for an appropriate upper limit to governmental funding of social programs is a difficult one, but the difficulty of creating a general theory should not force us to adopt the extreme claim that only rights justify governmental redistributive programs.

Saying this leads me to still another general point about moral theory that needs to be made at this point. There are some who would ground the social policy of providing health care to the indigent in a right to health care. There are others who would ground it in the general social utility of having such programs. I have provided two other types of moral considerations for grounding such programs, a respect for human health and life and a sense of social compassion. I suspect that still other groundings might be found. Any sound social policy, including the one that I have briefly outlined in Section III of this paper for the provision of care to the indigent, needs to fulfill the moral values found in all of these groundings.

The fabric of morality is far too complex to be the product of the simple weaving together of threads of human rights and social welfare.

In any case, we have seen that there are many significant moral bases for the provision of health care in the narrow sense and, I would suggest, for the provision of care in the broader sense discussed above, to the indigent. There is no reason to advocate a special place for the right to health care as opposed to any of these other moral bases. This then is one final reason why the right to health care is not all that important in policy debates, even if the question being looked at is simply the question of whether we should provide some health care for the indigent.

*Baylor College of Medicine*
*Houston, Texas*

BIBLIOGRAPHY

1. Annas, G.: 1985, 'Regulating Heart and Liver Transplants in Massachusetts', *Law, Medicine, and Health Care* **13**, 4–26.
2. Binkin, N. *et al.*: 1984, 'Preventing Neonatal Herpes', *Journal of the American Medical Association* **251**, 2816–2821.
3. Boyle, M.: 1983, 'Economic Evaluation of Neonatal Intensive Care of Very-Low-Birth-Weight Infants', *New England Journal of Medicine* **308**, 1330–1337.
4. Brody, B.: 1981, 'Health Care for the Haves and Have-Nots', in E. Shelp (ed.), *Justice and Health Care*, D. Reidel Publishing Co., Dordrecht, Holland, pp. 151–160.
5. Brody, B.: 1983, 'Redistribution Without Egalitarianism', *Social Philosophy and Policy* **1**, 71–87.
6. Brody, B.: 1986, 'Wholehearted and Halfhearted Care', in S.F. Spicker, S.R. Ingman, and I.R. Lawson (eds.) *Ethical Dimensions of Geriatric Care*, D. Reidel Publishing Co., Dordrecht, Holland, pp. 79–93.
7. Campion, F.: 1984, *The A.M.A. and U.S. Health Policy*, Chicago Review Press, Chicago.
8. Hubbell, F. *et al.*: 1985, 'The Impact of Routine Admission Chest X-Ray Films on Patient Care', *New England Journal of Medicine* **312**, 209–213.
9. President's Commission for the Study of Ethical Problems in Medicine: 1983, *Securing Access to Health Care*, Government Printing Office, Washington, D.C.
10. Starr, P.: 1982, *The Social Transformation of American Medicine*, Basic Books, New York.

SECTION III

# A QUALIFIED RIGHT TO HEALTH CARE: TOWARD A NOTION OF A DECENT MINIMUM

THOMAS HALPER

# RIGHTS, REFORMS, AND THE HEALTH CARE CRISIS: PROBLEMS AND PROSPECTS*

I am not sure that health care can profitably be discussed in the context of rights. For one thing, it is a term of many and shifting meanings, and thus virtually guarantees some degree of misunderstanding and confusion. For another, the very uttering of "rights" carries tones of ethical seriousness, if not outright superiority, that gives its advocates a substantial and unearned advantage in debate. Contemporary political scientists, as a result, have tended to feel more comfortable with "claims" (*cf.* [82]). Yet "rights" is what this book is about, and so "rights" must set the terms of the discussion.

What I hope to do in this paper is to discuss two kinds of rights to health care and how assertions of each has implications for reform proposals that, in turn, are tied to macroallocative resource allocations.

In one sense, this may suggest talk of apples and oranges, for rights and macroallocations would appear to occupy rather different planes of thought. Rights, after all, at least in the United States, normally are announced as attributes of individual persons (e.g., *Regents of the University of California v. Bakke* [98], while macroallocations concern large populations, perhaps entire societies. An instant's reflection, though, is enough to establish the connection: accepting or rejecting the principle of a right to health care may have important consequences regarding the allocation of resources necessary to enforce that right.

## I. THE RIGHT TO HEALTH CARE

What, then, is a right to health care? The answer necessarily is of two parts. It is an obligation on the part of society, negatively or procedurally, not to interfere with the individual's pursuit of health care and, positively or substantively, to provide that care when the individual demands it (*cf.* [34], pp. 243–244; [120]).

This view of rights is connected with the notion of obligation in three senses. First, a right is seen as simply an obligation perceived from the beneficiary's perspective; to switch metaphors, rights and obligations are ordinarily no more than opposite sides of the same coin. My assertion of a



*T.J. Bole III and W.B. Bondeson (eds.), Rights to Health Care*, 135–168.

right to a liver transplant, for example, is just another way of saying that some person or persons (or society itself) are obligated to obtain the organ, perform the surgery, and so on. As Radin declared of rights and obligations, "The two terms are as identical in what they seek to describe as the active and passive forms of indicating an act; '*A* was murdered by *B*'; or '*B* murdered *A*'" ([96], p. 1141).

Second, the enjoyment of a right is generally seen as creating a duty to meet one's own obligations. For rights do not exist in isolation but are part of a vast web of rules that, certainly in the liberal democracies we are most familiar with, presuppose some significant degree of reciprocity. Thus, I may not wish to fulfill my obligation to you, but I have an interest in your meeting your obligations to me and, indeed, in preserving a whole interlocking structure of mutual obligations, and this compels me to put my short term wishes aside and meet my obligation to you. (Perhaps, this merely restates the Golden Rule). This connection of right to obligation may not be composed of logic but merely prudence (or prudence masquerading as compassion). It is, however, no less strong for that.

Third, the right reflects a moral obligation on the part of the entire society. It is not an *in personam* right that one person may claim against another, but rather an *in rem* right that one may claim against the whole community.

## II. AN UNQUALIFIED RIGHT TO HEALTH CARE?

Now, I indicated that there were two kinds of rights to health care. The first I call an unqualified right, because the societal obligations are stated in more or less absolute form without mention of conditions or limitations (*cf.* [125], [86], [117], [76]). With perhaps a few peripheral exclusionary criteria, such as a residency requirement, an unqualified right operates quite irrespective of the personal qualities of the claimant. Age, race, ability to pay, moral worth, contribution to society – these characteristics and dozens more are simply irrelevant. The only decisive variable is medical need (I shall not consider the sometimes vexing question of how this is to be determined).

One argument for this position is that in order for an individual to utilize his political rights fully, he must first be educated, adequately fed, decently housed and clothed, and, perhaps above all, receive good health care. The very existence of political rights, therefore, is said to imply these correlative rights as a collection of necessary preconditions.

A second argument rests on an Aristotelian notion of justice that requires that like be treated as like. As Williams put it:

> Leaving aside preventive medicine, the proper ground of distribution [of health care] is ill health... [T]he situation of those whose needs are the same not receiving the same treatment, though the needs are the ground of the treatment,... is an irrational state of affairs ([125]; [69], pp. 105–106).

Similarly, Veatch contends that the essential equality of mankind entails the obligation of health care systems to offer "the amount of health care needed to provide a level of health equal, insofar as possible, to other persons' health" [117].

Legally, there is no constitutional right to health care (*Harris v. McRae* [52], p. 318; *Maher v. Roe* [68], p. 469), though there are a host of statutory rights, many deriving from Medicare or Medicaid legislation (e.g., *Schweiker v. Gray Panthers* [107], *Elder v. Beal* [28]). It is certainly not unthinkable, though, that later courts may devise such a constitutional right, and, in any case, the legal situation hardly forecloses ethical or prudential considerations.

There are a number of obvious problems attached to the assertion of an unqualified right to health care. First, it mistakes means for ends. It is not good health care that people desire, after all, but rather good health, and health care is only one of several means to that end. Indeed, societally, it is far from the most important means (see, e.g., [64], [78], [71]).

An exaggerated concern with health care, however, obscures the health contributions of other factors, attracts resources away from health promoting but non-health care activities (e.g., education, traffic safety, pollution reduction), and thus may actually impede progress toward better health. An unqualified right to health care, therefore, insures that health care providers will receive more money, and may even increase the quality and quantity of health care available, but it need not significantly promote better health.

Second, the assertion of an unqualified right to health care implies mutually conflicting positions on the issue of medical paternalism. Those emphasizing the individual's negative right to pursue his own health care may be troubled by the enormous, multi-tiered complex of regulations and institutions covering professional licensure, drug availability, and so on, all of which assume that health care is such a technical enterprise that the layman requires officially credentialled experts for his protection – even if he may not himself always recognize it. Those stressing the medical establishment's obligation to provide health care, on the other hand, will naturally seek to direct money and authority to exactly those forces whose paternalistic qualities may strike their neighbors as offensive. Moreover, by stressing the medical establishment's obligation to offer solutions, these advocates may eviscerate the patient's own role, seeing him as essentially passive and unable to contribute to his own health except by swallowing his pills at the appointed times.

Moreover, even those who dismiss the practical consequences of the collision between positive and negative rights may be forced to concede that the two are logically quite unrelated. Libertarians who stress the right of non-interference, therefore, find no right to well-being at the expense of other citizens [85].

Third, the assertion of an unqualified right to health care suggests that excellence and equality can be achieved simultaneously, when this is rarely the case. This is because excellence normally requires the concentration of resources (e.g., at a major teaching hospital), while equality entails their dispersion. But if equalizing "up" is typically too expensive to be realistic, equalizing "down" instantly meets stiff resistance from those advantaged elements who are asked to sacrifice merely for the greater good of the greater number.

Fourth, the assertion of an unqualified right to health care raises an endless stream of correlative questions. What health care, for example, is one entitled to? It is one thing to speak of emergency treatment of a potentially fatal trauma. But what of a buttocks tuck? or an abortion? The World Health Organization has defined health as "a state of complete physical, mental, and social well-being and not merely the absence of disease or infirmity". Does that imply that anyone not in a state of "complete well-being" (i.e., virtually all humankind) has a right to care that will elevate him to this condition without regard to limitations on resources or the evident absence of means to ensure this complete well-being? Is this perilously close to declaring that we have not only the Jeffersonian right to the pursuit of happiness, but to happiness itself?

And why limit the entitlement to health care? Why not add food, clothing, shelter, Club Med, everything? A frequently heard answer is that health care is special because ill health limits the normal opportunity range open to persons in a society [17]. This retort has the advantage of distinguishing between essential and non-essential care and, more broadly, between needs and wants. Abstract distinctions, however, do not apply themselves; Cyrano de Bergerac may perceive a right to cosmetic nose surgery that Shirley Temple may sniff at. Nor, again, does health care seem all that special; the lack of many goods and services, from heat in the winter to an air conditioner in the summer, may limit one's normal opportunity range. Empirically, moreover, these ranges would appear to vary from country to country, and within countries, from region to region and stratum to stratum. Shall we accept the highest? the lowest? the mean? the median? the one we know best? And what of those with incurable illnesses, whose conditions will never improve, let alone approach the normal opportunity range? Are those, the most helpless and vulnerable in our whole society, to be lumped with those seeking hair transplants as not worthy of a societal obgligation [50]? It may be possible to defend caring for this sad class on the basis of mercy, of course, but this is a far weaker principle for allocation than justice.

To what extent (if any) do the patient's desires or conduct shape his right to health care? Does intensity of demand reinforce his claim? Suppose he eschews medical orthodoxy for rolfing? Christian Science? diagnosis by divination from chicken entrails? Does he waive his right to health care by improvident health practices (like smoking or excessive drinking)? If he is unwilling to foreswear the pleasures of neglect, ought society, like some indulgent Daddy Warbucks, simply to pick up the tab? Is our obligation to the individual so powerful that it is unaffected by his own failure to look out for himself? And when, short of death, does the obligation end?

What of the impact of implementing an unqualified right to health care upon healthy people? Most of them, of course, would not freely choose to insure themselves and their families against all possible medical eventualities, because they implicitly calculate that the marginal utility of some of the coverage is not worth the price. (This is not mere speculation; a 1982 poll found that fully fifty-six percent of those with "hardly any" health worries preferred "cheaper and limited" coverage, while only thirty-seven percent preferred "full and expensive" coverage ([99], p. D23).) With an unqualified rights system, however, healthy people – indeed, virtually everyone – would in effect be coerced into purchasing this full and expensive coverage in the form of higher taxes. And driving up costs still further would be the increasing demand for health care brought about by severing the connection between consumption and payment.

And what of the consequences of encouraging consumption at the expense of production? Compassion, of course, has a fine, noble ring. It does not, however, create wealth, but merely uses it. Does this suggest that, if compassion (and other consumption oriented urges) is given too many rights at the expense of production, at some point there will be insufficient resources to support that compassion? It may be easy to agree that the only inequalities that should be allowed are those that improve the condition of the worst off [91], but if this requires an open-ended commitment to the seriously, frequently incurably ill, we will be forced to rearrange our priorities in ways that may seem neither prudent nor just. (In line with this, Sen ([109], p. 16) maintains that utilitarianism would dictate that fewer resources be directed toward the disabled than the normal, who would gain greater utility from each additional increment [but *cf.* [13], pp. 318–339].)

If the right to health care is truly a basic human right, does it necessarily apply equally to everyone? to Charles Manson no less than Amadeus Mozart or Francis of Assisi? to a demented ninety-year-old in need of dialysis as much as a nine-year-old? to Americans only, or to all humanity? Or does the sheer difficulty of ranking classes of persons as to their medical worthiness argue for a lottery system of selection (*cf.* [45], p. 7; [16], p. 138)?

And what of the health care providers? Society may be obligated to allocate sufficient resources for health care – whatever that means and however that is determined – but it is physicians and other health personnel who actually perform the work. Do they have rights – where to practice? how many patients to see? what tests and treatments to prescribe? what fees to charge? – that deserve protection, too? Or as one California dentist angrily put it:

> In current political parlance, the "right" to health care has come to mean *the right to health care at the expense of someone other than the recipient of the service....* To claim, then, that medical care is a right – that a man has a right to be cared for by somebody else – raises the question: What of that other somebody's rights? ([61], pp. 467–468, emphasis in original.)

Fifth (and to a political scientist, this must seem most egregious), the assertion of an unqualified right to health care represents a profoundly apolitical solution to a profoundly political problem. Politics, as the authoritative allocation of values in a society [25], entails a continual scramble among claimants for scarce resources. Denoting health care an unqualified right, however, places it outside the political process, assigning it sufficient – perhaps, in principle at least, unlimited –

resources without compelling its advocates to battle with competing claimants. As a kind of ultimate political victory, in short, winning acceptance of health care as an unqualified right raises it above politics, where life is safer and the rewards better.

What, however, it might be objected, is wrong with this? Can any American of voting age have failed to notice that politics is disreputable, if not selfish, dishonest, and demeaning? To which the political scientist can retort only that, stereotypes aside, politics cannot simply be banished like a party crasher by an upright butler. For politics derives not from the conduct of fools and knaves – though, of course, they can certainly make matters worse – but rather from the scarcity of resources and the incompatibility of human purposes, and more durable bases for a social phenomenon would be hard to discover ([51], chap. 1). To exempt any policy, even health policy, from this competition is to exempt it from effective accountability, as well.

Viewed from this perspective, it may simply be naive to assert an unqualified right to health care. Given such a right, of course, demand for health care would increase rapidly and indefinitely. This, however, would soon cause massive reallocations from other goods and services. How much of this reallocation, though, would the public, the bureaucracy, the elected officials, and the representatives of other suffering interests tolerate? After all, most of us view good health not as the goal of life but merely as a precondition to its full enjoyment. Indeed, when we are not ill or injured – and most of us are not – we may hardly think of it at all. Rather than giving health care a blank check, then, we may find ourselves assigning an equal or even higher priority to other things, some of them foolish or maybe unhealthy things. To acknowledge this is not to denigrate the importance of health care or the intelligence of society, but merely to face the fact that a free people given to focusing upon the short term is unlikely to give health care an unlimited call on its resources.

## III. A QUALIFIED RIGHT TO HEALTH CARE?

For all these reasons, most persons who speak of an unqualified right to health care do so only as a rhetorical device with which they hope to increase health care's share of available resources. Rights talk, in other words, tends to be more advocacy than philosophy. This is not to deny the practical impact of such declarations. Clearly, they sound a deep communitarian chord to which many citizens reflexively respond. Further, the very act of asserting an unqualified right to health care congratulates its supporters on their compassion and social commitment, and thus may be indispensable to sustaining and energizing many of the foot soldiers in the battle for money.

When we turn from sloganeering to thought, however, we are more likely to encounter advocates of a second kind of right to health care: a qualified right. Usually, this is construed to refer to an obligation on the part of society not to interfere unreasonably with the individual's pursuit of health care and to provide a reasonable level of that care when the individual demands it. Although it is often accomplished only implicitly, the insertion of the "reasonable-unreasonable" concept, of course, is the key.

Advocates of this position concede that an unqualified right to health care may be unworkable, counterproductive, absurd – in a word, unreasonable – and propose to snip off the offensive elements. Determining which elements should be pruned, however, is less a philosophical problem than a political one. As such, its solution is a function of the potency of competing claimants, developing standards in health care, general economic conditions, and numerous other transitory factors. Indeed, the essential vagueness of the core terms represents not so much an absence of intellectual clarity as a simple invitation to negotiate, compromise, and bargain.

In the field of education, for example, all of this is plain enough. Americans speak easily of the right of children to attend a tuition-free public school, but no one imagines that this requires all classes to be small, all teachers to be Nobel laureates, and all facilities to resemble the Stanford University campus. Instead, we put up with many departures from the ideal, being unwilling to divert tens of billions of dollars to this purpose and trusting that non-school factors (like parents) can at least partially compensate for the shortfalls. In health care, though, where life-or-death images operate with great effect, qualifying a right does not win quite so easy or complete an acceptance.

Sometimes, however, the right to health care may be qualified not so much by limitations on the services or payments offered as by limitations on the population to be served. In addition to the reasonable-unreasonable qualification, that is, a qualified right may refer to a societal obligation, which, though pretty thoroughgoing, applies only to a portion of that society. Medicare, for instance, singles out the elderly as meriting fairly comprehensive coverage, and among the millions suffering from various chronic conditions, Congress has selected those afflicted with end-stage renal disease (ESRD) for special attention. Thus, we can distinguish service or payment limited and population limited qualified rights.

Analytically, assertion of a qualified right to health care – at least as usually done – would seem to generate more questions than answers. For one thing, there is the old matter of line-drawing. It is not at all clear what the reasonable-unreasonable qualifiers actually mean, whatever the political utility of obfuscation. If the law of torts is any guide, such terms mean precious little *in vacuo*, acquiring significance only when applied in specific contexts. The problem is, however, that it took generations for narrow tort decisions to accrete into generally understood rules. Issues involving health care might well be resolved much faster today; after all, there *are* a vast array of pertinent legal precedents to draw on, not to mention gaggles of bureaucrats, insurance companies, and patients, all clamoring for answers. Exacerbating current difficulties, however, societal expectations of appropriate health care are constantly changing, and the medical technology that helps maintain the flux is always leading policy makers into uncharted areas. What is today's routine, after all, but yesterday's gamble?

Nor is it always plain why certain populations deserve more or less unqualified protection, while others do not. Why ESRD victims, for example, and not sufferers of multiple sclerosis or Alzheimer's disease or other expensive long term afflictions?

More basically, it is difficult to see how a vaguely qualified right differs from an ordinary claim. For viewed from one perspective, all a qualified right asserts is that an individual is entitled to as much health care as the society decides it can afford. The relevant portion of Britain's National Health Service Act of 1977, for instance, reads:

> It is the Secretary of State's duty to provide throughout England and Wales to such an extent as he considers necessary to meet all reasonable requirements... [for] medical... services.

It is certainly clear that this statute was not intended to give all the Secretary's unsupported subjective judgments the force of law; a full evaluation of relevant factors may be required. But it is even clearer that such a statute can hardly be the product of legislators intent upon creating an unqualified right to health care. Thus, when a senior government medical adviser recently observed that "there is no right to treatment" ([87], p. 113) for end-stage renal disease, he was merely repeating a commonplace. Hence, if ESRD treatment patterns in the UK are heavily skewed against patients over age fifty or with complicating diseases and vary greatly from one part of the country to another, no right to health care is being violated. On the contrary, these variations are perceived and defended by government as reflecting differing allocative regional resource schemes and differing physician patient selection patterns, all of which may be entirely legitimate [48, 49].

Similarly, Canada has a comprehensive medical insurance system that covers every citizen. As a matter of practice, though, the apparent unqualified right to health care is indeed qualified by a number of factors: quality differentials and queue rationing have resulted from chronic underfinancing; extra patient billing and doctors striking or leaving the country have resulted from physician dissatisfaction with fee scales; in short, resource constraints have made it impossible to guarantee a wholly unqualified right to health care. Although the former socialist premier of Saskatchewan may have declared a goal of providing "complete medical care without a price tag" (Douglas, in [72]), it has long since become plain that fees charged at the point of service are but one kind of price, and that free health care is no freer than the notorious free lunch.

In Britain and to a lesser extent in Canada, if "right" has any real meaning in this context, it is merely as a goal of providing minimum adequate levels of care for all [1]. Of course, these levels are constantly changing as a consequence of technological developments, relative economic prosperity, public opinion, and any number of other factors, and really resemble elevators more than floors. But however high they may be set, their advocates will lack the simplicity of message and majesty of the absolute of those speaking for unqualified rights.

## IV. RIGHTS AND HEALTH CARE REFORMS

How, then, do these rather abstract notions of unqualified and qualified rights connect to actual health care reforms? Four brief points must be made by way of preface. First, it is not always clear to outsiders whether a system features an unqualified or a qualified right to health care. Americans, for example, having learned that the National Health Service provides health care to all Britons at little or no charge at the point of service, usually conclude that the United Kingdom guarantees an unqualified right to health care. As we have seen, this impression is quite false.

Second, inasmuch as no system, actual or reasonably proposed, offers a totally unqualified right to health care, it may not be particularly useful to conceive unqualified and qualified rights as a dichotomy. Instead, they might better be thought of as poles on a continuum. Arguments against a purely unqualified right, therefore, are apt to seem like shots directed at a straw man.

Conceiving the rights issue in continuum terms also clarifies part of the policy maker's problem by placing it in a directional context. A generation ago, for example, when there was a broad consensus that older Americans were receiving far too little health care, policy makers could talk about the right of the elderly to health care and simply provide more of it. After proceeding in this direction for twenty-five years, however, the policy maker's situation has become much more difficult. By now, the elderly receive so much health care that there no longer is a consensus that they have a right to still more. Indeed, the question of where on the continuum the line ought to be drawn – that is, how much and what kinds of care a right to health care entails – has already become rather contentious. Some analysts, in fact, believe that too much care is being provided (or, at least that it is being provided at too little expense to the elderly), looking back at earlier policy decisions as naive and misguided and at today's aged as greedy and ungrateful.

Third, though it seems obvious *a priori* that an unqualified right to health care will seem more attractive to providers and consumers and a qualified right to those paying the bills, this indeed may not always be the case. This is true not simply because individuals may be activated by feelings other than material self-interest. In addition, complicating the picture is the fact that many taxpayers view themselves or members of their families as potential providers or, more likely, consumers, and this potential role may sometimes appear more salient to them than their actual role as taxpayers. By the same token, payers may not invariably seek to control costs. When their claims costs increase, for example, private insurance carriers can frequently simply raise their premiums, thus acquiring more funds to invest. (Moreover, private carriers may reasonably fear that aggressive cost containment efforts will alienate physicians, hospitals, or patients, and that ostensible cost limiting programs like preventive medicine may actually hike costs in the short or perhaps the long run.)

Fourth, though we may be repelled by Wildean characters who know the price of everything and the value of nothing, talk of rights and reforms compels us to try to price values. How much, to begin, is a life worth? Normally, what that question really means is how much compensation is due for wrongful loss of life, or how much should be spent to safeguard it from some particular risk. The answers, of course, vary widely, there being no generally accepted formula for determining such matters. Nor is it even clear how much the individual values his own life. Some would observe that, in the last analysis, life is all we have, and so each of us normally can be expected to view his own life as priceless. As one economist put it, "Death is an awesome and indivisible event that goes but once to a customer in a single large size" ([105], p. 158).

Others would reply that as individuals we take avoidable risks every day (consider the smoker, the jaywalker, the driver without a seatbelt), and as societies we routinely select policies that cost lives (whether deciding to build bridges, set highway speed limits, or fight wars). All that may be obscured by "society's humanitarian self-image" ([12], p. 233), which requires that life be described as priceless or sacred and that victims of storms, avalanches, or other catastrophes be rescued at immense cost and great risk. (This is why a few "identifiable" lives invariably have a vastly greater disproportionate call on public resources than a much larger number of "statistical" lives.)

The same kinds of pricing questions arise with injury and disease, and with far greater frequency.

In health care, these issues emerge most forcefully over expensive medical technology, such as that involved in heart and liver transplantation and hospital care for the dying. But when twenty billion dollars or over a quarter of the Medicare budget goes to maintain patients in their last year of life, most of that in their last month, platitudes about the "infinite value of life" begin to collapse under the sheer weight of accumulating medical bills. (Arguably, though, there is something useful about the prevailing confusion as to the dollar value of life, injury, or disease. For this very element of uncertainty makes it harder for government and business to look on these individual calamities as a predictable cost of doing business, and may induce a certain sense of caution.)

## V. THE HEALTH CARE CRISIS

The so-called crisis in health care, in the view of most policy analysts, is simply a matter of money: health care is seen as too expensive and as increasing in cost too rapidly. (Average Americans, on the other hand, do not rank cost high as a major health concern ([3], p. 53.) Although a "growth enterprise" for a century [5], health care really began to boom following the introduction of Medicare and Medicaid in 1965. Thus, relative to 1965, health expenditures now consume at twelve percent almost twice as large a share of the gross national product, average hospital cost per patient day has zoomed about 800 percent in constant dollars ([115], p. 240), and, perhaps most ominously, health expenditures have grown at an average annual rate of approximately twelve percent. Even though the rate of increase slackened substantially in the Reagan years, it still exceeds twice the overall consumer price index [90]. If no action were taken and current trends were to continue, personal health expenditures could be expected to rise from $393 billion in 1984 to more than $1.2 trillion in 1995 [34]. So stunning are these projections that no one imagines that policy makers will permit them to become reality.

Propelling this extraordinary growth has been a third party payment system perched upon fee for service for physicians and cost-based reimbursement for hospitals that leaves consumers paying directly less than a third of total personal health care expenditures [41]. The more complete the coverage, the less incentive the consumer has to search out lower cost providers, and thus the greater the providers' power to dictate prices. The longtime ban on provider advertising, of course, reinforced consumer ignorance for generations.

Initially promoted by hospitals trying to ensure payment for inpatient services [62], private third party coverage was joined by extensive public third party coverage with the enactment of Medicare and Medicaid [70]. From 1950 to 1976, some twenty-five to fifty percent of the increase in physician service expenditures was attributed to the growth in insurance coverage [24]. And by 1984, only nine percent of the total population was uninsured ([2], p. 1335).

(Since over three-fifths of the population's insurance coverage is employer provided, the percentage of uninsured is necessarily quite volatile, fluctuating with employment trends. Thus in 1983, the percentage of uninsured was fifteen [15]. Even this modest figure, in fact, may overstate the nature of the problem, for an earlier 1977 survey indicated that of this nine percent, more than a quarter were uninsured only temporarily (i.e., less than a year), nearly half were in the relatively healthy years from six to twenty-four (less than one percent was over sixty-five), and about a quarter belonged to the middle class or above and presumably were uninsured by choice. Even the difference between the proportions of uninsured and insured who rated their health as poor was so small that it may well have been explained by normal sampling error [123]. Of course, all this increase in coverage has been accompanied by ever higher charges for the insurance, but even this has not depressed service utilization, since these charges are not close enough in time for consumers to link the two [88].)

(As for physicians, so weak is their cost pressure that large numbers are quite uninformed as to the prices of the tests and treatments they order [22, 57, 58]; on a very different situation in Great Britain, see [48] and [49].)

What has resulted is a system that has effectively isolated many of the participants from the financial effects of their use of health care and rendered them unaccountable for their behavior: providers need not worry that high charges will lower demand for their services, and consumers need not worry about paying these high charges. There can be little doubt, in sum, that this system has encouraged both providers and consumers to act as if the right to health care were unqualified by resource considerations.

It is not necessary to inquire here as to how a society that in fact is committed to a qualified right approach saddled itself with so many unqualified right devices; perhaps, the matter was simply not well thought out when the various third party payment systems were devised and put in place; or perhaps, the short term gains, political and otherwise, appeared so tempting that a "let the future take care of itself" outlook naturally took hold. (For a fascinating analysis of the passage of ESRD legislation, see [101].) Whatever the reason, such systems develop enormous vested interests, become entrapped by their own rhetoric, and acquire a momentum that may prove extremely difficult to overcome.

Most of the major reform proposals addressing the cost problem can be separated into four major clusters. One pair, regulation and cost-sharing, features a payment- or service-limited qualified right; another pair, means tests and health maintenance organizations, relies on a population-limited qualified right.

## VI. REGULATION

One is regulation. By now, the medical community is thoroughly familiar with such measures as restricting hospital capital expenditures and mandatory state regulation of hospital cost increases. Recently, Medicare established a pre-admission review program, which allows peer review officials in each state to select surgical procedures they believe are performed too often and to require that physicians seeking government payment for these procedures receive prior approval from their review body. The Office of Health Care Review of the Health Care Financing Administration contends that, as a result of this program, in 1985–1986 approximately 1.4 million Americans were treated in doctor's offices or not at all for medical problems that until then would have involved a hospital admission.

Clearly, the best known and most significant regulatory device, however, has been the diagnostic related group (DRG), according to which Medicare sets prospectively determined prices on the basis of the average historic cost of caring for patients in different diagnostic categories. As with the pre-admission review program, the DRG concept springs from the conviction that a significant portion of the health care that is currently provided is medically unnecessary. But while the old "reasonable cost reimbursement" system facilitated such excesses and rewarded inefficiency – as the chief of the Health Care Financing Administration put it, "The more hospitals spent, the more they got" ([114], p. B20, col. 1) – DRGs are designed to provide hospitals with incentives to hold down expenses. For if costs exceed the predetermined payment, the hospital must assume the loss, but if it is less, the hospital receives a "profit". It is estimated that DRG rates will be set so that about seventy percent of hospitals will risk losses and thirty percent reap profits ([9], p. 5). (The twenty percent difference between seventy and fifty percent will represent savings to the government.)

Further, as doctor supply increases, raising competition for hospital privileges with it, the DRG system presumably will induce hospitals to select and retain admitting physicians partly on the basis of the cost of their style of practice. And this, in turn, it is hoped will lead physicians to cut down on testing, drugs, length of hospitalization, and so on. Another source of saving may simply be hospital closure, for perhaps as many as one hospital in five is unable to adapt to the DRG demands ([9], p. 5; but *cf.* [111]).

Since DRGs were not fully phased in until fiscal year 1987, it is still early to undertake a definitive evaluation, though there is some evidence that hospitals have substantially overstated their costs [39], and fears have been expressed that the system is too complicated to be implemented [9], that hospitals treating the poor will suffer [46], that clinical trials necessary for research will be discouraged [128], and that the financial savings will be short lived [27].

Certainly, earlier regulatory schemes do not conduce to optimism. The certificate of need (CON) program, for instance, may actually have driven up hospital costs by directing capital expansion from beds to high technology [67, 79, 104, 112], and professional standards review organizations quickly became known for quality control and high administrative costs, rather than cost containment ([5, 21], *cf.* [127]). Past price controls ([55], pp. 1, 5, 6) and lowered reimbursement rates ([102], pp. 67, 84), moreover, have not prevented physicians from increasing their Medicare income. On a state level, skeptics often point to New York's bizarre experience with trying to reduce excess bed capacity through regulation: first, it penalized hospitals with less than eighty-five percent bed occupancy; when hospitals predictably responded by filling beds with marginal patients, the state felt forced to change its regulations to reward hospitals with empty beds [113].

A change of rules always means a period of adjustment, many analysts believe, but before long provider ingenuity will again prevail, perhaps by transferring costs to other payers (Peterson in [93] and [103]) or trying to replace low profit DRGs with higher ones ([59]; Davis in [20]). As one cynic put it:

> When systems are pitted against people, the people will win in the long run because people are fluid, creative and changing, and they always learn how to beat the system. Prospective payment by diagnostic related groups will probably last from five to seven years. Then there will be a drastic change in direction or a complete replacement of the reimbursement concept with an entirely different system ([9], p. 2).

DRG critics also charge that the system fails to provide incentives to attract patients to more efficient hospitals. Instead, it merely bleeds the less efficient ones.

DRG advocates, on the other hand, have applauded the reform as a serious effort to contain costs, as well as one that raises provocative, important, and long overdue questions about the style of medicine that ought to be practiced in the United States. Already, the average hospital stay for Medicare patients has been shortened by 2.2 days to 7.4 days ([114], p. A1), and for all patients it has been cut to 6.7 days [56]. (Some have contended that this practice will merely increase later readmissions and not prove cost-effective (e.g., [54]), and many physicians claim that patients are simply being discharged "quicker and sicker". The Reagan administration, for its part, claimed that it had "seen no data showing an increased number of deaths or complications as a result of DRGs" ([92], p. D27, col. 3).) It is no wonder that the president of the Hospital Corporation of America declared, "We have gone from a cost-plus environment to a competitive market driven environment", in which "more efficient is better", rather than simply "more is better" (Bays in [95], p. A1). It is, as the manager of a health care mutual fund put it, "a phenomenal, profound change" (Hayes in [119]).

What is truly odd, of course, is that the same Reagan administration that so often attacked government regulation with evangelical fervor relied so heavily upon regulation as a health care cost containment technique. For many of the standard arguments made against regulation in general have been made against regulation of health care, too. Its rigidities are said to have weakened competitive pressure and slowed the adoption of new technologies and management innovations; winners and losers have too often been determined by mere political clout; and, most seriously, after an initial period of modest success, the regulations have generally failed to contain costs significantly (see, e.g., [19]).

Nonetheless, despite these contentions and in the face of its own ideological distaste, the Reagan administration continued to experiment with regulation. Why? One economist suggests a pragmatism born of desperation. They had "no overall philosophy about the structure of the health care industry or the distribution of health care resources", he indicated, "other than to cut public spending by whatever technique they can" (Enthoven in [90]).

## VII. COST SHARING

Second is cost sharing. This approach proceeds from a pair of assumptions. The first is that patients need protection against major outlays, and not minor ones. A plan, for example, might require an individual to pay a quarter of any covered service up to a maximum of ten percent of his household income. (A different set of figures might be set for the poor, or perhaps they might be excluded from the cost sharing completely.) However, the technical issues would be resolved, the goal would remain safeguarding persons against serious threats to their standard of living, while inducing them to share modest everyday costs and to do without unnecessary care.

The second assumption is that rising health care costs call for less, not more, third party payments. For as the consumer's share of the costs grows, he would presumably feel impelled to seek out less expensive providers and to forgo some utilization of services and, as the head of the HCFA put it, "to think about how you can keep yourself healthier" (Davis in [20], p. 40). All this would force providers into price competition, achieve greater efficiencies of operation, and thereby rein in costs.

Employers have tended to be particularly enthusiastic about sharing. Since employee health benefits typically account for five to nine percent of payroll costs – group health insurance costs consumed about ninety-one billion dollars in 1984 ([33], p. 138) – and have been growing much faster than other business expenses, companies have become highly sensitive to an issue that had once been barely noticed. The results have been predictable. One survey of health benefits at 250 large companies disclosed, for example, that the percentage of companies with health plans featuring deductibles rose from seventeen in 1982 to fifty-two in 1984, that the percentage providing full reimbursement for hospital room and board dropped from eighty-nine in 1979 to fifty in 1984, and that the percentage offering full reimbursement for surgery costs declined from forty-five in 1979 to twenty-seven in 1984 [89].

Whether and to what extent cost sharing would actually contain costs remains, of course, a key question. Setting higher deductibles theoretically ought to encourage patients to shop for providers of low-cost services, or perhaps do without them altogether. The deductible, however, is unlikely to be set very high; if it were, patients would be encouraged simply to purchase other insurance to cover it. Thus, the deductible probably will easily be exceeded, and once exceeded the patient's cost controlling incentives would be eliminated. Since it is the expensive items (chiefly, hospitalization) that some observers find most worrisome, the cost savings from raising deductibles may therefore turn out to be fairly modest. As for coinsurance, though it requires consumers to pay a percentage of their own health care bills, it also normally has a cutoff point and covers catastrophic conditions completely. It, too, then, is likely to encourage consumer shopping mainly for services that are already relatively cheap. And, again, if the cutoff point is set high, those who can afford to do so will simply supplement their regular coverage with a policy of their own, like most Medicare enrollees. (Medicare patients now pay over $500 for each hospitalization, more than double the 1981 charge.)

For one obstacle facing cost sharing plans is that many consumers approach the subject of health care with a strong instinct toward risk aversion. Barely an eighth of the participants in the Federal Employee Health Benefit Program, for example, select the substantially cheaper partial coverage in preference to more comprehensive [75], and to the extent that most people seem willing to accept cost sharing, they are more likely to do so for low cost items, like dental or vision care, where savings would be trifling when compared to inpatient hospital services ([42], p. 229). For serious or urgent matters, cost considerations are likely to recede into the background. Cost sharing, therefore, may not affect enough big ticket services to have much of an impact on cost containment. Furthermore, providers may devise ways of minimizing cost savings, such as by charging comparatively higher fees to patients with comprehensive coverage ([29], p. 90).

All the evidence, however, is not discouraging. A Rand Corporation study on cost sharing for physician visits and adult hospitalization found that total expenditures declined about nineteen percent per person [84], and there seems little doubt that sharing arrangements bring about lower utilization rates [10, 83, 108]. Nor are the savings from "little ticket" items to be so easily dismissed. In fact, one study found that the growing volume of such items (e.g., laboratory tests, X-ray studies) contributed much more to cost increases than did such low volume, big ticket items as CT scans or coronary artery bypass surgery [81]. Moreover, if government took steps to make cost sharing appear more attractive, its utilization might increase accordingly. The Treasury Department's initial 1985 tax simplification proposal, for example, among other things would have taxed as income employer contributions to employee health insurance in excess of seventy dollars a month for an individual or $175 for a family; subsequently, the Treasury suggested taxing the first twenty-five dollars a month of employee paid health insurance, climbing to the first sixty-three dollars a month by 1990. The earlier version, aiming at reducing public subsidization of generous health plans, arguably may have been more efficacious than the latter, which fails to distinguish as to lavishness of coverage. But either proposal, by placing health insurance on the tax agenda, may be counted on to help redirect employee and union efforts toward maximizing wages and salaries, not fringes. This probably will leave emplyees in coming years with higher pay and lower coverage in the form either of higher deductibles or increased coinsurance.

Another key question is the effect of cost sharing on health care quality. More health care, of course, need not mean better health care;

sometimes, in fact, it may even prove counter-productive. Waste involving diagnostic tests [18, 26], prescribed drugs [60, 110], and hospital acute care [40, 100] is well known, as is variability in surgery rates [116, 121]. One observer, in fact, conservatively estimated the cost of clearly excessive health care that delivers little or no benefit to the patient as greater than ten percent of the total health care budget ([80], p. 11–13); in 1990, this would amount to approximately sixty billion dollars. (A candid analyst, however, must admit that waste is frequently very difficult to define or agree upon, as it often seems merely a function of an individual's highly subjective value system; my need may be your waste. Nor is waste synonymous with failure; even reasonable and prudent research projects and medical procedures do not always meet with success. Nor is efficiency the only value we seek to maximize; equity, political harmony, and a dozen others also compete for resources. Nor, finally, is the elimination of waste invariably worth the cost, which may include loss of physician/scientist autonomy, stifling delays and paperwork, and fear of risks and preoccupation with bureaucratic procedures that may smother the spirit of inquiry and innovation. Some waste, then, may simply be a cost of doing business, and other "waste" may arguably really not be waste at all.)

On the other hand, the less expensive provider may not be cutting more waste but simply cutting more corners, and cost sensitive patients may not always be able to tell the difference. It is fine, after all, to eliminate unnecessary health care (as it is to eliminate unnecessary anything), but distinguishing what is unnecessary is not invariably an easy thing to do. The symptom that might, with first dollar or comprehensive coverage, have sent a patient to his physician may turn out to be not a transient ache or pain, but an early warning signal for some dread disease. Nor is the ordering of tests or the prescribing of drugs such an exact science that a physician can always be certain what is not needed, and as tests and drugs multiply this sense of uncertainty may well grow with them. Moreover, the physician surely knows that if a vital test or drug was omitted for its apparent lack of cost effectiveness, this explanation will hardly satisfy the patient, his family, or their malpractice attorney. To pretend otherwise, say the critics, is to claim that it is possible to get something – and, in health care cost containment, a very big something – for nothing. And that is rarely the case.

Both regulation and cost sharing take very much a payment or service limited view of health care rights. Although there is some talk that overutilized services may be medically counter-productive, advocates of these reforms plainly are far more concerned with containing costs than with improving care. Thus, they take it for granted that health care has only a limited call upon the nation's resources, that other claimants may sometimes take precedence, in short, that society's obligations to provide health care are clearly limited. Or, to assume the consumer's point of view, regulation and cost sharing guarantee him only restricted access to the health care system.

## VIII. MEANS TESTS

Third is the imposition of a means test. This would confine taxpayer supported health care services only to those able to demonstrate that they could not afford them. The Veterans Administration has already proposed a means test for veterans under age sixty-five [118], and some analysts have been urging means test for Medicare for a number of years.

The means tests' advocates' argument, stated simply, is that while the truly needy should be helped, it is moral and financial folly to help those who are already quite able to help themselves [47].

It is the elderly, of course, who would suffer most from the adoption of such a strategy, both because they generate a disproportionate share of health care expenses and because they benefit from numerous age-based programs. Yet, it has been replied, the popular stereotype of the destitute elderly is grotesquely overdrawn. Over the past two decades the aged as a class have done exceedingly well in terms of consuming public resources, certainly far better than have, say, the young [94].

The means test, then, is not seen as hostile to the poor, whose principal concern is surely not lack of health care but lack of money with which to purchase health care and all the other goods and services they desire. Rather, the test is said to be hostile to those to whom the law has granted an unmerited windfall, like the elderly earning $32,000 per year and over, who account for almost as much Medicare expenditures as those earning less than $10,000 ([122], p. 205).

Critics, for their part, have retorted that a means test would be mean-spirited and degrading, that its advocates would be so antagonistic to the poor and so intent on saving money that many needy persons would be denied care, that tendencies toward providing the poor inferior quality care would be accelerated, and that much of the savings that the test would create would be consumed by high administrative costs. The

Supplementary Security Income system, one of government's largest means tested programs, is frequently held up as a cautionary example, with its invasive twelve page application form, its pitifully miserly allowances, and its array of bullying functionaries ever on the scent of abuse.

## IX. HEALTH MAINTENANCE ORGANIZATIONS

Fourth is health maintenance organizations (HMOs). They provide health care to a voluntarily enrolled population for a fixed prepayment. They may take the form either of a prepaid group practice (PGP), where member physicians are salaried or paid on an enrollment basis and share facilities and ancillary personnel, or an individual practice association (IPA), in which participating physicians operate separate offices and are paid on a fee for service basis. A new type of individual practice association is the primary care network, in which primary care physicians maintain clinical and financial control of the total health service for each voluntarily enrolled member. The PGP, however, remains the dominant form of HMO, and as such merits the closest attention.

Until a few years ago, HMOs faced a number of potent legal obstacles: some states required them to maintain reserves, like insurance companies, or to have boards of directors dominated by physicians, for example, and the American Medical Association had a rule forbidding the contract practice of medicine. With the Supreme Court's ruling that the learned professions are subject to antitrust laws (*Goldfarb v. Virginia State Bar* [44]), a decline in the political potency of organized medicine, and the vast pressures for cost containment brought about by surging health care expenditures, most restrictions have been swept aside [30]. Today about fifteen million persons are enrolled in HMOs, a figure expected to double by the end of the decade [35]. A new Medicare ruling that will pay HMOs directly for memberships for the elderly and the disabled has also opened a thirty million person market, though HMOs' wariness about future Medicare cost containment rulings has seriously tempered the urge to expand in that area. (Moreover, Medicare HMO demonstration projects in Florida produced decidedly mixed results [38].) The giant HMOs, like the Kaiser Health Plan in California and the Health Insurance Plan in New York, have always been nonprofits, but for-profit HMOs – many of them formerly nonprofit that converted to attract capital more easily – are beginning to multiply ([73], p. 22). Already hospital chains like Humana have begun to organize their own HMOs, and HealthAmerica has become the first HMO chain to be listed on the New York Stock Exchange.

The great argument for PGPs is lower health care costs. Indeed, PGP savings run between ten and forty percent of conventional insurance plans ([66], p. 511). How are these savings to be explained? The obvious answer is that the provider has a powerful incentive to hold down costs, since he operates on a prepaid and not a cost-plus basis. This is reflected in managerial cost cutting innovations, increased use of nonphysicians and generic drugs, greater willingness to purchase services from other providers when it is economically advantageous to do so, more emphasis upon preventive care, and by far most important, lower hospital admission rates.

More broadly, what PGPs reward is a less elaborate, more conservative style of practice. This entails not only a lower rate of surgical procedures and laboratory tests [7, 125], but also a less prominent role for specialists. While the United States is unique in its reliance upon specialists [106], who tend to adopt and use high technology because of their training and socialization [8], HMOs offer a much greater role to generalists, who have a more parsimonious style and serve as gatekeepers to prevent precisely those specialists from becoming overutilized. Where regulations that change the rules tend after a while to be undermined by evasive behavior, HMOs, it is said, encourage a rather impressive level of discipline. (It is also possible, however, that HMO savings are partly a consequence of physician self-selection [75]. That is, HMOs may not so much produce a conservative style of practice as attract physicians who already adhere to that style. To the extent that this is the case, HMOs cost saving potential for the society as a whole would be undercut.).

Indeed, substantial HMO membership may help to contain costs for nonmembers as well, for as HMO membership costs become widely known, other providers may feel compelled to adopt a more competitive price position [44]. Price dispersion, therefore, should narrow even as price increases are restrained.

Meanwhile, from the patient's point of view, he has acquired comprehensive coverage, and need not fear the cost implications of health problems nor even fret over paperwork. Indeed, the simplicity of the arrangement from the patient's perspective is one of its chief advantages, for families find simple programs far easier to understand [71], making it more likely that consumers will play their expected role. Thus, though provider decisions tend to hold down costs (mainly by a reluctance to recommend hospitalization), consumer decisions go in the opposite direction (i.e., patients are more likely to see their doctors). Because the consumer decisions are far less costly than the providers', however, consumers can be accommodated without raising total health care costs.

HMOs also facilitate consumer choice by greatly simplifying his options and thus requiring less information from him. (This is not a trivial advantage, for the process of becoming informed is so costly in terms of time, effort, rejected alternatives, and so forth that it is a potent disincentive to acquiring knowledge.) Instead of feeling that he must search out information on a large number of physicians, he can obtain literature from the local HMOs, talk to a couple of subscribers and compare their offerings with those of the physicians he has learned about. If HMO membership is offered by the consumer's employer – along, perhaps, with a cost sharing option – the individual's choice is simpler still.

Can all this be accomplished without sacrificing quality? The evidence is incomplete, but the signs are hopeful ([66], pp. 52, 57). If patients in conventional insurance plans seem somewhat happier with their relationship with their physicians, HMO enrollers express greater satisfaction with the technical aspects of their medical care ([126], p. 544), and most observers would rate that variable as more important. For just as providers have incentives to lower costs, they also have incentives to maintain quality: a dissatisfied patient may be able to terminate his membership; a good reputation is essential to long term prosperity; peer pressure is enhanced by the group practice setting; standards of external quality review or accrediting boards must be met; the cost and publicity of malpractice suits must be avoided; and so on. With HMOs, the care costs less chiefly because the patient receives less. Health care, however, is not the true output of a health care system; health is. And less health care need not – and, normally, in the case of HMOs does not – mean poorer health.

Of course, HMOs are not without their disadvantages: the patient's choice of primary physicians is confined to those working for the HMO (unless he chooses to go outside the system and absorb all the costs himself); patients selecting HMOs may not always understand that they are choosing not only insurance coverage but a particular style of medical practice; HMOs may not be conveniently located, attract a prestigious clientele, or feature many amenities [11]. Still, sizable percentages of employees choose HMOs when given the option ([32], p. 284; [31]), and as more employers see these as a cost containment device, they are likely to become considerably more numerous. They need entrepreneurial talent, capital, and time to hire physicians, develop facilities, and enroll patients, and so their enrollment may not be able to grow faster than about ten percent per year ([42], p. 233; but *cf.* [35], where a twenty percent per year growth estimate is reported.) In many regions, though, they will play a very important role; in some, like Minneapolis and Seattle, they already do.

A variation on the HMO is the preferred provider organization consisting of a group of hospitals and physicians that contract with third party payers to provide comprehensive health care coverage. These providers practice a conservative style of medicine, and in order to induce patients to join, they are freed from cost sharing or deductible arrangements that would apply if other providers were used.

The HMO approach proceeds from the assumption that health care costs can best be contained by increasing efficiency, and that this involves restructuring incentives for both providers and patients. It also assumes that the process will not be easy or entail sacrifices only from ancillary personnel, but will call for a major shift in the style of practicing medicine that, in turn, will require fewer physicians and hospital beds than Americans have grown accustomed to.

A recent conservative prediction foresees a surplus of 51,800 by 2000 [53]. In this context, the only way for physician incomes to rise is for physicians to practice more elaborate medicine. Technology certainly will make that possible. But HMOs – and the societal dissatisfaction with the rising health care costs that lies behind them – will just as certainly struggle against rising costs. And it may be hard to justify a ratio of 2.9 physicians per thousand persons, when HMOs frequently get by with half that figure. Similarly, hospitals seeking to maintain a usage rate of 1200 patient days per thousand persons will find HMOs with rates of approximately 800, and there will be no obvious way to make up the difference. (Gaus and his colleagues reported rates of 356 for Medicaid patients using HMOs, as contrasted with 936 for those relying upon fee for service physicians [31].) Complicating matters further, physicians will doubtlessly feel forced to compete with hospitals through ambulatory surgery centers, home care organizations, and various forms of group practice, while large employers will be pressing for discounts from the providers who will care for their employees. All of these are evident today; the future will merely accentuate these developments.

Both means tests and HMOs take a population-limited view of health care rights. That is, instead of limiting payment to providers (like DRGs) or consumers (like cost sharing arrangements) or limiting services (like Medicare's pre-admission review program), means tests and HMOs focus on limiting the population to be served. It is not that these populations are offered only partial medical coverage but instead that they are rather narrowly defined, either in terms of need (as with means tests) or formal enrollment (as with HMOs). For it is precisely the opportunity to provide comprehensive coverage to its limited population that offers these reform proponents the opportunity to practice a more conservative style of medicine on a large scale.

## X. CONCLUSIONS

To Americans, our ears filled with noisy exhortation and mumbled advice, there is something ineffably attractive about rights talk. It is a homage to our Great Past; it is a retort to those who smirk at our capacity for Deep Thought; it is a call to arms for those straining to Fight the Good Fight. Thus, even though "rights" carries with it a certain Enlightenment or Victorian aura, it is no surprise that it is quite fashionable today. Perhaps, it has always been in.

What is fairly new, though, is talk not of such generally accepted rights as the right to freedom of speech or the right to be safe from unreasonable searches and seizures, but such rights as the right to health care. This had to wait for affluence to make it seem possible and the triumph of welfare state mentality to make it seem desirable.

The right to health care is, on the one hand, a wholly unqualified obligation on the part of society not to interfere with the individual's pursuit of his own health care, but also, on the other hand, the obligation to provide that care when the individual demands it. The former may be viewed as a negative, procedural right, and the latter as positive and substantive. As one moves across the spectrum, the right becomes steadily more qualified, as "reasonable" grounds are found for defending a limitation here or a restraint there. Normally, the pressure for these qualifications comes from resource scarcity, though it is, of course, not always acknowledged as such.

There are a series of major problems associated with the assertion of an unqualified right to health care, aside from the fact that policy makers have never believed that health care merits an unlimited call on the nation's resources. It mistakes the means of health care for the end of health; it implies mutually contradictory positions on the issue of medical paternalism; it foresees no conflict in the simultaneous pursuit of excellence and equality; it represents a profoundly apolitical solution to a profoundly political problem; and it creates a mob of practical and often controversial sub-issues. Indeed, even if investing in an unqualified right to health care were demonstrated to guarantee a concomitant improvement in health, the unqualified right would not be established to everyone's satisfaction. To some, the coercion involved in the massive resource redistribution to the health care industry would simply seem excessive. To others, the neglect of competing goods that this transference would entail would appear offensive. And to still others, the downplaying of the individual's responsibility for his own health would seem absurd, if not actually immoral. Absent a consensus on the unqualified right to health care, these dissenters would maintain, it is certainly premature to elevate it to that lofty position.

These defects in and dissatisfactions with the unqualified rights approach, which lead many to dismiss it as mere special interest pleading, may well constitute the strongest argument for the qualified rights approach. For whether qualified in terms of services or payments offered or population to be served, the problem of line drawing retains its vexing and, to some irreducible degree, arbitrary nature. In fact, a qualified right may seem indistinguishable from an ordinary claim, except maybe as a statement of a social policy goal.

In an era of a widely perceived health care crisis – that is, a time when health care costs appear on the verge of becoming intolerably high – the qualified right approach has an insurmountable advantage. This becomes even clearer when the four chief health care reform ideas are sketched. One pair, regulation and cost sharing, qualifies the right to health care with payment or service limitations, seeking to restrict the consumer's access to care. A second pair, mean tests and health maintainance organizations, qualifies the right by confining it to a well defined population served (in the case of HMOs) by providers given incentives to contain costs. Each of these reforms, like most human contrivances, has its costs/risks and benefits, though the decisions as to which to adopt and how far to commit may be made less on technical grounds than on political ones. What does seem sure, though, is that rights talk will contribute more heat than light to the discussion.

*Baruch College*
*City University of New York*
*New York, U.S.A.*

## NOTE

* I am grateful to the Center for the Study of Business and Government of Baruch College for its generous support.

## BIBLIOGRAPHY

1. Abel-Smith, B.: 1978, 'Minimum Adequate Levels of Personal Health Care: History and Justification', *Milbank Memorial Fund Quarterly* **56**, 7–21.
2. Aday, L. and Andersen, R.: 1984, 'The National Profile of Access to Medical Care: Where Do We Stand?', *American Journal of Public Health*, **74**, 1331–1339.
3. American Board of Family Practice: 1985, *Rights and Responsibilities: A National Survey of Health Care Opinions*, American Board of Family Practice, Lexington, Kentucky.
4. Andersen, O.W.: 1984, 'Health Services in the United States: A Growth Enterprise for a Hundred Years', in T.J. Litman and L.S. Robins (eds.), *Health Politics and Policy*, Wiley, New York, pp. 67–80.
5. Andersen, O.W.: 1976, 'PSROs, the Medical Profession, and the Public Interest', *Milbank Memorial Fund Quarterly*, **54**, 379–388.
6. Annas, G.J.: 'Regulating the Introduction of Heart and Liver Transplantation', *American Journal of Public Health*, **75** 93–95.
7. Arnould, R.J., Debrock, L.W., and Pollard, J.W.: 1984, 'Do HMOs Provide Specific Services More Efficiently?', *Inquiry*, **21**, 243–253.
8. Banta, H.B., Behney, C.B., and Willems, J.S.: *Toward Rational Technology in Medicine*, Springer Publishing, New York.
9. Beck, D.F.: 1985, 'The Hospital's Financial Future: DRGs and Beyond', *Health Care Supervisor*, **3**, 1–10.
10. Beck, R.G.: 1974, 'The Effects of Co-Payment on the Poor', *Journal of Human Resources*, **9**, 128–142.
11. Berki, S.E. and Ashcraft, M.L.F.: 1980, 'HMO Enrollment: Who Joins What and Why: A Review of the Literature', *Milbank Memoral Fund Quarterly* **58**, 581–603.
12. Blumstein, J.F.: 1976, 'Constitutional Perspectives on Governmental Decisions Affecting Human Life and Health', *Law and Contemporary Problems* **40**, 231–305.
13. Brandt, R.: 1979, *A Theory of the Right and the Good*, Clarendon Press, Oxford.
14. Brook, R.H., Ware, J.E., Rogers, W.H. *et al.*: 1983, 'Does Free Care Improve Adults' Health?', *New England Journal of Medicine* **309**, 1426–1434.
15. Census Bureau: 1985, *Survey of Income and Program Participation.*

16. Childress, J.F.: 1979, 'A Right to Health Care?', *Journal of Medicine and Philosophy* **4**, 132–147.
17. Daniels, N.: 1981, 'Health Care Needs and Distributive Justice', *Philosophy and Public Affairs* **10**, 146–179.
18. Dixon, R.H. and Laszlo, J.: 1974, 'Utilization of Clinical Chemistry Services by Medical House Staff', *Archives of Internal Medicine* **134**, 1064–1067.
19. Dobson, A., Greer, J.G., and Carlson, R.H. *et al.*: 1978, 'PSROs: Their Current Status and Their Impact to Date', *Inquiry* **15**, 113–128.
20. Donlan, T.G.: 1985, 'Carolyne Who? Carolyne K. Davis Talks Softly, Controls Big Purse', *Barron's*, March 11, pp. 36, 40.
21. Donabedian, A.: 1978, 'The Quality of Medical Care', *Science* **200**, 856–864.
22. Dresnick, S.J., Roth, W.I., Linn, B.S. *et al.*: 'The Physician's Role in the Cost-Containment Program', *Journal of the American Medical Association* **2418**, 1606–1609.
23. Dyck, F.J., Murphy, J.K., Road, D.A. *et al.*: 1977, 'Effects of Surveillance on The Number of Hysterectomies in the Province of Saskatchewan', *New England Journal of Medicine* **296**, 1326–1328.
24. Dyckman, Z.Y.: 1978, *A Study of Physicians' Fees.* Government Printing Office, Washington.
25. Easton, D.: 1953, *The Political System: An Inquiry into Political Science*, Knopf, New York.
26. Edwards, L.D., Levin, S., and Balagatas, R.: 1973, 'Ordering Patterns and Utilization of Bacteriologic Culture Reports', *Archives of Internal Medicine* **132**, 678–682.
27. Egelman, A.: 1984, 'HMOs: Incentive for Cost-Efficiency', *New York Times*, December 27, p.A20, col. 1.
28. *Elder v. Beal*: 609 F, 2d 695 (3rd Cir.).
29. Enthoven, A.C.: 1980, *Health Plan*, Addison-Wesley, Reading, Mass.
30. Falkson, J.: 1980, *HMOs and the Politics of Health System Reform*, American Hospital Association, Chicago.
31. Farley, P.J. and Wilensky, G.R.: 1983, 'Options, Incentives and Employment-Related Health Insurance', in R.M. Schefler and L.F. Rossiter (eds.), *Advances in Health Economics and Health Services Research*, JAI Press, Greenwich, Connecticut, pp. 57–82.
32. Feldstein, P.J.: 1979. *Health Care Economics*, Wiley, New York.
33. Fisher, A.B.: 1985, 'The New Game in Health Care: Who Will Profit', *Fortune* 4 March, **111**, 138–143.
34. Freeland, M.S. and Schendler, C.E.S.: 1983, 'National Health Expenditure Growth in the 1980s: An Aging Population, New Technologies, and Increased Competition', *Health Care Financing Review* **4**, 1–58.
35. Freudenheim, M.: 1985, 'H.M.O. Growth Displays Vigor', *New York Times*, April 16, p. D2, col. 1.
36. Fried, C.: 1975, 'Rights and Health Care – Beyond Equity and Efficiency', *New England Journal of Medicine* **293**, 241–245.
37. Gaus, C.R., Cooper, B.S., and Hirschman, C.G.: 1975, 'Contracts in HMO and Fee-for-Service Performance', *Social Security Bulletin*, May 29, 3–14.

38. General Accounting Office: 1985, Problems in Administering Medicare, Health Maintenance Organization Demonstration Projects in Florida, GAO/HRD–85–48.
39. General Accounting Office: 1984, Excessive Respiratory Therapy Cost and Utilization Data used in Setting Medicare's Prospective Payment Rates GAO/HRD–84–90.
40. Gertman, P.M. and Restuccia, J.D.: 1981, 'The Appropriateness Evaluation Protocol: A Technique for Assessing Unnecessary Days of Hospital Care', *Medical Care* **19**, 855–871.
41. Gibson, R.M. and Waldo, D.R.: 1982, 'National Health Expenditures, 1981', *Health Care Financing Review* **4**, 1–55.
42. Ginsburg, P.B.: 1981, 'Altering the Tax Treatment of Employment-Based Health Plans', *Milbank Memorial Fund Quarterly* **59**, 224–255.
43. Goldberg, L.G. and Greenberg, W.: 1980, 'The Competitive Response of Blue Cross to the Health Maintenance Organization', *Economic Inquiry* **18**, 55–68.
44. *Goldfarb v. Virginia State Bar*: 1975, 421 U.S. 773.
45. Gorovitz, S.: 1966, 'Ethics and the Allocation of Medical Resources', *Medical Research in Engineering* **5**, 5–7.
46. Griffith, J.R.: 1984, 'DRGs – What's Next?', *Two Views*, Mount Sinai School of Medicine, New York.
47. Halper, T.: 1984, 'Aging Policy in the Eighties: Second Thoughts on a Strategy That Has Worked', in S.F. Spicker and S.R. Ingman (eds.), *Vitalizing Long-Term Care*, Springer Publishing, New York, pp. 3–13.
48. Halper, T.: 1985, 'End-Stage Renal Failure and the Aged in the United Kingdom', *International Journal of Technology Assessment in Health Care* **1**, 41–52.
49. Halper, T.: 1985, 'Life and Death in a Welfare State', *Milbank Memorial Fund Quarterly* **63**, 52–93.
50. Halper, T.: 1979, 'On Death, Dying and Terminality: Today, Yesterday and Tomorrow', *Journal of Health Politics, Policy, and Law* **4**, 11–29.
51. Halper, T.: 1981, *Power, Politics, and American Democracy*, Scott, Foresman, Glencoe.
52. *Harris v. McRae*: 1980, 448 U.S. 297.
53. Health and Human Services, Bureau of Health Professions: 1984, *Report on Physician Supply.*
54. Herron, C.: 1985, 'Hospital Readmission Is Not Cost-Effective', *New York Times*, May 2, p. A26, col. 6.
55. Holahan, J. and Scanlon, W.: 1978 'Physician Pricing in California: Price Controls, Physician Fees, and Physician Incomes from Medicare and Medicaid', Health Care Financing Growth and Contracts Report prepared pursuant to contract SSA 600–76–0054.
56. 'Hospital Costs Rose by 4.6% in '84, Lowest Rate Since '63': 1985, *New York Times*, April 14, I, p. 46, col. 4.
57. Kelly, S.P.: 1978, 'Physicians' Knowledge of Hospital Costs', *Journal of Family Practice* **6**, 171–172.

58. Kirkland, L.R.: 1979, 'The Physician and Cost Containment', *Journal of the American Medical Association* **242**, 1032.
59. Kohlman, H.A.: 1984, 'Determining a Contribution Margin for DRG Profitability', *Healthcare Financial Management*, April, 108–110.
60. Kunin, C.M., Tupasi, T., and Craig, W.A.: 1973, 'Use of Antibiotics: A Brief Exposition of the Problem and Some Tentative Solutions', *Archives of Internal Medicine* **79**, 555–560.
61. Kurskey, G.F.: 1973, 'Health Care, Human Rights and Government Intervention: A Critical Appraisal', *California Dental Association Journal*, in R. Hunt and J. Arras (eds.): 1977, *Ethical Ideas in Modern Medicine*, Mayfield Publishing, Palo Alto, pp. 465–471.
62. Law, S.: 1974, *Blue Cross: What Went Wrong?*, Yale University Press, New Haven.
63. Levin, S. and Balagatas, R.: 1973, 'Ordering Patterns and Utilization of Bacteriologic Culture Reports', *Archives of Internal Medicine* **132**, 678–682.
64. Levine, S., Feldman, J.J., and Elinson, J.: 1983, 'Does Medical Care Do Any Good?', in D. Mechanic (ed.), *Handbook of Health, Health Care, and the Health Professions*, Free Press, New York, pp. 394–404.
65. Luft, H.S.: 1980, 'Assessing the Evidence on HMO Performance', *Milbank Memorial Fund Quarterly* **58**, 501–536.
66. Luft, H.S.: 1981, 'The Operations and Performance of Health Maintenance Organizations: A Synthesis of Findings from Health Service Research', prepared for the National Center for Health Services Research.
67. Luft, H.S. and Frisvold, G.A.: 1979, 'Decisionmaking in Regulatory Health Planning Agencies', *Journal of Health Politics, Policy, and Law* **4**, 250–272.
68. *Maher v. Roe*: 1977, 432 U.S. 464.
69. Marmor, T.: 1985, 'A Political Scientist's View', *Bulletin of the New York Academy of Medicine* **61**, 101–106.
70. Marmor, T. and Marmor, J.: 1973, *The Politics of Medicare*, Aldine Publishing Co., Chicago.
71. Marquis, M.S.: 1983, 'Consumers' Knowledge about Their Health Insurance Coverage', *Health Care Financing Review* **5**, 65–80.
72. Martin, D.: 1983, 'Health Care in Canada: Popular System Now Rocked by Criticism', *New York Times*, February 15, p. C1, col.1.
73. Masso, A.R.: 1985, 'HMOs in Transition: What the Future Holds', *Business Health*, January-February, **2**, 21–24.
74. McClure, W.: 1982, 'Implementing a Competitive Medical Care System through Public Policy', *Journal of Health Politics, Policy, and Law* **7**, 2–43.
75. McClure, W.: 1982a, 'Toward Development and Application of a Qualitative Theory of Hospital Utilization', *Inquiry* **9**, 117–135.
76. McCulloch, L.B.: 1979, 'The Right to Health Care', *Ethics in Science and Medicine*, **6**, 1–9.
77. McKeown, T.: 1976, *The Role of Medicine: Dream, Mirage or Nemesis.* Nuffield Provincial Hospitals Trust, London.
78. McKinlay, J.B. and McKinlay, S.M.: 1977, 'The Questionable Contribution of Medical Measures to the Decline of Mortality in the U.S. in the Twentieth

Century', *Milbank Memorial Fund Quarterly* **55**, 405–428.
79. Melnick, G.A., Wheeler, J.R.C., and Feldstein, P.J.: 1981, 'Effects of Rate Regulation on Selected Components of Hospital Expenses', *Inquiry* **18**, 240–246.
80. Menzel, P.T.: 1983, *Medical Costs, Moral Choices*, Yale University Press, New Haven.
81. Moloney, T.W. and Rogers, D.E.: 1979, 'Medical Technology – A Different View of the Contentions Debate over Costs', *New England Journal of Medicine* **301**, 1413–1419.
82. Morgan, R.E.: 1985, *Disabling America: The "Rights Industry" in Our Time*, Basic Books, New York.
83. Newhouse, J.P., Manning, W.G., Morris, C.N. *et al.*: 1981, 'Some Interim Results from a Controlled Trial of Cost Sharing in Health Insurance', *New England Journal of Medicine* **305**, 1501–1507.
84. Newhouse, J.P., Manning, W.G., Morris, C.N. *et al.*: 1982, *Some Interim Results from a Controlled Trial of Cost Sharing Health Insurance*, Rand Corporation (publication R–2847–HHS), Santa Monica.
85. Nozick, R.: 1974, *Anarchy, State, and Utopia*. Basic Books, New York.
86. Outka, G.: 1974, 'Social Justice and the Equal Access to Health Care', *Journal of Religious Ethics* **2**, 11–32.
87. Parsons, F.M. and Ogg, C.S. (eds.): 1983, *Renal Failure – Who Cares?* MTP Press, Lancaster.
88. Pauley, M.V. and Langwell, K.M.: 1982, *Research on Competition in the Market for Health Services: Problems and Prospects*. Applied Management Systems, Silver Spring, Maryland.
89. Pear, R.: 1985a, 'Companies Tackle Health Costs', *New York Times*, 3 March, III, p. 11, col. 1.
90. Pear, R.: 1985b, 'Health Care Regulation: Bane? Balm? Accident?' *New York Times*, January 15, p. A16, col. 3.
91. Pear, R.: 1985c, 'Medical Inflation Slackens Slightly', *New York Times*, January 28, p. A1, col. 1.
92. Pear, R.: 1985d, 'U.S. Plans to Freeze Medicare Hospital Payments', *New York Times*, May 29, p. A1, col. 4.
93. Pear, R.: 1985e, 'Washington Drives a Hard Bargain on Medical Costs', *New York Times*, January 20, IV, p. 3, col. 1.
94. Preston, S.H.: 1984, 'Children and the Elderly in the U.S.', *Scientific American* **251**, December, 44–49.
95. Purdum, T.S.: 1985, 'Hospital Company and No. 1 Supplier Plan Huge Merger', *New York Times*, p. A1, col. 1.
96. Radin, M.: 1900, 'A Restatement of Hohfeld', *Harvard Law Review* **51**, (1928), 1141–64.
97. Rawls, J.: 1971, *A Theory of Justice*. Belknap Press of the Harvard University Press, Cambridge.
98. *Regents of the University of California v. Bakke*: 1978, 438 U.S. 265.
99. Reinhold, R.: 1982, 'Competition Held Key to Lower Medical Costs', *New York Times*, 1 April, p. A1, col. 2.

100. Restuccia, J.D. and Holloway, D.C.: 1976, 'Barriers to Appropriate Utilization of an Acute Facility', *Medical Care* **14**, 559–573.
101. Rettig, R.A.: 1976, 'The Policy Debate on Patient Care Financing for Victims of End-Stage Renal Disease', *Law and Contemporary Problems*, Autumn, 196–230.
102. Rice, T. and McCall, N.: 1982, 'Changes in Medicare Reimbursement in Colorado: Impact on Physicians' Economic Behavior', *Health Care Finance Review* **3**, 67–84.
103. Rosko, M.D. and Boyles, R.W.: 1985. 'Unintended Consequences of Prospective Payment: Erosion of Hospital Financial Position and Cost Shifting', *Health Care Management Review* **9**, 35–44.
104. Salkever, D.S. and Bice, T.W.: 1978, 'Certificate-of-Need Legislation and Hospital Costs', in M. Zubkoff, I.E. Raskin, and R.S. Hart (eds.), *Hospital Cost Containment: Selected Notes for Future Policy*, Prodist, New York, pp. 429–460.
105. Schelling, T.C.: 1968, 'The Life You Save May Be Your Own', in S.B. Chas (ed.), *Problems in Public Expenditure Analysis*, Brookings Institution, Washington, pp. 127–162.
106. Schroeder, S.A.: 1984, 'Western European Responses to Physician Oversupply – Lessons from the United States', *Journal of the American Medical Association* **252**, 373–384.
107. *Schweiker v. Gray Panthers*: 1981, 453 U.S. 1.
108. Scitovsky, A. and McCall, N.: 1977, 'Coinsurance and the Demand for Physicians Services: Four Years Later', *Social Security Bulletin* **40**, 19–27.
109. Sen, A.K.: 1973, *On Economic Inequality*, Norton, New York.
110. Shapiro, M., Townsend, T.R., Rosner, B., Kass, E.H.: 1979, 'Use of Antimicrobial Drugs in General Hospitals: Patterns of Prophylaxis', *New England Journal of Medicine* **301**, 351–355.
111. Shepard, D.S.: 1983, 'Estimating the Effects of Hospital Closure on Areawide Impatient Hospital Costs: A Preliminary Model and Application', *Health Services Research* **18**, 513–550.
112. Sloan, F.A.: 1981, 'Regulation and the Rising Cost of Hospital Care', *Review of Economics and Statistics* **63**, 479–487.
113. Sullivan, R.: 1979, 'New York's Hospitals Will Gain by Filling Fewer Beds', *New York Times*, December 31, p. A13.
114. Sullivan, R.: 1985, 'Decline in Hospital Use Tied to New U.S. Policies', *New York Times*, April 16, p. A1, col. 1.
115. Taylor, A.K.: 1984, 'Hospital Cost Inflation, Health Insurance and Market Incentives in the Hospital Industry', R.M. Scheffler and L.F. Rossiter (eds.), *Advances in Health Economics and Health Services Research*, Vol. 5, JAI Press, Greenwich, Connecticut, pp. 237–276.
116. Vayda, E.: 1973, 'A Comparison of Surgical Rates in Canada and in England and Wales', *New England Journal of Medicine* **289**, 1224–1229.
117. Veatch, R.M.: 1976, 'What Is a 'Just' Health Care Delivery?', in R.M. Veatch and R. Branson (eds.), *Ethics and Health Policy*, Ballinger, Cambridge, pp. 127–153.

118. Veterans Administration: 1984, *Caring for the Older Veteran*, Government Printing Office, Washington.
119. Wallace, A.C.: 1985, 'Finding the Right Niche in Health Care', *New York Times*, April 7, III, p. 10, col. 2.
120. Warner, R.: 1980, *Morality in Medicine*. Alfred Publishing, Sherman Oaks.
121. Wennberg, J.E., Barnes, B.A., and Zubkoff, M.: 1982, 'Professional Uncertainty and the Problem of Supplier Induced Demand', *Social Science and Medicine* **16**, 811–824.
122. Wilensky, G.R.: 1982, 'Government and the Financing of Health Care', *American Economic Review* **72**, 202–207.
123. Wilensky, G.R. and Walden, D.C.: 1981, 'Minorities, Poverty, and the Uninsured', National Center for Health Services Research, Department of Health and Human Services, Hyattsville, Maryland.
124. Willems, J.S.: 1979, 'The Relationship between the Diffusion of Medical Technology and the Organization and Economics of Health Care Delivery', in J.L. Wagner (ed.), *Medical Technology*, National Center for Health Services Research, Hyattsville, Maryland.
125. Williams, B.: 1962, 'The Idea of Equality', in P. Laslett and W.G. Runciman (eds.), *Philosophy, Politics, and Society, 2d ser.*, Blackwell, Oxford.
126. Wolinsky, F.D.: 1980, 'The Performance of Health Maintenance Organizations: An Analytic View', *Milbank Memorial Fund Quarterly* **58**, 537–587.
127. Worthington, N.L. and Piro, P.A.: 1982, 'The Effects of Hospital Rate-Setting Programs on Volumes of Hospital Services: A Preliminary Analysis', *Health Care Financing Review* **4**, 47–66.
128. Yarbro, J.W. and Mortenson, L.E.: 1985, 'The Need for Diagnosis-Related Group 471', *Journal of the American Medical Association* **253**, 684–685.

ALLEN E. BUCHANAN

# RIGHTS, OBLIGATIONS, AND THE SPECIAL IMPORTANCE OF HEALTH CARE

## I. INTRODUCTION

My aim in this essay is to develop a more rigorous moral case than has hitherto been made for establishing a general right to a "decent minimum" or "adequate level" of health care.[1] The most distinctive characteristic of the approach I shall pursue is that it does not attempt to support the establishment of such a general legal right by appealing to a general moral right to health care. Since attempts to justify the claim that there is a general moral right to health care have been extensively criticized by myself and others, I will not here rehearse the difficulties of the general moral right approach [5]. Nor will I review the equally familiar and compelling reasons for maintaining that if there is a sound moral justification for a general legal right to health care, then it must be a limited right, a right to a "decent minimum" or "adequate level" of health care or "health care floor", rather than a right to all beneficial care, a right of all to equal dollar amounts for care, or a right to a level of care equal to that which others receive ([10], Vol. I, pp. 18–21). Instead, I shall concentrate on developing a case for a limited general legal right to health care based on the thesis that there is a general moral obligation to provide such care to those who cannot provide it for themselves, even if there is no general moral right to health care of any kind. In doing so, I shall offer answers to four questions. (1) To what extent is the obligation to provide health care for the needy limited by their responsibility for their health status? (2) If the general moral obligation to provide a decent minimum of care for the needy is an obligation for which there is no correlative moral right, how can it be viewed as an enforceable "societal" or "collective" obligation to provide care for all the needy rather than as an unenforceable individual duty of charity which the individual may discharge by contributing to the health needs of only those persons whom he or she chooses to aid? In other words, if there is no general moral right to some level of health care, how is it possible to justify the use of government's coercive power to ensure that *everyone* in need has access to some health care? (3) If there is



*T.J. Bole III and W.B. Bondeson (eds.), Rights to Health Care*, 169–184.

an enforceable moral obligation to provide a decent minimum of health care for all, is there also an enforceable moral obligation to provide minimal levels of other important goods such as food and shelter? (4) What does the argument for a general moral obligation to provide a decent minimum of health care for all imply about the morally acceptable means of seeing that this obligation is fulfilled? In particular, what is the proper role for the Federal government?[2]

## II. THE SPECIAL VALUE OF HEALTH CARE AND THE GENERAL MORAL OBLIGATION OF CHARITY

The President's Commission Report, *Securing Access to Health Care*, (Vol. 1) [10], lists four features of health care that make it especially valuable. Health care can promote personal well-being, broaden an individual's range of opportunities, provide information that relieves worry or enables a person to plan how to cope with his or her situation, and can serve to affirm a sense of community in the face of the suffering and death to which we are all subject.

The Commission's view here requires qualification. First, even if all forms of health care are intended to serve one or more of these functions, not all of them are actually efficacious in any of these ways. Second, some of the functions that are said to make health care of special importance are not unique to health care. For instance, education widens opportunity, and food and shelter contribute to well-being. So even if some of the features listed are unique to some forms of health care and even if all of the forms of health care that possess these features are of great value, it only follows that some forms of health care are especially important relative to some other goods. It does not follow that health care is *uniquely* valuable, if this means that it is preeminently valuable among all goods. The President's Commission Report carefully (though perhaps not convincingly) skirted the issue of whether the special importance of other goods such as food, shelter, and basic education, also grounds a "societal" obligation to provide some minimal level of these for all. In order to answer this question, we must first set out more carefully the connection between the special importance of health care and a general moral obligation to ensure a decent minimum of health care, before addressing the question of whether the conclusion can be generalized to an obligation to provide a more comprehensive welfare floor.

All of the major traditions of religious ethics and all major philosophical normative ethical theories include a general obligation of beneficence or charity – a duty to help others in need, even when one stands in no special relationship to them and even though they have no right to one's aid. Indeed, it would be quite surprising if they did not, since morality is centrally, though not exclusively, concerned with human well-being. It would be very strange, in particular, if morality contained strict and definite injunctions against harming others or infringing their autonomy, but was utterly silent on whether we are to aid them when we can do so without significant cost to our own well-being or autonomy. After all, at least part of the point of not harming is concern for well-being, and at least part of the value of autonomy is that a competent person who is allowed to make his or her own choices is, at least in general, and in the long-run, more likely to promote his well-being effectively than if he were ruled by another (especially if that other is the state). So if, as I shall assume without further argument, any plausible morality will include a general obligation to help the needy, and if health care, or at least some forms of it, makes contributions to well-being that are especially important and that at least in some cases cannot be secured by other means, then it follows that the general obligation to help the needy includes an obligation to provide some types of health care to those who need it but who cannot secure it for themselves.

This obligation, like any obligation of charity, is always limited, and in some circumstances may be suspended altogether. The chief limitation is that one is not required to render aid to the needy if doing so imposes excessive or unreasonable costs on oneself. The great difficulty, of course, is the vagueness of this limitation. However, the problem does not seem insurmountable if several points are kept in mind. First, equally vague "reasonableness" standards are applied, with apparently generally acceptable results, as limitations on other obligations, including negative legal duties in the criminal and civil law. For example, a successful plea of self-defense against the charge of murder requires that the killer's action was that which a "reasonable man" would perform in the circumstances; and in the law of torts a determination that a standard of reasonable care was not met is required for proof of negligence [1]. Second, even if the borderline between charity which involves "unreasonable" or "excessive" costs and that which does not is fuzzy, some acts of aiding are clearly on one side of the line or the other. For example, simply lifting a drowning child's head from the bathtub or contributing a small percentage of one's very high income to the poor does not involve excessive costs. Whether or not *requiring* someone to render aid in these ways is morally justified or would involve a violation of his or her rights despite the fact that the costs are trivial is a different issue, which will be addressed below. At present, my point is simply that it is implausible to maintain that an obligation of easy rescue or an obligation to contribute a small portion of one's income to the most needy, and then only if one's own income exceeds some generous threshold, imposes excessive costs. Consequently, if enforcing such obligations violates individuals rights or is otherwise morally unjustifiable, it cannot be *because* the costs of fulfilling such obligations are excessive.

Some have suggested that even where the costs of aiding would not be excessive, obligations of charity may be reduced in scope, or even suspended altogether, if the individual in need is responsible for his plight. Whether or not an individual's free choice to engage in extremely risky behavior or to squander his means of support suspends entirely obligations of charity toward him, is perhaps not clear. But this much seems uncontroversial: What counts as an excessive cost in aiding an individual in need, and hence the scope of our obligation to aid, may be determined in part by the extent to which that individual is responsible for being in a condition of need. For example, I might be justified in refusing to delay my own vacation by even an hour in order to aid a foolish motorist who willfully ignores signs warning that the road is impassable except to four-wheel drive vehicles. Yet the same delay – of one hour – might not count as excessive costs if someone were in need of my help through no fault of his own. In general, it seems that the more serious a person's need, the more certain we must be that he was fully or at least substantially responsible for his plight, if we are to be justified in concluding that the obligation to aid him is suspended (or even significantly reduced in scope).

The relevance of these reflections to the obligation to provide some minimal level of health care is clear. Even if, as seems highly likely, individuals are often at least *in part* responsible for their ill-health, this is not sufficient to suspend (or even perhaps significantly reduce the scope of) the obligation to provide some minimal level of health care for them when they lack the resources to do so themselves, at least in cases in which the care in question is very likely to make a significant impact on their well-being or opportunity. For in order to show that the obligation to provide uncontroversially important types of care is suspended or significantly reduced in scope, it is not sufficient to show that individuals are *in part* responsible for their health care needs. Instead, at least where their need is great, it would have to be established that they are fully or at least largely responsible. Granted our current ignorance of the etiology of many diseases, and granted that chance and heredity undoubtedly play a contributing role in many of the more serious health problems, this burden of proof would be very hard to bear.

It might be argued that the grounds for an obligation to aid the medically needy are still incomplete. Even if the costs of aid were not excessive and even if the need were great and the individual were in no way responsible for his ill-health, we still might have no significant obligation to aid him if he could have taken steps to protect himself financially against ill-health. In other words, even if the individual is not responsible for his ill-health, he may be responsible for failing to take reasonable steps to cope with ill-health, and this failure may reduce or suspend our obligation to him. This point is well taken. Nevertheless, since the need for health care is extremely unpredictable (relative to the need for other important goods such as food and shelter), and since some of the most important types of modern health care are extremely expensive, only a small minority of persons could be reasonably expected to provide for all their important health care needs without aid from others.

The social response in the United States to the unpredictability and high cost of health care needs has been the development of a market in health care insurance. The problem is that those most in need of protection are either uninsurable or cannot be insured without exorbitant costs. A market for health care insurance does not make it possible for everyone to provide for his or her own health care.

We may conclude that if there is a general moral obligation of charity, then there is a general moral obligation to provide some minimal level of health care to the needy, even if people are in part responsible for their health status, and even if we are not obligated to aid those who could have themselves provided for their health care needs.

## III. THE CASE FOR COLLECTIVE, AS OPPOSED TO INDIVIDUAL, EFFORTS TO DISCHARGE THE OBLIGATION TO PROVIDE A DECENT MINIMUM OF HEALTH CARE

It is often said that the obligation to ensure an adequate level or decent minimum of health care is a "*societal*", as distinct from an individual, obligation. This view is explicit in the President's Commission Report on access.

> Society has a moral obligation to ensure that everyone has access to adequate care without being subject to excessive burdens. In speaking of a societal obligation the Commission makes reference to society in the broadest sense – the collective American community. The community is made up of individuals, who are in turn members of many other, overlapping groups, both public and private: local, state, regional, and national units; professional and workplace organizations; religious, educational, and charitable organizations; and family, kinship, and ethnic groups. All these entities play a role in discharging societal obligations ([10], p. 22).

The assumption that there are collective, and more specifically, institutional as well as societal, obligations to provide care for the indigent has also surfaced in recent criticisms of for-profit hospitals. Non-profit hospitals, and in particular, public hospitals, have complained that for-profit hospitals have "skimmed the cream" of the patient population, attracting paying patients away from non-profit hospitals while refusing to provide their "fair share" of indigent patient care.

Both the President's Commission Report and the critics of for-profit hospitals assume that sense can be made in this context of the notions of *collective* and *institutional* obligations. Yet that this is so far from clear. The clearest cases of societal obligations are those in which a society is said to incur an obligation as a result of its collective actions or as a result of the authorized action of its agents. For example, the German nation, meaning Germany's citizens collectively, incurred an obligation to provide financial "compensation" to survivors of the holocaust and the families of its victims. Similarly, a society may be obligated to provide health care for those of its citizens who were injured in a war which it waged. These societal obligations, however, are special obligations, generated by specific undertakings of individuals or societies, and as such they shed no light on the notion of a societal general moral obligation to ensure a decent minimum of health care.

Nor is it clear that particular health care institutions, such as hospitals, have obligations to provide health care independently of their contractual obligations to particular patients or groups of patients. An exception, of course, is the legal obligation to treat a percentage of indigent patients under the terms of the Hill-Burton Act, which made Federal funds for hospital construction contingent on the performance of these services. Absent such special arrangements, it is difficult to see why hospitals or other health care institutions can be said to have obligations to contribute to care for the needy, apart from their general tax obligations to contribute to government revenues, some of which will in fact be used to finance indigent care under Medicare, Medicaid, and the Veteran's Administration.

It is sometimes said that health care institutions have special obligations (quite independently of any general obligations they may have as taxpayers) to provide access to care for the poor because they are heavily subsidized, directly or indirectly, by public funds for medical education and research as well as by billions of dollars of Medicare and Medicaid reimbursements. Indeed, it is no exaggeration to say that, without this massive infusion of public funds, the health care system as we know it would not even exist. It is not at all obvious, however, that these subsidies ground a special institutional obligation to provide health care for the needy. It is true that the education of physicians is heavily subsidized by public funds, but since physicians own the human capital in which these social resources have been invested and can sell their services on the market, it is physicians (and indirectly their patients), not health care institutions, who are the principal beneficiaries. Consequently, the existence of subsidies for medical education is a stronger argument for special obligations on the part of physicians as individuals than for special institutional obligations.

Similarly, the fact that medical research is heavily subsidized by public funds does not itself show that health care institutions have special obligations to provide care for the indigent, since it is patients, rather than those institutions, who are the principal beneficiaries of those subsidies. It is true that health care institutions do not bear the full research costs of these treatments as part of their delivery costs. But if, as is increasingly the case, health care institutions operate in a competitive environment, they will be forced to pass on these subsidies to consumers or patients. (And if they operate in a largely non-competitive environment, there will be a strong case for some form of regulation of their rates.) The price that patients pay for health care treatments whose research costs were subsidized by the government will not include those research costs and so will not reflect true costs. It is then the consumers of health care, not the institutions, who principally benefit from research subsidies, and any obligation arising from this subsidy presumably lies on them.

Finally, consider the large public subsidy represented by Medicare and Medicaid. These programs created a vast expansion in the market for health care which many hospitals and other institutions serve. This is new health care business which heretofore did not exist and from which they benefit. But does it follow that this benefit grounds a special institutional obligation to provide subsidized care for the poor? The most obvious difficulty with such a view is that the subsidized health care consumers, not the deliverers of the health care, are by far the principal beneficiaries of Medicare and Medicaid. Any profit or revenue surplus that health care institutions receive from serving Medicare and Medicaid patients is only a small proportion of the overall cost of their care. It must be granted, nevertheless, that health care institutions do benefit from these subsidized patients. But it is difficult to see how this fact by itself is sufficient to ground a special institutional obligation to subsidize free care for the poor. In no other cases of government-generated business is it held that merely earning a profit (or revenue surplus) from such business grounds such a special obligation. Virtually no one holds that defense contractors, supermarkets who sell to food stamp recipients, highway builders, and so forth, have any analogous special obligation based on the fact that their business is created by government funds. It seems, then, that further grounds would have to be adduced to support the assumption that health care institutions, as opposed to individuals, have obligations to provide care for the indigent, distinct from whatever general tax obligations they may have and distinct from explicit contractual arrangements or specific legislation (such as Hill-Burton) [1].

It is somewhat more plausible to maintain that society *could* make the granting of tax-exempt status to (non-profit) health care institutions contingent upon their providing some specified amount of care for the needy. However, non-profit health care institutions are not currently under any such legal obligation by virtue of their tax-exempt status. Absent any explicit special legal obligation of non-profit institutions, and absent any more comprehensive system for distributing the costs of indigent care across individuals and institutions in society at large, no determinate content can be given to the alleged moral obligation that non-profit institutions are said to have.

Talk of societal or institutional obligations here is neither illuminating nor necessary. Instead, what is needed is a clear account of why the general moral obligation, which each of us has as an individual, can best he discharged by *collective* arrangements. The case for collective, as opposed to merely individual, charity in health care is straightforward, resting partly on considerations of efficiency and partly on moral grounds. The most obvious consideration in favor of collective charity toward the medically needy is the relatively large scale of investment needed for modern, high-technology medical equipment and services. Equally important is the need to coordinate contributions to avoid duplication of effort while seeing that gaps in services are avoided. Thus, valuable types of health care will not be provided unless there is an effective mechanism for *combining* and *coordinating* the contributions of many individuals. Unless effective mechanisms for collective charity are available, the individual may reluctantly conclude that her contribution will do more good if expended in an individual, uncoordinated act of charity, even though she knows that more good could be done if appropriate collective arrangements did exist. Where appropriate collective arrangements are lacking, the cumulative result of uncoordinated, individual acts of charity may be less than optimal even if each individual acts so as to do the most good she can with her resources [5].

A mechanism for collective charity can play another important role: it can specify obligations to contribute in such a way as to achieve a fair distribution of the costs of caring for the needy. In some cases it may not even be possible to specify the scope of any individual's obligation without specifying the obligations of others. At least in an environment of strong competition for resources, whether a given level of contribution by a particular individual would constitute his "fair share" may depend upon the level of contribution provided by others. For all of these reasons, those who acknowledge their obligations as individuals to provide for the medically needy should seek to develop collective arrangements for discharging their individual obligations to provide care for the needy efficiently and fairly.

## IV. THE CASE FOR ENFORCEMENT OF INDIVIDUAL CONTRIBUTIONS TO COLLECTIVE CHARITY IN HEALTH CARE

Collective charitable efforts, like other collective ventures, may founder if contribution is left strictly voluntary. The most familiar difficulty is the free-rider problem. Even if each potential contributor to a good whose production requires a collective effort recognizes the importance of producing it, so long as contribution is not enforced, each may elect not to contribute, since contributing is a cost to him, if he believes that either enough others will contribute to achieve the good or they will not, regardless of whether he contributes.

The free-rider problem threatens to block voluntary collective action, of course, only if non-contributors will not be excluded from "partaking" of the good if it is produced. Whether this condition is satisfied in the case of the collective good of providing health care for the needy will depend upon whether the individual views the good to be attained simply as *the provision of care for the needy* or as *the efficient discharging of my obligation to help provide health care for the needy*. If a sufficient number of individuals are concerned primarily with the former good – that is, if their main concern is that the needy be provided for, rather than with their providing for the needy – then the collective good may not be produced. On the other hand, if what each individual wants is to act charitably and in the most efficient way, and does not simply want charity to be done (by others), then he will not attempt to be a free-rider, but will contribute, at least if he has assurance that enough others will do so to achieve the greater efficiencies of collective as opposed to individual charity.

The extent to which extent people are motivated to help the medically needy out of a desire to act charitably rather than simply out of a desire that someone or the other provide for the needy, is an empirical question. However, anecdotal evidence suggests that many people are motivated by the latter. For example, it is often said that at least emergency care ought to be provided even for helmetless motorcyclists and smokers with lung cancer or others whom we are not obligated to help because they have (allegedly) voluntarily ruined their health, simply because it would be too painful for us to see them dying in the streets unattended. Clearly, the goal of those who hold this view is not to discharge an obligation of charity, since they deny that they are obligated to help these people. Instead, their goal is that these people should be cared for. Many people may desire that the needy be provided for because they believe that we all benefit, if only indirectly, from a more productive labor force, or that a country with a healthy citizenry is more secure and able to defend itself. Finally, some of the better off may believe that a system of health care for the needy is sound insurance against social unrest or revolution. To the extent that people are motivated by the desire to achieve the goal of charity rather than the desire to be charitable, the free-rider problem may thwart voluntary collective charity in health care. A standard response to the free-rider problem, here as with other important collective goods, is to use the threat of coercion to enforce contribution. If it is acknowledged that enforcement is sometimes justified when necessary to secure important collective goods in other areas, it is difficult to see why enforcement of contributions to collective charity in health care is not sometimes justifiable as well.

Yet, even if individuals are motivated by the desire to act charitably rather than simply a desire that the goals of charity should be achieved, they may, as was mentioned earlier, be unwilling to contribute, or to contribute as much as they otherwise would, if they lack assurance that others will contribute as well. Their reluctance to contribute without assurance of reciprocity may stem from either or both of two sources. They may think that their obligation to contribute to the collective charity is suspended unless they can be assured that others are doing their fair share, or they may simply be unwilling to divert their contribution into a collective enterprise that may not reach the required level of contributions to be effective when they could instead use the same resources for some individual act of charity whose efficiency does not depend on the actions of others. In either case, lack of assurance that others will contribute may block or reduce the scope of collective charity, and in some cases enforcement may be required to provide the necessary assurance. Finally, even if collective charity in health care can be initially established without enforcement, the continued effectiveness of the scheme may require enforcement in order to provide assurance that each is continuing to do her fair share and that some are not taking a free ride on the contributions of others.

My aim here has not been to show that the success of collective efforts to provide a decent minimum of health care for the needy always requires enforcement, nor even to try to specify precisely the conditions under which coercion will be necessary for success. Instead, I have only tried to make clear that arguments for enforcement to avoid failures of collective action that are familiar and widely accepted in other areas also apply to collective charity in health care.

It might be tempting to reply that if such arguments for enforcing obligations to aid the needy succeed, they show not that charitable obligations may sometimes be enforced, but that if the obligations in question may be enforced, then they *must* have corresponding rights. This reply either assumes that (a) it is a necessary truth that obligations of charity may not be enforced, or that (b) only obligations that have correlative rights may be enforced.

I am not confident that anything can be settled here by appealing to alleged analytic truths about 'charity'. But even if assumption (a) is granted, then all that follows is that the obligation to contribute to collective aid to the medically needy is not an obligation of *charity*. The conclusion of my argument for enforced collective aid remains: there is a strong *prima facie* justification for enforcing obligations to aid even if there is no right to aid. Assumption (b) simply begs the question. The public goods argument for enforcing obligations to collective aid does not depend upon any assumption that individuals have a *right* to the good in question.

The argument for enforcing contributions to collective aid in health care can be generalized to an argument for enforcing contributions to collective arrangements for providing a decent minimum of food, shelter, and income, if, for each of these welfare goods, the following conditions are satisfied. (i) We as individuals do have a moral obligation to help those in need of these goods. (ii) The most fair and effective way to discharge these obligations is through collective, coordinated efforts. (iii) Strictly voluntary collective efforts would fail. Though I will make no attempt here to work out the details of the argument for a more comprehensive welfare floor, I submit that conditions (ii) and (iii) will often be satisfied in a society like ours. Instead I will concentrate on condition (i), since it provides the most direct challenge to the assumption, implicit in the President's Commission's position that the obligation to provide health care is in some sense unique. Assuming that there is a general moral obligation to help the needy, there is an obligation to provide them with some minimal level of food and shelter, unless the obligation is suspended because they are fully or largely responsible for being without these goods.

If needs for food and shelter are generally both more predictable and less costly to satisfy than needs for health care, we will generally be more confident in holding an individual responsible for satisfying these needs by his own efforts than for satisfying his health care needs. Nevertheless, the difference in responsibility, and hence the difference in the scope of our obligation to aid, will be a matter of degree. Especially in a society in which full equality of opportunity has not yet been achieved, it will not be plausible to say that our obligation to provide a minimum of food and shelter for all is suspended because needy individuals are fully or largely responsible for their plight. I conclude, then, that even if our obligations to provide health care to the needy are especially strong, they are not unique, and that the argument for enforcing these obligations does have more general implications.

It is one thing to show that enforcement may sometimes be necessary to achieve the goals of collective charity, and that this need provides a *prima facie* justification for enforcement. It is quite another to show that enforcement in such cases is morally justified all things considered. In the next section I argue for this stronger indirectly, by attempting to rebut arguments against the enforcement of obligations of charity.

## V. ARGUMENTS AGAINST ENFORCING OBLIGATIONS OF CHARITY

There are three main objections that might be raised against the attempt to enforce obligations of charity (or beneficence). Since I have attempted to rebut these objections in detail elsewhere, I will only summarize that discussion here ([3], ch. 3). The first objection is that obligations of charity may not be enforced because they are imperfect duties. Obligations of charity are said to be imperfect in two senses: both the amount and type of aid and the choice of a beneficiary are left to the discretion of the benefactor. Because duties of charity are indeterminate in this sense, it would be inappropriate to try to enforce them since efforts at enforcement would be too arbitrary and subject to abuse.

The first objection loses its force once we cease thinking of charity as a matter of independent, private beneficence. If there are collective arrangements for specifying and fairly distributing obligations to provide for the needy, then the indeterminateness which makes enforcement inappropriate is eliminated. To argue that obligations of charity may not be enforced because they are indeterminate is circular, if institutions are available which can make them determinate and enforce them.

The second objection to the enforcement of obligations of charity is that obligations of charity have no correlative rights, and that coercion may only be used to enforce rights. Presumably, the latter claim is presented as a necessary truth, analytic upon the concept of a right or the concept of justified coercion. To reject all arguments for enforced charity on the basis of this alleged analytic truth is unconvincing. There is nothing *logically contradictory* about the proposal that enforcement is sometimes justified if it is necessary for securing certain important collective goods, including collective arrangements for charity, even in the absence of a right to those goods. Simply stipulating that only rights may be enforced begs the question. To think that we can determine the sole conditions under which coercion is morally justified simply by analyzing terms is naive. What is needed is a *theory* of the moral uses of coercion.

The third objection is that, in the absence of a right to aid, enforcing charitable obligations necessarily violates individual rights, either a general moral right to negative liberty, i.e., a right against coercion, or a moral right to private property. It is important to emphasize that this objection can succeed only if a sound justification is given for the claim that there *are* such rights, and that they are such virtually *unlimited* rights that respecting them rules out in principle all attempts to enforce charitable obligations. To my knowledge such a justification is not presently available.[3] Until an adequate justification for such an extreme libertarian theory of rights is produced, enforced charity cannot be ruled out in principle.

## VI. THE ROLE OF THE FEDERAL GOVERNMENT

So far, a case has been made for enforcing obligations to provide a decent minimum of health care for the needy, but nothing has been said about who the enforcer should be or how broad the collective arrangements should be. We often assume, too quickly, that if obligations are to be enforced, they must be enforced by the government. Similarly, when we think of collective arrangements for specifying and distributing obligations to contribute, we tend to assume that these tasks can only be discharged by the government. In particular, we may be tempted to assume that if my argument for enforced collective aid to the medically needy succeeds, it shows that the Federal government should establish and enforce a general legal right to a decent minimum of health care.

This conclusion does not follow – at least not without the addition of a rather complex set of contingent assumptions. For one thing even if government is to be the ultimate enforcer of the obligations in question, enforcement might be provided without the establishment of a general legal right to health care. Instead, groups of individuals and health care institutions might voluntarily make legally binding contracts with one another to share the burden of providing care for the indigent in their region. Like all legal contracts, such agreements would be enforced by the state, even in the absence of any legal right to health care. Whether a collection of such local or regional contractual arrangements for collective charity would be as effective as a nationwide system in which the Federal government specifies and enforces a legal right to health care, is largely an empirical issue for political economy to decide, not a fundamental question for ethical theory.

*University of Arizona*
*Tucson, Arizona, U.S.A.*

## NOTES

[1] In this essay I do not address the problem of how to determine what a 'decent minimum' or 'adequate level' of health care is. For a beginning of this difficult task, see *Securing Access to Health Care*, Report of the President's Commissions for the Study of Ethical Problems in Medicine and Biomedical and Behavioral Research ([10], Vol. I, pp. 35–41).

[2] The approach followed in this paper is similar to that in Chapter One of *Securing Access to Health Care* ([10], Vol. 1). The general strategy that the Commission Report follows of arguing in terms of a moral obligation to provide access to care rather than in terms of a moral right drew upon an earlier paper by this author that was prepared for the Commission prior to the writing of the Access Report, Vol. I [4]. The present essay is an attempt to work out this general strategy in a more rigorous way by answering questions (1)–(4), which in my opinion were not fully addressed in the Commission Report, *Securing Access to Health Care* {10}.

[3] Robert Nozick's frank admission that he can provide no systematic justification for the libertarian rights he espouses is representative of libertarian views generally. Nozick does offer indirect support for his list of exclusively negative rights by attempting to show that the establishment of positive (general) rights inevitably leads to unacceptable results [9, part II]. However, these attempts to rule out all positive rights seem unconvincing. See A. Buchanan ([2]; [3], Ch. 3). See also G.A. Cohen [6] and T. Nagel [8].

## BIBLIOGRAPHY


1. Brock, D. and Buchanan, A.: 1986, 'Ethical Issues in For-Profit Health Care', in B. Gray (ed.), *For Profit Enterprise in Health Care*, Institute of Medicine, Washington, D.C., pp. 229–249.
2. Buchanan, A.: 1982, 'Deriving Welfare Rights From Libertarian Rights', in P.G. Brown *et al.*, *Income Support*, Rowman and Allanheld, Totowa, New Jersey, pp. 233–246.
3. Buchanan, A.: 1985, *Ethics, Efficiency, and the Market*, Rowman and Allanheld, Totowa, New Jersey.
4. Buchanan, A.: 1983, 'Is There A Right to a "Decent Minimum" of Health Care?', in President's Commission, *Securing Access to Health Care*, Appendices, Vol. 2, U.S. Government Printing Office, Washington, D.C., pp. 207–238.
5. Buchanan, A.: 1984, 'The Right to a "Decent Minimum" of Health Care?', *Philosophy and Public Affairs* **13** (1), 18–21.
6. Cohen, G.A.: 1978, 'Robert Nozick and Wilt Chamberlain: How Patterns Preserve Liberty', in J. Arthur and W.H. Shaw (eds.), *Justice and Economic Distribution*. Prentice-Hall, Englewood Cliffs, New Jersey, pp. 246–262.
7. Grey, T.C.: 1983, *The Legal Enforcement of Morality*, Alfred A. Knopf, New York.
8. Nagel, T.: 1975, 'Libertarianism Without Foundations', *Yale Law Journal* **85**, 136–149.
9. Nozick, R.: 1971, *Anarchy, State, and Utopia*, Basic Books, New York.

10. President's Commission for the Protection of Human Subjects in Medicine and Biomedical and Behavioral Research: 1983, *Securing Access to Health Care*, Vols. I and II, U.S. Government Printing Office, Washington, D.C.

GEORGE J. AGICH

# ACCESS TO HEALTH CARE: CHARITY AND RIGHTS

There are many problems associated with access to health care in the United States. Some of these problems are conceptual and definitional. The empirical literature on access, for example, lacks agreement even regarding the nature of the problem. It is not settled whether inequity of access is to be measured by utilization rates, by "process" variables such as travel or waiting time, or by market considerations [1–3, 6, 9, 13, 15, 18]. Paul Starr has shown that the problem of inequity of access has large-scale historical and sociological features which imply that historically there are several different problems of access to health care.[1] Despite these problems, discussion of health care access in recent years, especially in political contexts, has tended to be expressed in terms of the language of rights.

The addition of rights language to the problems of access or distribution of health care has met with at least three competing responses. The first response rejects as unfounded any right to health care on general libertarian grounds, namely, that such a right would violate liberty, particularly the freedom of providers to participate in the marketplace [17]. The second response sees the ambiguity in talk of equity and equality of health care access as a challenge calling for a philosophical justification of a right to health care based on a theory of social justice [11, 12]. Such a theory would provide a principled defense of a right to health care while correcting some of the excesses of strong right-claims to health care, claims which lend support to the libertarian critique. A third response is that the right to health care and its associated social justice foundation is problematic and should be replaced by a theory based on beneficence or charity [7]. The social justice and enforced beneficence approaches constitute the focus of mainstream discussion; they share three basic points of agreement which are summarized by the phrase "a decent minimum or basic minimum" of health care. First, it is assumed that entitlements and access to health care should be seen in a society-relative sense. Second, if there is a right to health care, it must be a limited as opposed to an unqualified right. And third, the entitlement to health care should be defined in terms of need rather than wants or



*T.J. Bole III and W.B. Bondeson (eds.), Rights to Health Care*, 185–198.

desires; hence, there should be a decent or basic minimum of health care. The last two features allegedly avoid some of the excesses associated with a strong equal access principle such as waste and burgeoning health care expenditures.

In the present essay I consider two contributions to this ongoing discussion: Thomas Halper's 'Rights, Reforms, and the Health Care Crises: Problems and Prospects' [14], and Allen Buchanan's 'Rights, Obligations, and the Special Importance of Health Care' [8]. Both Buchanan and Halper offer admonitions regarding an unqualified right to health care. Buchanan sets aside talk of moral rights to take up the task of providing a moral foundation for a legal right to health care. This argument turns on there being an enforceable obligation of charity rather than the needy individual's having a valid moral right to health care. Halper seeks to clarify the concept of a legal right to health care by arguing, first, that an unqualified right is, in effect, a rhetorical ploy without substantive foundation, second, that talk of a right to health care must be understood in historical and political terms, and third, that a right to health care in the political context must be a qualified right at best. These papers, then, share a criticism of unqualified right claims to health care and view the right to health care as a limited legal entitlement. Whereas Halper is concerned to demonstrate how the right has been limited by the political process in recent years, Buchanan is intent to establish a moral foundation for a general limited legal entitlement by showing that health care is an enforceable collective obligation of charity. In discussing these essays, I focus primarily on three questions: First, is Halper's taxonomy of how rights to health care are limited sufficiently broad to support a political analysis of health care rights and problems of access to health care in the United States? Second, is Buchanan's account of the moral foundation of the collective nature of the obligation to aid the needy adequate? And third, does Buchanan establish the enforceability of moral obligations to care for the medically indigent?

For Halper a right-claim to health care comprises a vague demand that is more a matter for politics than philosophy to address:

> As such, its solution is a function of the potency of competing claimants, developing standards in health care, general economic conditions, and numerous other transitory factors. Indeed, the essential vagueness of the core terms represents not so much the absence of intellectual clarity as a simple invitation to negotiate, compromise, and bargain ([14], p. 142).

Given this orientation, Halper outlines some of the ways in which an absolute right-claim to health care is pragmatically or politically qualified. In effect, Halper retains talk of an unqualified right-claim as an anchor or pivot from which he lets out line along which political accommodations based on, for example, nging economic conditions, developments in biomedical science and health care delivery, and changing public expectations of health care qualify the absolute right. The problem that Halper sees with talk of an absolute claim is that it places health care "outside the political process, assigning it sufficient – perhaps, even in principle unlimited – resources without compelling its advocates to battle with competing claimants..." ([14], pp. 141–142). Instead, Halper wants to view the question of entitlement to health care as a continuum which "clarifies part of the policy maker's problem by placing it in a directional context" ([14], p. 145). One of Halper's concerns is to categorize some of the main ways in which the right to health care is qualified in recent proposals to reform health care. My primary concern is whether Halper's taxonomy provides an adequate framework from which to interprete the political process whereby entitlements to health care are created and modified.

The reforms which Halper discusses are directed at the widely perceived "crisis in health care", which seems to be the result of a system in which both consumers and providers of health care are insulated from the financial effects of health care decisions and, consequently, are made unaccountable economically for their behavior. The crisis in health care is thus mainly economic or fiscal. Health care costs are growing at an alarming rate, now consuming over eleven percent of GNP. To be sure, demographics, an increasingly aging population, and development of new and expensive medical technologies, such as renal dialysis, organ transplantation, specialized coronary care units, and their associated diagnostic and treatment modalities, have complexly contributed to this burgeoning growth. Past government policies, such as Hill-Burton, which has resulted in significant over-building of hospital beds, and Medicare and Medicaid, which effectively gave health care providers carte blanche control over public expenditures for medical care, have also contribute significantly to spiralling health care costs, especially hospital-based patient care. But what has this set of problems, referred to in the phrase "the crisis in health care", to do with rights to health care? Was the rhetoric of rights crucial in the development of our system of health care since the 1950s, or did talk of an unqualified right to health care emerge as policy makers established entitlements to health care without adequate restraint or reflection on the fiscal and policy implications of their commitments?

These are important questions for social and political historians. I raise them only to stress a point made by Halper, namely, that the rhetorical ideal of an unqualified or absolute right to health care should be seen in practical and political terms, that is, in terms of its persuasive power in political negotiation and compromise, rather than theoretical or philsophical terms. But is this to say that the present crisis in health care can be causally attributed to policy makers having taken seriously a general unqualified right to health care? Or, is complaining about the rhetoric of the right to health care an elliptical way of referring to the powerful interests that have benefitted from programs – perhaps naively intended to achieve greater equity and equality of access and distribution of health care in the United States? I am not sure what Halper's views are regarding these questions, because his analysis is more taxonomic and descriptive than analytical or critical.

The only substantive legal rights to health care that exist in the United States are privately created, legally enforceable rights to certain medical care or to insurance money to pay for it ([19], p. 42). Entitlements to health care, like entitlement to other goods, depend either upon one's present ability to pay or (through prudent financial planning) on an insurer's contractual obligation to pay. Public programs, to be sure, have in a piecemeal fashion attempted to modify inequities and inequalities in the distribution of health care services in the United States, but it is not clear that they have been implemented with sufficient commitment to establish a clear entitlement to care. Legal rights are entitlements not simply because they are in principle enforceable or "on the books", but because there is the political will to enforce them. Public policy in health care has until very recently seemed unwilling to deal with powerful medical interests effectively to enforce even limited entitlements to health care and so inequity and inequality have persisted in health care despite ever-growing public expenditures.

Hill-Burton and Medicare and Medicaid are prime examples. These public policies did not establish a general legal right to health care, but only a right qualified or limited in specific ways.[2] The provisions of the Hill-Burton program, for example, a program which built approximately one-third of the hospital beds in the United States with Federal funds since 1946, requires that "a reasonable volume of services" of the facility built with such funds be made available to all persons residing in the facility's territorial area" (as a "community service"). These provisions, however, were virtually unenforced during the first twenty-five years after Hill-Burton was enacted ([14], p. 40). In fact, it was not until lawsuits were brought on behalf of impoverished patients who had been denied service at various hospitals that DHEW issued regulations interpreting the above provisions ([19], p. 40). Further lawsuits were required to force some liberalization of the regulations after they were promulgated [10]. By 1974, Congressional hearings on the implementation of the "free service" and "community service" provisions made clear that they were still in the stage of infancy [16].

The long record of evasion was background to a new Federal law, enacted in 1974, which authorized lawsuits against Hill-Burton grantee hospitals and facilities for specific performance of their "free service" and "community service" obligations ([19], pp. 40–41; [22]). Even assuming a vigorous enforcement of Hill-Burton, how much of a right to hospitalized medical care was created by the program? The "free service" provision means only that the hospital is required to give a certain amount of service (up to ten percent of the size of the Hill-Burton grant or three percent of its operating costs in any one year) for a period of twenty years after the grant to persons unable to pay. The economic definition of the person who is unable to pay is left up to the state and the hospital. While the "community service" provision means that the hospital cannot discriminate against whole groups of people who reside in the community, for instance Medicaid recipients, it does not mean that any individual, even one with great need for hospitalization, has to be accepted ([19], p. 41).

Under Hill-Burton a person could be denied hospital admission for any reason other than his ability to pay; he can also be denied hospital admission on the ground that he is unable to pay and the hospital has met its "reasonable volume" quota of those unable to pay. While under the "community service" provision, a person applying for admission as a Medicaid recipient can nonetheless be denied admission on the ground that the hospital's affiliated physicians refuse to take him and other Medicaid patients (if the hospital has no house staff service for Medicaid patients). All of these qualifications hardly indicate that anything like an unqualified right to health care motivated the writers of Hill-Burton. And if it did, then one can only conclude that their effort failed miserably. The important point is that the limitations of Hill-Burton indicate the need for a framework which looks beyond the language of rights in order to see the ways that competing interest groups and social and political forces contribute to the present problems of access.

On Halper's view, reforms such as cost-sharing and Certificate of Need are seen as qualifying a right to health care. To be sure, they may have partly had that effect, but it is not entirely clear that such was the primary intent. One might reasonably argue that an alternative hypothesis needs to be explored as well, namely, that such approaches were designed mainly to regulate health care providers who systematically turned even limited entitlements to health care services to their own interests and, hence, created ever-increasing claims on government resources. Were these reforms designed to qualify or restrict an established right to health care, which Halper seems to imply, or were they attempts to introduce rational planning and accountability into public spending for health care? After all, if rights or entitlements are constrained by the process of debate, compromise, and negotiation – as Halper describes the political process – then reform of health care financing may be misleadingly characterized as qualifying an absolute right to health care. Certainly, that is one effect, but this way of interpreting the present concern over health care delivery and finance tends to oversimplify the complexities of both the fiscal crisis of health care in America and the problems of access [20].

At best, then, a talk of an unqualified right must be regarded as a fiction. It is myth rather than reality. The important point, however, is that the myth or rhetoric of a right to health care has played a role politically. The question is why? Public policy makers indeed may have derived and may continue to derive comfort from the belief that the ship of state rides out political storms tethered to an unqualified right to health care, but that comfort must be seen as illusory if we take seriously Halper's criticism of rights to health care. I, for one, propose that this criticism be taken seriously. I also think, however, that Halper's observations regarding rights to health care are preliminary to the really interesting and important work. That work needs to provide a critical analysis of the development of legal entitlements to health care in the United States. It should focus on the rise of health insurance, the social and economic, and political forces which contributed to that development. It will clarify the relationship between private and public insurance programs, as well as show how entitlements to health care are related to broader social and political developments in America such as social security, unemployment compensation, and disability compensation. Halper's discussion certainly whets our appetite for just such an analysis, but it only begins the task of identifying the key research questions and topics.

If the anchor of an unqualified (moral) right to health care is illusory, then even qualified legal rights to health care will lack a foundation, a fact that lends a philosophical significance to the phrase "the crisis of health care". That is where Buchanan's contribution is central. Buchanan simply assumes both the criticism of the general moral rights approach to health care and a limited legal right to health care. He offers a substantive argument for "a general legal right to health care based on the thesis that there is a general moral obligation to provide such care to those who cannot provide it for themselves...." ([8], p. 169). Thus, even if there is no general moral right to health care of any kind, Buchanan argues for a moral foundation for a limited legal right to health care. In other words, Buchanan's project is to forge theoretically a new anchor for a limited legal right to health care to replace the insubstantial concept of an unqualified moral right.

In pursuit of this goal, Buchanan argues that it makes sense to think of health care as an enforceable obligation to care for the needy; this obligation is a duty of charity, and not a matter of a right to aid on the part of the needy. Thus, while preserving the language of a right to health care at the legal level, Buchanan wants to replace obligations founded on moral rights with obligations of a different sort, namely, the more general moral obligations of charity. Two features of this argument deserve particular attention: the collective nature of the obligation to aid the needy, and the enforceability of the obligation. Neither of these features seems necessary to establish a general moral obligation of charity, but rather are needed to justify the special and enforceable obligation to provide some types of health care.

Buchanan's argument, that the general moral obligation to aid the needy, which every individual has, can best be discharged by collective arrangements, mixes considerations of psychological motivation based on efficiency and considerations of morality. He cites the relatively large scale of investment needed for modern, high-technology medical equipment and services as the most obvious reason in favor of collective charity as opposed to individual charity. This requires that there be effective institutional mechanisms for combining and coordinating the contributions of many individuals. Buchanan argues for this point by noting that individuals may be reluctant to contribute to a collective effort unless they can be assured that collective arrangements will enhance the effectiveness of their contributions. While such considerations probably are correct as regards psychological motivation, it does not seem that this establishes why collective arrangements are required for *moral* reasons. To develop an account in moral terms, Buchanan would have to show – and I think it highly plausible to argue – that the obligation of charity is an obligation not simply to act in order to do something for the needy, but to intend to actually aid the needy, to provide *effective* aid. Perhaps that is what Buchanan means, but the point needs to be made explicitly.

The moral obligation of charity calls forth not just any response to the needy, but a response that is appropriate and effective to the needs in question. Hence, knowledge and judgment are required in the moral exercise of the obligation. Effective action requires knowledge, and knowledge involves at least minimal social arrangement for its dissemination, if not development. So even with respect to the actions of individuals to aid the needy, there is already a reference to a social or collective stock of knowledge regarding what should be done, how it should be done, and so on. Such a stock of knowledge might include recipes for action or simple folk remedies, but they are nonetheless social in nature. Anything beyond such informal interventions, however, obviously requires greater institutional organization and, hence, the need for collective action becomes more apparent. The point here is that it is not simply for reasons of aggregate efficiency that collective arrangements are necessary, but effective interventions even of an individual sort already assume a socially determined base of knowledge or skills. Because the obligation to care for the needy is itself an obligation to provide *effective* care, a social and possibly collective dimension is implicated [4]. The implication is for logical reasons internal to the very notion of an obligation of charity. These points clarify the collective character of the moral obligation to aid the needy, but do they establish a case for the enforceability of individual contributions to collective charity?

The connection between charity and collective arrangement is crucial for Buchanan's position, because he wants to justify an *enforceable* legal entitlement to health care. If enforceability were not justified, the free-rider problem would considerably lessen the plausibility of the charity-based justification of limited rights. Legal rights, after all, logically entail enforcement, whereas moral obligations do not. Curiously, Buchanan's arguments for enforceability turn on consideration of the free-rider problem. But that already assumes just what is in question, namely, whether the obligation of charity is enforceable. If such obligations were not enforceable, then the free-rider problem and associated questions of fairness would not apply with the same pertinence.

Buchanan's account, to be sure, focuses less on fairness than on psychological motivations that individuals might have in making contributions to discharge their general obligations to aid the needy. Unless individuals can be assured that their "contributions" will make a difference and that the general goal of aiding the needy is achieved, the account will fail to justify enforceable rights to health care. It is an empirical question, Buchanan acknowledges, regarding how individuals actually are motivated. Nonetheless, he thinks there is anecdotal evidence to suggest that many people are motivated by the desire to see that the needy are efficiently cared for rather than by the desire to act charitably simply out of a sense of obligation. Examples of the former motivation might be a belief that utility will be maximized if the needy receive care or the desire to avoid the embarrassement or pain of seeing them die in the streets unattended. These examples differ from cases of acting from the obligation of charity where the motivation is simply the rational desire to care for the needy.

Even if Buchanan is correct regarding motivation, it is not clear that this point helps to establish the enforceability of the obligation on *moral* grounds. Buchanan seems to provide an argument that shows that many people are motivated to act not from the moral obligation of charity, but according to some other consideration such as utility or convenience. The argument is not moral, but prudential.

Perhaps Buchanan's position might be interpreted in the following fashion. Those who are motivated to act from the obligation of charity to care for the needy do not have to be forced to act; they are, after all, already so motivated on moral grounds. Enforceability of the obligation is necessary not for such individuals, but for those – perhaps even the majority – who are motivated by other considerations. Does this reading salvage the moral argument for enforceability of a limited legal right to health care?

One might be inclined to think that it does if certain unstated moral assumptions are made. For example, if one assumed that the obligation of charity as regards health care *morally* requires a social organization in which individuals – at least those individuals untrained or uneducated in effective health care techniques – might be made to discharge their obligation indirectly through, perhaps, monetary contributions raised by taxation rather than in-kind effort. This assumption also would preclude even morally motivated individuals from thwarting collective social efforts by offering direct help when such help is less-than-fully effective. For example, my charitable urge to care for a neighbor who has suffered a heart attack is morally laudible, but does nothing to help build an appropriate, effective medical care system with trained medical professionals in coronary care centers. I have, in fact, already supplied a suggestion that might be helpful here, namely, that the obligation of charity should be seen as an obligation not simply to offer aid, but to offer *effective* aid. Hence, charitable individuals, when properly educated, would not have to be *forced* to desist from individual efforts or forced to channel their efforts into effective avenues; they would be motivated willingly to participate in the most effective ways. Thus, morally motivated individuals, at least, would not need to be forced to contribute to collective arrangements to provide care *if* these arrangements are singularly more effective than individual or alternative efforts. After all, Buchanan's goal is to provide a moral foundation for a *limited* legal right to health care or, as others have termed it, a decent or basic minimum. Limitation is central to Buchanan's concept of legal right; hence, there is no *prima facie* reason why effectiveness should not be a criterion for helping to set the limits.

Effectiveness, however, is an empirical matter. Given the literature on variation in utilization rates, or what John Wennberg terms "practice style variables", it is unclear that there is sufficient consensus on which to base enforceability [23–26]. Also, given rational disagreements regarding preventive versus treatment-oriented care or aggressive treatment of some tumors, it is hard to see why even morally motivated individuals *should* concur and blindly discharge the moral obligation of charity through contributions to collective arrangements that stress one approach over another. If they do not concur for morally valid reasons, then it is harder to justify enforceability on moral grounds. The important point here is that while effectiveness helps to explain and justify collective arrangements, it helps to justify enforceability possibly at the cost of restricting it to uncontroversial services. This restriction need not be seen as a serious impediment to Buchanan's account, however, because his intent, as noted earlier, is to justify a *limited* right to health care. Unless one can show that the restriction would be so severe as to constitute a *reductio ad absurdum* of the right as a right to *health care* – based on some unanalyzed intuitive notion of what health care or medical care really is – the point does not invalidate the position. When viewed in terms of the mainstream concern for a decent or basic minimum of health care, however, it is worth noting that this argument, like others designed to establish legal rights to a basic minimum as a matter of social justice, might just lead to a more radical set of implications for the health care delivery system than is apparent at first glance ([5], pp. 624–627).

What about those individuals who are not motivated by the moral obligation of charity? Even if there is a moral obligation to aid the needy that entails collective action because of some consideration such as effectiveness, it is still hard to see why the non-moral motivations of individuals, even the majority of individuals, should be at all relevant to *moral* argument? Buchanan thinks that such considerations are central because he wants to justify the *enforceability* of the moral obligation. Enforceability is even more pressing a problem for those individuals who are not motivated by moral considerations. Has Buchanan succeeded?

Enforceability is tacitly understood in terms of a limited legal right, presumably a legal right which has reference to national political realities, otherwise it is not clear what sense can be given to the term legal right. However, the moral argument establishes a *general* or universal obligation to aid the needy. It is not an obligation to aid the needy limited to a neighborhood, city, state, or nation. Such limitations, however, are assumed by the mainstream discussion of the basic or minimum of health care standard. The concerns are historically conditioned by the perceived inequities in the distribution of health care in the United Staes. It is hard, however, to see such a focus as *morally* required. Ironically, Buchanan's *moral* argument is designed to establish a universal obligation to aid the needy, whereas the argument for enforceability seems to be limited to legal rights to health care in the United States rather than political mechanisms whereby the needy world-wide are cared for. Undoubtedly, legal rights are central to our political system, but it is not clear why moral analysis should be constrained by local political circumstances.

This observation is not a criticism of Buchanan's argument as a *moral* analysis, but a reminder based on Halper's allusion to the historical and political context of discussions of the right to health care that the context of the discussion of access to health care be made clear.

In other words, it is a reminder that Buchanan's effort to justify a limited legal right to health care is part of an on-going public policy discussion regarding access to health care in the United States. More specifically, it is an effort to justify a basic or decent minimum standard for health care. Nonetheless, the importance of Buchanan's stress on the moral obligation of charity is precisely that it broadens moral inquiry and shifts attention both from a narrow focus on health care to other human needs and to the needy outside the political boundaries of the United States.

Understandably, then, the context that gives rise to Buchanan's analysis is the problem of a right to health care and the problem of access in the United States. Buchanan's moral argument, however, establishes a *universal* obligation to aid the needy which implies collective arrangements to discharge it. While the argument has significant implications for the political debate over access to health care in the United States, it has even more important implications for national policy regarding the needy world-wide. As Buchanan points out, his argument says nothing about either who the enforcer should be or about the scope of the collective arrangements.

*Department of Medical Humanities*
*Southern Illinois University*
*Springfield, Illinois, U.S.A.*

## NOTES

[1] Starr distinguishes *mass inequality* that characterized the period of 1870–1940 and *marginal inequality* in the period since 1940 [21]. Mass inequality of utilization of medical care was influenced by geographical inequalities in hospital and physician availability between, for example, the north and south and rich and poor counties, and by access variables such as availability of telephone services, automobiles, public transportation, and money or insurance coverage. The development and distribution of health insurance, unemployment compensation, and Hill-Burton hospital construction contributed to equalizing access and availability, which in turn affected utilization and helped to transform mass inequality into marginal inequality. Key to this development was private health insurance – subsidized through the Internal Revenue Service's exemption of employer contribution to health benefit plans from federal income tax – which had absorbed costs somewhat to provide care for the poor, but gradually, by the 1960s, abandoned this redistributive practice, leaving the poor, unemployed, and chronically ill dependent upon direct government support.

[2] Medicaid is limited by state-determined eligibility requirements, which vary considerably from state to state. Diagnosis-related group (DRG) based reimbursement for Medicare is designed to restrain hospital costs than to assure that patients receive any specific set of entitled services.

## BIBLIOGRAPHY


1. Aday, L.A.: 1975, 'Economic and Non-Economic Barriers to the Use of Needed Medical Services', *Medical Care* **13**, 447–456.
2. Aday, L.A. and Andersen, R.: 1974, 'A Framework for the Study of Access to Medical Care', *Health Service Research* **9**, 208–220.
3. Aday, L.A. *et al.*: 1980, *Health Care in the U.S.: Equitable for Whom?*, Sage Publications, Beverly Hills, California.
4. Agich, G.J.: 1981, 'The Question of Technology in Medicine', in Stephen Skousgaard (ed.), *Phenomenology and Understanding Human Destiny*, The Center for Advanced Research in Phenomenology and University Press of America, Washington, D.C., pp. 81–92.
5. Agich, G.J. and Begley, E.C.: 1985, 'Some Problems with Pro-Competition Reforms', *Social Science and Medicine* **21**, 623–630.
6. Andersen, R. *et al.*: 1975, *Equity in Health Services: Empirical Analysis in Social Policy*, Ballinger Publishing Company, Cambridge, Massachusetts.
7. Buchanan, A.: 1984, 'The Right to a Decent Minimum of Health Care', *Philosophy & Public Affairs* **13**, 55–78.
8. Buchanan, A.: 1990, 'Rights, Obligations, and the Special Importance of Health Care', in this volume, pp. 169–184.
9. Carpenter, E.S.: 1983, 'Concepts of Medical Underservice: A Review and Critique', in President's Commission for the Study of Ethical Problems in Medicine and Biomedical and Behavioral Research, *Securing Access to Health Care, Volume Three: Appendices*, U.S. Government Printing Office, Washington, D.C., pp. 189–222.
10. *Corum v. Beth Israel Medical Center*, 373 F Supp. 550, 557–8 (1974).
11. Daniels, N.: 1981, 'Health Care Needs and Distributive Justice', *Philosophy & Public Affairs* **10**, 146–179.
12. Daniels, N.: 1986, *Just Health Care*, Cambridge University Press, Cambridge.
13. Enthoven, A.: 1980, *Health Care Plan: The Only Practical Solution to the Soaring Costs of Medical Care*, Addison-Wesley Publishing Co., Reading, Massachusetts.
14. Halper, T.: 1990, 'Rights, Reforms, and the Health Care Crises: Problem and Prospects', in this volume, pp. 135–168.
15. Havighurst, C.: 1977, 'Health Care Cost-Containment Regulation: Prospects and an Alternative', *American Journal of Law and Medicine* **3**, 309–322.
16. Hearings before the Sub-Committee on Health of the Senate Committee on Labor and Public Welfare (November 25, 1974): 1974, *Implementation of Hill-Burton Amendment*, 93rd Congress, 2nd Session.
17. Sade, R.M.: 1971, 'Medical Care as a Right: A Refutation', *New England Journal of Medicine* **285**, 288–292.
18. Sloan, S. and Bentkover, J.D.: 1979, *Access to Ambulatory Care and the U.S. Economy*, Lexington Books, Lexington, Massachusetts.
19. Sparer, E.V.: 1976, 'The Legal Right to Health Care: Public Policy and Equal Access', *Hastings Center Report* **6**, 39–46.
20. Starr, P.: 1982, *The Social Transformation of American Medicine*, Basic Books,

Inc., Publishers, New York.
21. Starr, P.: 1983, 'Medical Care and the Pursuit of Equality in America', in President's Commission for the Study of Ethical Problems in Medicine and Biomedical Research, *Securing Access to Health Care, Volume Two: Appendices*, U.S. Government Printing Office, Washington, D.C., pp. 3–22.
22. Title XVI, The National Health Planning and Resources Development Act of 1974 (P.L. 93–641).
23. Wennberg, J.E. and Gittelsohn, A.:1975, 'Health Care Delivery in Maine I, Patterns of Use of Common Surgical Procedures', *Journal of Maine Medical Association* **66**, 123–130, 149.
24. Wennberg, J.E. *et al.*: 1980, 'The Need for Assessing the Outcome of Common Clinical Practices', *Annual Review of Public Health* **1**, 277–295.
25. Wennberg, J.E. and Gittelsohn, A.: 1982, 'Variations in Medical Care Among Small Areas', *Scientific American* **246**, 120ff.
26. Wennberg, J.E.: 1984, 'Dealing with Medical Practice Variations: A Proposal for Action', *Health Affairs* **6**, 32.

SECTION IV

# EQUALITY, FREE MARKETS, AND THE ELDERLY

NORMAN DANIELS

# EQUAL OPPORTUNITY AND HEALTH CARE RIGHTS FOR THE ELDERLY

## I. RIGHTS AND THEORIES OF JUSTICE

The literature on rights to health care – and this volume is no exception – is not shy about pointing to the difficulties talk about such rights brings with it, including problems of specifying their scope, limits, and bases. In this essay, I will try to give some plausible, if abstract, content to the notion of a health care right, indeed to the health care rights of the elderly. But my working assumption in this discussion is that the appeal to a right to health care is not an appropriate starting point for an inquiry into what just health care involves. Rights are not moral fruits that spring up from bare earth, fully ripened without cultivation. Rather, we are justified in claiming a right to health care only if it can be harvested from an acceptable general theory of justice for health care. The theory tells us which kinds of right claims are legitimately viewed as rights. Such a theory also helps us specify the scope and limits of justified right claims. I will appeal to the fair equality of opportunity account of just health care [7] to specify the content of these derivative health care rights.

Before turning to the fair equality of opportunity account, however, it will be useful to make some important if not novel distinctions about health care rights (*cf.* [3]). I take the expression 'right to health' to be elliptical for the expression, 'right to health care', but this needs some comment. Some suggest that a 'right to health' is a negative right – a claim that others refrain from actions that threaten health. One might then try to see how much health care – I quite broadly use the term to include personal medical services and public health and preventive measures – might be required on the basis of this negative right. But advocates of a right to health often treat it both as a negative and positive right, requiring people to take positive steps to improve conditions that threaten health. So forcing the expression into the narrow mold of a negative right does not capture what is being said. Some philosophers have remarked that the expression 'right to health' embodies a confusion about the kind of thing which can be the object of a right claim. Health is an inappropriate object,



*T.J. Bole III and W.B. Bondeson (eds.), Rights to Health Care*, 201–212.

but health care, action which promotes health, is appropriate. The point is that, if my poor health is not the result of anyone's doing, or failing to do, something for or to me that might have prevented, or might cure my condition, then it is hard to see how any right of mine is violated.

To avoid these objections, I will understand someone who claims a 'right to health' to be claiming that certain individuals or groups (or society as a whole) are obliged to perform certain actions which promote or maintain his good health and are obliged to refrain from actions which interfere with it. The reference to 'health' should be construed as a handy way to characterize *functionally* the category of actions about which one is making a claim. This classification allows us to see why some advocates have insisted on a 'right to health' and not just on a 'right to (some) health care services'. They want their right claim to include a broad range of actions that affect health, e.g., protection of the environment, even if these actions are not normally construed as health care services.

Those who claim a right to health care often gloss over an important distinction. They may intend only a system-relative claim to health care: whatever health care services are available to any within the given health care system should be accessible to all. Such a claim may be met by removing from the system services accessible only to a privileged few. Thus this equality of access demand is not a demand for an independently determined level of health care, only for equality relative to whatever level of services the relevant system provides. Contrast this right claim with one that requires that some specifiable range of health care services be made available to all (and perhaps that any additional services be made available to some only if they are available to all). Such a substantive demand might require specific expansion or contraction of the existing health care system, not just in terms of *who* is treated, but in terms of *what services* are offered. The two right claims may have vastly different implications for reform of a given system, nor is it obvious which demand is more radical or conservative.

Quite different theoretical issues may underlie justification of the two different right claims, and this would be true even if they happen to require exactly the same reform of a given health care system. The system-relative right claim to equal access may ultimately best be defended by considerations that depend, not on the special nature of health or health care, but rather on more general considerations of equality itself. For example, it might be argued that the public nature of health care institutions – their dependency on public subsidies for training and facilities – requires that each person is due 'equal protection'. This argument is compatible with there being no basic right to any particular level of health services (as with the legal right to public education, where 'due process' requires equality in its provision, but not its provision). In contrast, a right claim to some particular level of health care may require for its justification a theory of basic needs – which is the type of account I develop. Moreover, a right claim to equality of health care at a level that exceeds satisfaction of such basic needs may require a different type of justification. In general, right claims to different kinds of health services may require different justifications.

In what follows I develop an account of just health care that allows us to characterize health care rights in a non-system relative fashion. Actual entitlements to health care, the content of such rights, will, however, turn out to be system-relative in a way that will be explained.

## II. THE FAIR EQUALITY OF OPPORTUNITY ACCOUNT

We can begin with the question, "Is health care 'special'?" Is it a social good that we should distinguish from other goods, say video recorders, because of its special importance? Does it have a special moral importance? And does that moral importance mean there are social obligations to distribute it in particular ways, ways that might not coincide with the results of market distribution? I believe the answer to each of these questions is "yes".

Health care – I mean the term quite broadly – does many important things for people. Some extends lives, some reduces pain and suffering, some merely gives important information about one's condition, and much health care affects the quality of life in other ways. Yet, we do not think all things that improve quality of life are comparable in importance: the *way* quality is improved seems critical. I have argued (*cf.* [6], [8]) that a central, unifying function of health care is to maintain and restore functional organization, let us say 'functioning', that is typical or normal for our species. This central function of health care derives its moral importance from the following fact: normal functioning has a central effect on the opportunity open to an individual. This claim can be made more precise.

The *normal opportunity range* for a given society is the array of life plans reasonable persons in it are likely to construct for themselves. The normal range is thus dependent on key features of the society, such as its stage of historical development, level of technological development and wealth, and other cultural facts. In this way, the notion of normal opportunity range is socially relative. Facts about social organization, including the conception of justice regulating its basic institutions, will also determine how the total normal range is distributed in the population. The share of the normal range open to an individual is also *determined in a fundamental way by his talents and skills.* Fair equality of opportunity does not require opportunity to be equal for all persons. It only requires that it be equal for persons with similar skills and talents. Thus individual shares of the normal range will not in general be *equal*, even when they are *fair* to an individual. This means that the general principle of fair equality of opportunity does not imply 'levelling' individual differences.

I can now state a fact central to my approach: impairment of normal functioning through disease and disability restricts an individual's opportunity *relative to that portion of the normal range his skills and talents would have made available to him were he healthy.* If an individual's fair share of the normal range is the array of life plans he may reasonably choose, given his talents and skills, then disease and disability shrinks his share from what is fair. Restoring normal functioning through health care has a particular and limited effect on the individual's share of the normal range. It lets him enjoy that portion of the normal range to which his full range of skills and talents would give him access, assuming these too are not impaired by special social disadvantages. Again, there is no presumption that we should eliminate or level individual differences, which act as a baseline constraint on the degree to which individuals enjoy the normal range. Only where differences in talents and skills are the results of disease and disability, not merely normal variation, is some effort required to correct for the effects of the "natural lottery". The suggestion that emerges from this account is that we should use impairment of the normal opportunity range as a fairly crude measure of the relative moral importance of health care needs at the macro-level.

Some general theories of justice, most notably Rawls's, provide foundations for a principle protecting fair equality of opportunity. If such a principle is indeed a requirement of an acceptable general theory of justice, then I believe we have a natural way to extend such general theories to govern the distribution of health care. We should include health care institutions among those basic institutions of a society which are governed by the fair equality of opportunity principle.[1] If this approach to a theory of just health care is correct, it means that there are social obligations to provide health care services that protect and restore normal functioning. In short, the principle of justice that should govern the design of health care institutions is a principle that calls for guaranteeing fair equality of opportunity.

This principle of justice has implications for both access and resource allocation. It implies that there should be no financial, geographical, or discriminatory barriers to a level of care which promotes normal functioning. It also implies that resources be allocated in ways that are effective in promoting normal functioning. That is, since we can use the effect on normal opportunity range as a crude way of ranking the moral importance of health care services, we can guide hard public policy choices about which services are more important to provide. Thus the principle does not imply that any technology which might have a positive impact on normal functioning for some individuals should be introduced: we must weigh new technologies against alternatives to judge the overall impact of introducing them on fair equality of opportunity – this gives a slightly new sense to the term 'opportunity cost'. The point is that social obligations to provide just health care must be met within the conditions of moderate scarcity that we face. This is not an approach which gives individuals a basic right to have all their health care needs met. There are social obligations to provide individuals only with those services which are part of the design of a system which, on the whole, protects equal opportunity.

If social obligations to provide appropriate health care are not met, then individuals are definitely wronged; injustice is done to them. These obligations are thus not merely like imperfect duties of beneficence, even though decisions have to be made about how best to protect opportunity. The case is similar to individuals who have injustice done to them because they are discriminated against in hiring or promotion practices on a job. In both cases, we can translate the specific sort of injustice done, which involves acts which impair or fail to protect opportunity, into a claim about individual rights. That is, the principle of justice guaranteeing fair equality of opportunity shows that individuals have legitimate right claims when their opportunity is impaired in particular ways – against a background of institutions and practices which protect equal opportunity. Health care rights on this view are thus a species of rights to equal opportunity.

The account I have been sketching thus provides grounds for right claims to health care, but more has to be said about how the scope and limits of these rights are specified. So far I have characterized these rights in a non-system relative way: they are rights to have opportunity protected in a particular fashion, through health care services. But the specification of the scope and limits of these rights – the entitlements they actually carry with them – requires a system-relative account. We must design a health care system that protects opportunity within the limits imposed by resource scarcity and technological development for a given society. The point can be put as follows. We cannot make a direct inference from the fact that an individual has a right to health care to the conclusion that he is entitled to some specific health care service, which in fact might meet his health care need. Rather, the individual is entitled to a service only if it is or ought to be part of a system which appropriately protects fair equality of opportunity. I will later return to consider objections to a health care right whose scope is only so indirectly determined, but first it will be important to see that complex decisions about institutional design must be made before health care rights can be made determinate.

## III. HEALTH CARE RIGHTS FOR THE ELDERLY

I would like to try illustrate the sense in which the scope and limits of health care rights must be specified in a system-relative manner by considering what is involved in meeting the needs of the elderly. Doing so will require elaborating the fair equality of opportunity account slightly. First, I will stipulate what I argue for elsewhere (*cf.* [4], [5], [9]), namely, that in designing our health care institutions, we must protect individual shares of the normal opportunity range *at each stage of life*. That is, the fair equality of opportunity principle requires that we protect the age-relative normal opportunity range for each stage of life.

Second, we must think of a health care system as a set of institutions distributing services over our lifetimes – as we age, we pass through it at different stages of life, with actuarially different needs at these stages. For example, when we are over age sixty-five, we will consume in the aggregate 3.5 times the health care services we consume at earlier stage of life. If we are prudent, we will want to design health care institutions so that they act as a savings scheme which defers the use of resources at a prudent rate. This means that we may not want to allocate to ourselves at each stage of life the same amounts of resources or the same types of resources: a prudent allocation will allocate to each stage the amounts and types of resources appropriate to protecting the age-relative normal opportunity range. The effect of these elaborations of my account is the proposal that just health care for the elderly will involve the prudent allocation of resources over a lifetime in ways that protect opportunity at each stage of life.

Consider some implications of this approach. Imagine we are prudent deliberators who do not know certain facts about ourselves, such as our age or medical history. How would we assess the importance of various personal care and social support services for the partially disabled and frail elderly as compared to personal medical services? From the perspective of these deliberators, both types of care would have the same rationale and the same general importance. Personal medical services restore normal functioning and thus have great impact on an individual's share of the normal opportunity range. But so too do personal care and support services for the partially disabled and frail elderly. They compensate for losses of normal functioning in ways that preserve an individual's fair share of the normal range.

The frequency of partial disability increases for the elderly, especially for those over age seventy-five. These disabilities are in general not life-threatening, and people usually live for many years with them. But they can have a dramatic impact on an individual's opportunity to carry out otherwise reasonable parts of his life plan, and 'quality of life' may be sharply reduced – that is, if there are no personal care and social support services that promote independent living. It is not prudent to design a system which ignores these health care needs, since they affect such a substantial part of the later stages of life, and which pays attention only to acute crises. This form of imprudence involves either not transferring enough resources to the later stages of life or transferring resources in the wrong form, as claims on acute services instead of personal care and social support services.

A major criticism of the U.S. health care system, that it encourages premature and inappropriate institutionalization of the elderly, should be assessed in this light. The issue becomes not just one of costs and the relative cost-effectiveness of institutionalization versus home care. Rather, opportunity range for many disabled persons will be enhanced if they are helped to function normally outside institutions. They will have more opportunity to complete projects and pursue relationships of great importance to them, or even to modify the remaining stage of their life plans within a greater range of options. Often this issue is discussed in terms of the loss of dignity and self-respect that accompanies premature institutionalization or inappropriate levels of care. My suggestion is that the underlying issue is loss of opportunity range, which obviously has an effect on autonomy, dignity, and self-respect.

Viewed in this light, the British health care system, in which extensive home care services exist, far more respects the importance of normal opportunity range for the elderly than does our system. They put their resources into improving opportunity range for the substantial number of elderly disabled over significant periods of the late stages of life. In contrast, we put our resources into marginally extending life when it is threatened in old age by acute episodes – a major chunk of our hospital bill is spent marginally extending the life of the dying elderly. Their approach is more prudent because it better protects age-relative opportunity range than ours.

The comparison with the British National Health Service (BNHS) is open to the objection that the British age-ration life extending resources, whereas we do not. The case is well documented for dialysis and some other technologies (*cf.* [1]). The objection is important to our discussion of 'rights to health care' because it might be thought that any account that recognized such rights could not tolerate different treatment of the elderly from the young. Such 'inequality' in treatment would seem to grant rights to the young that are denied to the elderly.

This objection is especially instructive to consider, for it forces clarification of some central claims made early. First, if health care rights are a species of rights to equal opportunity, as I have argued, then the actual entitlements such rights give rise to will depend on comparing technologies and considering the different impact they will have on preserving equal opportunity as a whole. Much attention has recently been focused on the sponsorship by the Humana hospital chain of artificial heart implantation, a technology which benefits a relatively small group of patients at considerable cost. (Indeed, the distributive effects of this technology were the object of concern some years ago.) Humana would have received far less attention had it announced that comparable money was being spent reestablishing prenatal matrenal care programs which were victims of budget cuts in 1980. Yet these programs can save far more lives at much lower cost than the heart implantation program. A recent report [11] indicated a significant increase in infant mortality in Massachusetts between 1980 and 1983 – which was attributed to the termination of the prenatal maternal care programs. Under conditions of real resource scarcity, we must carefully consider which technologies contribute most effectively to the protection of opportunity – and it is not obvious that many of the very high cost, high-tech services on which we lavish public attention contribute most to that goal. So the first point to make is that health care rights, in a generally just system, will give us claims only to services which promote the opportunity-enhancing task of that system. We will not necessarily have legitimate right claims to the introduction and dissemination of all feasible life-extending technologies.

The second point is directly relevant to the question of age-discrimination: Our health care rights might give us legitimate claims to services at one stage of our life but not give us such claims at another. Under certain conditions, it would be prudent – or at least not imprudent – to design a health care system so that it denied certain life extending technologies to people over some particular age (say seventy-five) in order to make those or other services available to people younger than that age. The effect of such age-rationing might be to make it more likely that the young will reach a normal life span and slightly less likely that the old will live to be even older. But such a rationing scheme operates over the whole course of an individual's life. Each individual will benefit from the prudence of the scheme and bear its burdens at one stage or another in his life. In this way, the scheme treats people 'equally' from the perspective of their whole life, though it *appears* to treat the young and old 'unequally'. I have defended the moral acceptability of age-rationing under specified conditions in some point further here. But I do want to return to its implication for the objection about rights to health care.

If age-rationing proves just under some conditions, for the reasons I mention, than our health care rights may indeed entitle us to different services at different ages. My rights to health care will not be 'unequal rights' compared to yours if I am denied dialysis at age seventy-five and you are given it at age forty, provided I also would have been entitled at forty and you will no longer be at seventy-five. The qualifier is that the differential treatment here must be part of a system which is prudent and which protects age-relative opportunity range at least as well as alternative, feasible arrangements. Any piecemeal implementation of age-rationing, for example, to shave hospital costs here and there, would thus violate health care rights of the elderly. (It might also be pointed out that the rationing scheme would have to be publicly understood, or publicity requirements on principles of justice would be violated. Age-rationing in the BNHS may be faultable on these grounds [*cf.* [1] and [7]).

## IV. OBJECTIONS AND REPLIES

I have characterized health care rights for the elderly as a species of rights to equal opportunity. More precisely, they are rights to health care which protect the age-relative normal opportunity range. But the entitlements which such rights carry with them, I have insisted, can be specified only relative to a health care system, for the design of a just health care system will involve many judgments about which types of services will most effectively function to protect fair equality of opportunity. Moreover, on this account health care services will compete with other institutions, such as education and job training and placement, which have an impact on fair equality of opportunity. Because we cannot read our entitlements to health care services off directly from the characterization of the right itself, some have claimed that the account "falls short of establishing health care as a universal human right" ([12], p. 335). In contrast, I think my account shows the limits that must be imposed on any such right claim to health care.

A more positive response can be made as well. I think the notion of health care rights that emerges from the fair equality of opportunity account can play the critical and progressive role that appeals to rights have traditionally played in the search for social justice. Consider the problem of access to personal medical services. In the U.S., for example, over twenty-five million people lack any health care insurance coverage at all, and many others have inadequate coverage. There is evidence that gaps in utilization rates between the rich and the poor and between blacks and whites, which had been closing from the mid-1960s until the late 1970s, are beginning to widen again (*cf.* [2], [10]). The health care rights that follow from the fair equality of opportunity account are violated by such grievous inequalities in access to care. There is no way the denial of services to the many who lack financing for them can be defended on the grounds that existing entitlements constitute an arrangement that best protects fair equality of opportunity, or that the inequalities of opportunity the system permits work to the advantage of those with the least opportunity.[2] It is far more plausible to think that granting access to basic care for the poor, even if it meant rationing or eliminating some high-technology services, would yield a system that better protected equal opportunity.

The account I have offered of health care rights for the elderly suggests another way these rights may be appealed to in attempts to reform the

U.S. system. An advocate of reform of our long-term-care system, which prematurely institutionalizes the frail elderly since it provides inadequate financing for home care services, could reasonably claim that the health care rights of the elderly are violated. His claim would be based on the contention that resources in our system fail to protect the normal opportunity range of the elderly because excessive resources are drained into prolonging death and diverted from providing less restrictive environments for the frail elderly. Here the argument would have to be supported by evidence about the effects of our system on the opportunity of the elderly compared to the effects of alternative arrangements and allocations of resources. But if the argument is sound, as I believe it is, then our system does violate the health care rights of the elderly because it denies them many services they need and lavishes the wrong services on them instead. The result is that it imprudently impairs opportunity.

I might add, by way of conclusion, that the fact that rights to health care can be asserted on the basis of the fair equality of opportunity account makes it a more concrete and effective way of urging reform of our health care system than the more guarded appeal to 'social obligations', which is found in the President's Commission Report *Securing Access to Health Care* [13]. The Commissioners eschew any appeal to health care rights, a stand which I endorse as a starting point for thinking about just health care, but which is not necessary when the job of constructing a theory is complete. The root of the difference between my account and the Commission's lies in their more eclectic and broader answer to the question, Is health care special? (*cf.* [8], Ch. 4). The Commission account fails to make it clear whether, or in just what way, the social obligations are obligations of justice. The fair equality of opportunity account does address just that point, with the result that it provides an account of health care rights.

## ACKNOWLEDGEMENTS

Research for this paper was funded by the National Endowment for the Humanities (NEH) Grant No. RO20456 and by the Retirement Research Foundation.

*Department of Philosophy*
*Tufts University*
*Boston, Massachusetts, U.S.A.*

## NOTES

[1] This requires modifications of Rawls's equal opportunity principle, however. Cf. Daniels [5].
[2] This last clause is included to indicate how the fair equality of opportunity principle would have to be applied within the framework of Rawls's theory.

## BIBLIOGRAPHY


1. Aaron, H. and Schwartz, W.: 1984, *The Painful Prescription: Rationing Hospital Care*, Brookings Institute, Washington, D.C.
2. Aday, L.A. and Andersen, R.M.: 1983, 'Equity of Access to Medical Care: A Conceptual and Empirical Overview', in President's Commission for the Study of Ethical Problems in Medicine and Biomedical and Behavioral Research, *Securing Access to Health Care*, Volume Three: Appendices. U.S. Government Printing Office, Washington, D.C.
3. Daniels, N.: 1979, 'Rights to Health Care: Programmatic Worries', *Journal of Medicine and Philosophy* **4** (2), 174–191.
4. Daniels, N.: 1982, 'Am I My Parents' Keeper?', *Midwest Studies in Philosophy* **7**, 517–540.
5. Daniels, N.: 1985b, 'Equal Opportunity, Justice, and Health Care for the Elderly: A Prudential Account', in H.T. Engelhardt, Jr. and S. Spicker (eds.), *Geriatrics: Ethical and Economic Conflict for the 21st Century*, D. Reidel, Dordrecht, Holland.
6. Daniels, N.: 1981, 'Health Care Needs and Distributive Justice', *Philosophy and Public Affairs* **10** (2), 146–179.
7. Daniels, N.: 1985c, 'Is Age-Rationing Just', invited paper delivered to the Pacific Division, March, American Philosophical Association, San Francisco, California.
8. Daniels, N.: 1985a, *Just Health Care*, Cambridge, Cambridge University Press.
9. Daniels, N.: 1983, 'Justice Between Age Groups: Am I My Parents' Keeper?', *Milbank Memorial Fund Quarterly/ Health and Society* **61** (3), 489–522.
10. Davis, K. and Rowland, D., 1983: 'Uninsured and Underserved: Inequities in Health Care in the U.S.', in President's Commission for the Study of Ethical Problems in Medicine and Biomedical and Behavioral Research, *Securing Access to Health Care* Volume Three: Appendices, U.S. Government Printing Office, Washington, D.C.
11. Knox, R.: 1984, 'Fund Cuts Linked to Infant Death Rise', *Boston Globe* **225** (145) (May 20), 1.
12. Moskop, J.: 1983, 'Rawlsian Justice and a Human Right to Health Care', *Journal of Medicine and Philosophy* **8** (4), 329–338.
13. President's Commission for the Study of Ethical Problems in Medicine and Biomedical and Behavioral Research, 1983: *Securing Access to Health Care* Volume One: Report, U.S. Government Printing Office, Washington, D.C.

NANCY K. RHODEN

# FREE MARKETS, CONSUMER CHOICE, AND THE POOR: SOME REASONS FOR CAUTION

## I. INTRODUCTION

"Consumer choice" is the rallying cry for health policy analysts who espouse an essentially free, competitive market for health care. Some, of course, would make their market freer than others. On the one extreme are the libertarians, who would allow health care to be distributed precisely as the market allocates it, with no independent evaluation of the outcome. On the other are modified market theorists, such as Alain Enthoven, who would give the poor vouchers for health care and would restrict the sorts of health plans that competing providers could offer. Despite the significant differences in their views, however, both pure market advocates and voucher proponents share an underlying faith in not only the market, but also the ability of health care consumers to make their demands effectively heard in this market.

In this article I explore some of the reasons why a competitive health care market, even with the safeguards built in by theorists such as Enthoven, might not work well for all. My focus is not on the economics of markets, or even primarily on the many peculiarities of the health care market. It is instead on those people whose health depends on consumer effectiveness. Consideration of some of the correlates of poverty suggests that even with vouchers, the poor may fail to obtain needed medical care in a system that expects providers to respond to consumer vigilance. I argue that this is because poverty, contrary to the assumption of all market theorists, pure and modified alike, often means more than merely a relative lack of money.

Because the basic assumptions about health care as a market commodity and individuals as cost-conscious health consumers are exhibited in their most pristine form in the writings of pure market proponents, I first analyze them in this largely libertarian context. This should help in recognizing the extent to which, despite their modifications, voucher proposals share these basic assumptions and hence are vulnerable to the same criticisms. I will then look to the voucher proposals themselves,



*T.J. Bole III and W.B. Bondeson (eds.), Rights to Health Care*, 213–241.

criticizing not the proposals in general but, specifically, their underlying assumption that what the poor (and other vulnerable groups, such as the disabled, infirm elderly, mentally ill, etc.) need is access to the health care market, rather than also requiring the sorts of outreach and education about their health needs that a system that sees health care as something more than a consumer commodity might provide.

## II. THE PURE MARKET APPROACH

Health care in this country unquestionably has some ailments, not least among them its cost. One solution, often proposed by persons more or less in the libertarian camp, is to scrap our complex and often counterproductive set of health care regulations, forget national health insurance (NHI), and simply let the market – the standard distributor of goods and services – handle health care distribution [4, 32]. When critics object that the poor then will have *no* access to health care, pure market proponents tend to suggest giving the poor cash ([4], pp. 151–153; [32], pp. 73–75). Cash, they argue, will respect the autonomy of the poor by letting them choose to purchase whatever goods they prefer, rather than denying them choice by offering only in-kind aid.

While a complex set of beliefs underlies this pure market approach, including, of course, a robust faith in the free market, certain specific assumptions are necessary in order to embrace fully a market approach to health care. I want to spell out these assumptions, and then show why they are wrong. One could reject this libertarian approach on the pragmatic ground that, since government officials want aid to the needy to buy necessities like health care rather than luxuries, the cash aid needed to render a market system even minimally fair will not be forthcoming. However, inasmuch as market proponents could simply respond that the government must overcome its paternalistic predelictions and provide cash, this line of argument does not take us very far. Instead, since market-oriented conclusions seem to flow so nicely from their proponents' assumptions, it seems more fruitful to scrutinize their underlying assumptions.

The first assumption is that health care is a consumer good like any other. If it were a public good, like clean air, a straightforward market approach simply would not work. But people privately purchase doctors' services, hospital care, drugs, etc., much as they purchase car parts and auto repair services. This suggests that health care is not a truly public good, and market proponents conclude that it must therefore be a private consumer good like any other. They admit that the desire for health care is sometimes more pressing than for other goods, like fine wine, but maintain that this alone provides no reason to treat it differently.

Another primary assumption is the fundamental rationality of individual consumers. As David Friedman puts it, the first assumption of economics is that people are rational and will tend to make choices that serve their ends. Closely related to this assumption is the primacy of autonomy and indvidual choice. This emphasis explains why market advocates would favor giving the poor cash over in-kind assistance. Since people will rationally pursue their ends, those people so economically disadvantaged that they cannot buy health care (or whatever) could be given money enough that they could do so. But their rights of autonomous choice should not be preempted by insisting that health care, rather than something else, be purchased.

If health care be a consumer good, markets generally the way to distribute such goods, and all people, if given money, able to participate rationally and effectively in such markets, it follows that a free market approach to health care is ideal. But what if the end result seems somehow unfair – i.e., what if the poor, even when given cash, get very little health care? Another important tenet of the market approach is that the proper focus of evaluation is not outcome. After all, outcome is not important with other consumer goods: it is not considered unjust if middle-aged women buy fewer computers than young men, or Yuppies buy more Perrier than other groups. These disparities merely reflect people's rational choices – as would the choice of the poor to use their cash supplements for things more pressing (or more fun) than health insurance.

These assumptions of health care as just another consumer good, consumer rationality, and irrelevance of outcome will be questioned here. While it is undoubtedly true that this libertarian approach is not a serious contender for public policy implementation, it is nonetheless somewhat influential. Its influence is perhaps most keenly felt in certain modified free market proposals, which will be considered later on. These proposals share, to some extent, some of the assumptions found in their strong form in the pure market proposals. Criticizing these assumptions in their strong form will thus help us to see the flaws in the weaker, but related, assumptions underlying the modified free market plans.

*A. The Irrelevance of Outcome*

First, let us consider the idea that health care can be viewed as just another consumer good, with its distributional outcome irrelevant if the market is functioning properly. Market proponents agree that for a market to yield results that are acceptable, such that independent scrutiny is unwarranted, the base-line for which consumers start should not be grossly inequitable ([32], p. 91). This does not mean that persons' starting shares must be equal, but merely that they should not be so disproportionate that some people are debarred from even participating in the market. Market proponents would support giving the poor cash so that the starting points are less unequal. Were this approach to be taken seriously, of course, there would be the thorny problems of *how much cash*, and how to know what amount is enough. When would the least well off have a sufficient starting base that the outcome of the market would be irrelevant, and would there be a problem of circularity, in that outcome would have to be evaluated to determine whether the starting point was sufficiently equitable?

For example, suppose the government terminates in-kind assistance, assesses the general cost of living, and decides to give up to $7,000 to each United States family of four. Suppose further that subsequent assessment of this policy shows that, although ninety-five percent of working Americans making $10,000 or more have health insurance, only forty percent of persons receiving cash supplements have insurance. The other sixty percent have no health insurance and, therefore, no access to the health care market (unless they can pay cash for needed care on the spot). Suppose the study also shows that persons without insurance are four times as likely as others to die preventable deaths, live an average of fiteen years less than the general population, and have an infant mortality rate thrice that of the rest of society. Would this indicate the the $7,000 was not enough, or, if the amount were properly chosen according to cost of living indices, should these results simply be ignored?

While it is fairly easy to dismiss quite significant disparities in purchases of fine wine, home computers, etc., between rich and poor, this hypothesized disparity in health status seems harder to ignore. I would surmise that most people would find such an extreme disparity distressing – although the consistent libertarian may have a strong enough stomach (or hard enough heart) to ignore it. When statistics about our society (with both welfare and in-kind health care assistance) are given that, for example, show that blacks have an infant mortality rate twice that of whites ([2], p. xiv), the implication is that these figures suggest something is wrong. It seems it would be hard for even the most thoroughgoing libertarian to contemplate a truly *enormous* disparity in health status and not at least wonder if perhaps the initial background for this "market justice" were not askew. What this suggests is that the notion that outcome can ever be completely ignored when health is at issue is extremely problematic: even with great faith in the market, an inordinate disparity could indicate that the poor's cash supplement should be increased yet again. The end result could be a pure market health care system, but huge cash subsidies to the poor.

### *B. Health Care as Just Another Consumer Commodity*

The disparities focused upon were not ones of health care consumption – e.g., number of doctor visits or hernia operations – but rather were typical health status indicia like infant mortality, life span, and preventable deaths. What would be the equivalent indices were we asking whether the distribution of fine wines or home computers were fair? Would it be the number of low income members of wine tasting clubs or subscribers to magazines such as *Personal Computing*? Somehow these latter indicia lack the same sort of force. This suggests a problem with the premise that health care is like any other consumer commodity. Few consumer goods (food and housing being notable exceptions) have such a close connection with basic human well-being. For most goods, the only relevant statistics are the number of items purchased – like the number of doctor visits or gallbladder operations. With health care, however, the indices of health status seem quite important: at least, a system where the poor had as many operations as the rich, but had the sort of gruesome health statistic hypothesized earlier, would likely be deemed unsatisfactory. This is not to suggest that health status of rich and poor must be equal: poverty, nutrition, living conditions, genetics, etc., influence health, and ill health itself may lead to poverty. Nonetheless, as well as indicating a high degree of disparity between social classes, such statistics could suggest that the health care system is not even partially ameliorating these disparities.

This focus on health status (and the absence of a corresponding "fine wine status" or "computer status") suggests that perhaps health care differs in a significant way from most consumer goods. Norman Daniels has nicely delineated how people, although they may desire many consumer items, often *need* health care ([10], pp. 6–7). A desire for fine wine is just that – a desire (though a desire for basic nutrition is also an expression of a need). But a person with acute appendicitis *needs* an appendectomy. As Daniels emphasizes, in moral contexts we tend to appeal to some objective criteria of well-being, and hence some objective measures of need ([10], pp. 6–7). At least in regard to certain types of health care, the concepts of objective need and of clearly diminished well-being if the need is not met, seem peculiarly appropriate.

Daniels has also provided a promising explanation for just why health care is special, and hence why satisfying health care needs seems more important than satisfying desires for most other consumer goods. He takes the view that because health is a necessary condition for equality of opportunity, under a Rawlsian theory of justice health care institutions would be among the background institutions involved in providing for fair equality of opportunity ([10], p. 24). Health care helps to minimize the obvious disadvantages caused by ill-health, and thus enhances equality of opportunity. This analysis, which takes into account the fact that needs are distinguishable from desires and their fulfillment more important in a just society, makes health care seem, as Daniels suggests, analogous to education in many important respects. According to Daniels, both education and health care address needs that are unequally distributed among individuals. Moreover,

> Various social factors – such as race, class, and family background – may produce special learning needs; so too many natural factors, such as the broad class of learning disabilities. To the extent that education is aimed at providing fair equality of opportunity, special provision must be made to meet these special needs. Hence educational needs, like health care needs, differ from other basic needs – such as the need for food and clothing – which are more equally distributed between persons. The combination of unequal distribution and the great strategic importance of the opportunity to obtain both health care and education puts these needs in a separate category from those basic needs we can expect people to purchase from their fair income share ([10], pp. 26–27).

Taking this analysis seriously should strongly undercut the notion that all private goods can be treated alike and simply distributed by the market. While health care is not like clean air or the national defense, it differs from most other consumer goods in that the need for it is more urgent, more unevenly distributed, and more important in allowing people an equal opportunity to pursue their vision of a good life. The car accident victim does not merely want a doctor, she needs one – and in a different way from how her distraught husband needs a drink or will need a new car. The suffering and severe inequality caused by denying such a need does not necessarily show that health care should not be distributed via the market, but does suggest that its special nature must be given its due, and that market proponents cannot merely invoke the efficiency of markets and say that a pure market approach should prevail because health care is a consumer good.

### *C. Consumers as Rational Health Care Decisionmakers*

Thus far we have seen that even if the background of income distribution were fair, health care, unlike consumer goods, would still warrant independent examination of outcome, largely in terms of health status. There is no correlate to health status for most consumer goods, because there can be an objective need for health care that differs from the subjective desires for most other goods. The assumptions that outcome can ever properly be ignored, and that health care can be treated just like any ordinary consumer good (rather than like, say, education) are thereby undercut. The other fundamental assumption of this individualistic market approach is that adult consumers are rational and autonomous and hence able to make the health care choices that best promote their ends and their vision of the good life. A problem, however, with making assumptions and then deriving policies from them is the tendency to forget that the original assumption is merely a convenient assumption and not a proven fact. The statement that most persons are rational is undoubtedly far closer to the truth than its opposite. But such a blanket statement is not exactly attentive to the many nuances of psychology, especially as it may be affected by illness.

Section IV will deal with the question of whether the poor can act as effectively and autonomously in the health care marketplace as the middle class, so the discussion of the impact of social conditions on choice will wait until then. But here I should briefly note that illness engenders irrationality and dependency in us all, and that our attitude toward risk when remote is far different from our view of it when close and threatening. It may seem rational one day to use funds for other things and not to insure against the seemingly slim chance of catastrophic illness. But the next day, when catastrophe strikes, it would be virtually astonishing not to seek care anyway, or at least not to desperately want it. And it would be inhuman for others not to wish to provide it given such dire need. Paul Menzel says this is mere emotion, the logical inconsistency of weak humans. It would, he suggests, respect the autonomy of the uninsured to harden our hearts to their plights. He imagines an uninsured and sick person responding to the anguish that others may feel denying him care as follows:

> *[Y]our* anguish is *your* problem. ... It surely isn't *my* responsibility as a welfare recipient who didn't insure myself. Out of respect for my decision, in fact, you ought not to indulge in your anguish! Unless I consent somehow to the restrictions of choice which in-kind aid involves, why should you feel guilty? ([32], p. 99)

But if our hearts so clash with our minds, perhaps it is our minds that are wrong on this one.

One last point on rationality: proponents of this assumption in its strong form (i.e., people are rational and their health care choices should not be influenced, aided, constricted, etc.) seem to imply that the only alternatives are that people are rational or that they are not (much as the notion of health care as merely another consumer commodity seems to imply that goods either are public or are private, consumer ones, with no hybrids such as education). Since we generally think of humans as more rational than not, any caveats, conditions, fluctuations within a person or variations among groups of persons tend to be ignored. The only real exception is children. Since children cannot make their own health care choices and will suffer if denied care, Menzel, for example, would favor universal maternal and child health ([32], p. 78). Unquestionably, children (at least young ones) are less capable of appreciating their needs and competently setting out to meet them. But human rationality, or effectiveness at matching needs with courses of action, is much more a continuum than an either/or proposition. Consumers in the health care marketplace often get little chance to shop around, comparing price, effectiveness, etc. They may come in acutely ill and anxious; they may have had their discriminating powers affected by chronic mental or physical illness; they may know precious little about the "product" to be "consumed"; they may have denied the possibility of illness (and therefore not insured for it); and, as will be discussed in Part IV, some of them may be predictably affected by severe social, economic and educational disadvantages. No assumption should be allowed to mask these significant and sorrowful features of human existence.

The assumptions that underlie this pure market approach to health care

are thus extremely problematic and oversimplified. Outcome *is* important, and if health care is to be viewed as a consumer good, it certainly must be recognized to be an unusual one. Nor can people simply be assumed to make rational decisions that aid their self-interest in an area as complex, problematic and perilous as health care – the variations in human capacity and need for assistance in decision-making must likewise be recognized. Having rejected the libertarian's assumptions in their strong form, let us now turn to proposals for a modified, voucher-enhanced market system, and consider the validity of the somewhat modified assumptions that underlie these proposals.

## III. VOUCHER PROPOSALS FOR MARKET ACCESS

Realistically, the chances of this country embracing the sort of libertarianism exemplified by the pure market approach are practically nil. Unless very large cash supplements are given, it has been noted that: "Since the richest 20% of all U.S. households have eleven times as much income as the poorest 20%, any efficient market mechanism will end up giving eleven times as much medical care to the top 20% as it gives to the bottom 20%" ([48], pp. 1570–1571). Most Americans do not really want their fellow citizens to die because they cannot afford medical care, even if they foolishly squandered their cash allotment on something else. Hence at least some market modifications are necessary.

### *A. The Voucher Proposals*

The modifications proposed in recent years by some highly influential theorists are to provide health care via a rejuvenated market system, with vouchers for the poor to enter the competitive health care marketplace. One of the best known of these proposals is Alain Enthoven's Consumer Choice Health Plan (CCHP). Professor Enthoven proposes a system comprised predominantly of competing organized health systems that would provide comprehensive services, frequently on a prepaid basis, to a population that has voluntarily enrolled with them. Under the CCHP, all citizens would be able to enroll (with open enrollment once a year) in any health plan in their area – a fee-for-service plan, a health maintenance organization (HMO), etc. – and would be able to remain continuously enrolled regardless of changes in job status ([16], p. 117). Enrollment would be financed through a system of refundable tax credits. Because the credit would equal only sixty percent of the individual's or family's actuarial cost (the average per capita cost for covered services for persons in a particular actuarial category), consumers would have a strong incentive to economize in choosing health plans.

The poor would receive vouchers with which to enroll in the health plan of their choice. For the most needy, these vouchers would cover 100 percent of actuarial costs. As family income increased, the amount covered by the voucher would decline. The poor would likewise have an incentive to be economical in choosing plans. To the extent that the voucher exceeded the cost of the plan chosen, the extra money would purchase additional benefits such as dentistry or cover any direct medical expenses such as co-payments ([16], p. 123). According to Enthoven, competition among health care groups would favor prepaid plans such as HMOs, which provide doctors with incentives to keep costs down and avoid unnecessary tests and hospitalizations. Groups could not economize by screening out high risk customers, however, because providers would be required to allow open enrollment and would be prohibited from screening applicants for health status.

Enthoven argues that, so organized, a market system would efficiently reduce costs by rewarding frugality in the provision of health care. It would also fill in the gaps in our current system, under which more than twenty million people lack health insurance and in which loss of employment often terminates health care coverage. Enthoven's proposal likewise incorporates certain safeguards to ensure that cost concerns do not adversely affect the health care coverage of the poor. These are: (1) that all policies cover a set of basic services; (2) that vouchers be set at a high enough level to ensure access to a plan with adequate benefits; and (3) that there is an upper limit on the amount that an individual ever has to pay for medical expenses ([16], pp. 128, 139). Enthoven suggests that with these safeguards a market system, while not providing full equality in health care, will meet the demands of justice by providing a "decent miminum" of health care to all. In considering whether the disparity between the plans selected by rich and poor would become too great, Enthoven states: "I would expect competition to keep the difference in premiums among the plans within fairly narrow limits, so that low-income families would not be limited in their effective choices to the least costly plans" ([16], p. 24).

Clark Havighurst's proposal for a consumer choice system (CCS) bears much similarity to Enthoven's. It would, however, incorporate somewhat fewer safeguards. For example, Havighurst would allow voucher recipients who choose an inexpensive plan to reap the rewards of their frugality in the form of a cash rebate representing the difference between the voucher and the chosen plan. He also would not mandate a minimum benefit package. According to Havighurst, the government should not judge plan acceptability on an *a priori* basis.[1] Rather, consumer protection "might take the form of a requirement that no plan could be offered that did not have the minimum number and percentage of premium payers enrolled otherwise than through the CCS" ([20], p. 405). (Havighurst would, however, consider restricting low income persons from selecting plans whose coverage gaps and cost-sharing provisions made them suitable only for the well-to-do ([20], p. 405).) He would also allow providers some discretion in accepting customers: open enrollment (without screening for health status) and community ratings (without actuarial adjustments) would not be required ([20], p. 405). Any limitation on the amount an individual should have to pay for catastrophic illness would differ as well: while not rejecting the idea that providers should at least be strongly encouraged to provide adequate catastrophic illness coverage, Havighurst says that such coverage should not necessarily provide for everything medically possible. Rather, he would explore a way of requiring a minimum coverage of catastrophic costs in such a way that cost-benefit trade-offs could still be made, and in this regard would rely to some extent upon hospitals' charitable impulses toward the severely ill ([20], pp. 422–424).

Despite these differences, Enthoven and Havighurst are in full agreement on the basic proposition that once the poor are given vouchers to enter the health care market, the competitive market can serve their needs. In other words, vouchers plus free competition (and certain safeguards) can yield a happy congruence of economic efficiency and a sufficient degree of equity in health care. Because certain psychological and philosophical assumptions underlie this conclusion, our task is to spell them out and then ask, as with the libertarian assumptions, whether they are adequately justified.

### *B. Suppositions of Voucher Proposals*

An important assumption underlying voucher proposals is that health care is primarily a consumer commodity, albeit a rather special one, and that it is suited to the language and values of the marketplace. This general assumption will not be scrutinized here, because, despite its importance, such scrutiny is beyond the scope of this critique. I will merely note that it is an assumption that is neither universally shared – witness the many countries with national health insurance or socialized medicine, and the proponents of each of these in the United States – nor unproblematic. According to one writer:

> the patient is increasingly becoming the consumer in the language of public debate ... A word taken from the vocabulary of the market place is displacing a term appropriate to an expert-dominated system. The language of demands is threatening to replace the language of needs ([26], p. 90).

And, as Alan Buchanan astutely notes:

> There is a relationship between the increasing acceptance of the market model as the appropriate form of explanation for a certain sphere of human activity and the growing tendency of that sphere of activity to fit the market model.... In other words, to the extent that we come to view our interactions as market transactions they may actually come more closely to approximate the model by which we seek to explain them ([6], p. 160).

Whatever the validity of the idea that health care is a consumer commodity, voucher proposals, unlike pure market ones, reflect its special and unusually important nature. While market outcome is ordinarily irrelevant, voucher proponents recognize that an outcome in which the poor lack health care violates notions of social justice. Havighurst's plan affords fewer safeguards against this; for example, a cash rebate could overly influence the poor to choose the very cheapest plan, whatever their real needs.[2] Havighurst shows a correspondingly lower degree of independent evaluation of outcome, stating:

> There is ... great resistance to the idea that the competitive process, if it works reasonably well, validates the outcome. Many observers want to be told in advance what the result of competition would be so that they can decide whether they approve ([20], p. 112).

Nonetheless, within at least some broad range of outcomes, Havighurst, like Enthoven, does profess to care what the outcome is, and does believe that the poor should have a decent minimum of health care.

Perhaps the most important tenet of market proposals is that consumers can make rational choices that will serve their ends. It has been remarked that: "societies allow market mechanisms to work when buyers are knowledgeable or willing to live with their mistakes and when society is willing to distribute goods and services in accordance with the market distribution of income. In the case of health care, neither of these two necessary conditions exist" ([47], p. 1571). Having taken care of the problem of unequal distribution of income by providing vouchers, voucher proponents would argue that the chance of mistakes through poor choices could be minimized (with Enthoven's proposal minimizing them more than Havighurst's), and that consumers under such a system can usually choose knowledgeably. The reason they can choose knowledgeably is that, because all they have to decide about is an insurance plan, cost-benefit analysis need not be made at each episode of illness. Havighurst emphasizes how consumer choices about insurance are usually made with the benefit of professional advice or employment groups ([20], p. 43). He says that a consumer's primary problem connected with the lack of information about health services is identifying a trustworthy physician and choosing a hospital and specialist. He confidently predicts that, "perceiving the importance of a correct choice, the consumer can be expected to invest considerable time and effort in the search" ([20], p. 78).

Enthoven likewise emphasizes that while individual episodes of medical care are not usually subject to rational economic calculation, because an urgent need for care renders the transaction not fully voluntary ([16], p. 34), consumers can choose far more rationally when picking a plan in advance. A consumer, of course, knows his own values and preferences, as Havighurst emphasizes ([20], p. 43). But, equally importantly, he may be able to anticipate his health needs. According to Enthoven, much of health care is elective, rather than being urgent or necessary at the time. He states:

> I have not seen a systematic study of this point, but I believe that well over half of all medical care spending is for the treatment of chronic conditions, for care postponable for a year without great risk of suffering, or is elective with respect to timing (for example, a decision to become pregnant or undergo corrective orthopedic surgery) ([16], p. 8).

This of course resembles the pure market notion that health care satisfies wants as much as needs. It does not completely deny needs – health care is, after all, treated as special, and vouchers for other consumer goods are not supplied. But market proponents seem to emphasize the elective aspects of health care more than do analysts less enamoured of the market.

One other general assumption which will not be assessed here is that, as consumers make their desires known (and rationally seek to economize), the institutional sector, including health insurance plans, hospitals and professionals, will respond. Some writers feel that there is little evidence that the responsiveness envisioned could actually occur. According to David Mechanic, to believe in this, "one must assume that consumers would make informed choices on the basis of economic interests rather than habit, inertia, or psychological considerations, and that large providers would feel pressured to compete in offering more economical and efficient plans" ([30], p. 11). Likewise, Lawrence Brown questions the degree of faith in both market incentives in health care and other areas:

> Some policy analysts have apparently persuaded themselves that the merest flick of an incentive system can, like Sumner's mores, make anything right. Right incentives, one is assured, will lead businesses back into central cities (enterprise zone proposals); make companies produce and consumers buy less gas (decontrol of gas prices); lead polluting firms to pollute 'optimally' (pollution taxes, fees and 'rights'), etc. ([5], p. 154).

While it seems that a bit more skepticism than voucher proponents muster is warranted, the issue of market responsiveness in general will not be considered here, especially because this is an empirical issue about which, inasmuch as such a system is not in force, we can have little or no evidence.

Part IV will, however, ask whether providers will respond as well to the desires of consumers who are poor. This requires elucidating the assumptions about poverty implicit in market proposals. Such proposals largely share the assumption of the pure market advocates that the only problem the poor have is that they lack money. While libertarians would just give the poor money, voucher proponents realistically recognize that the cash provided might not be enough, or, given the poor's other needs, might not go toward health care. But voucher proponents believe that nothing more than vouchers, which provide market access, is necessary, because the poor are considered as capable of being rational, effective, cost-conscious consumers as anyone else. This sounds reasonable – as both Brody and Friedman put it, "the poor lack money, but they are not stupid".

A corollary of this seemingly reasonable viewpoint that the poor suffer only from a relative lack of funds is the view that the poor's health needs are as elective as those of anyone else, and hence, are equally amenable to advance planning. In other words, the supposition is that the poor, like anyone else, will know what sorts of health care they will want, and will be able to seek out those plans and medical groups that will be most likely to provide it. In other words (again quite reasonably), it is assumed that once the poor have access to the market, they will treat their health needs (or wants) just like everyone else treats theirs. Providers, in turn, will respond to the demands of the poor like they respond to the demands of any other consumer group, now that the poor have purchasing power. Thus, what the poor want will be, for the most part, both what they need and what the competitive system gives them. The combination of these various assumptions leads inexorably to the conclusion that once the poor have the vouchers to buy into the system, a competitive system will work as well for them as for anyone else. Hence, it is these underlying assumptions which Part IV will explore.

## IV. CONSUMER BEHAVIOR OF THE POOR – LIMITATIONS ON EFFECTIVENESS

Some of the poor already have a voucher program of sorts. Medicaid was enacted to allow poverty stricken individuals to buy their way into mainstream medicine, and Medicaid recipients unquestionably have obtained access to health care through this program. Indeed, one of the most often cited problems with our health care system is the Medicaid gap – individuals not poor enough to be eligible for Medicaid, but not sufficiently solvent to have any (or adequate) health insurance ([36], pp. 92–100). Health care for Medicaid recipients is not, however, provided in exactly the manner health care is ordinarily provided. Poorer, less adequately staffed public hospitals, foreign physicians, long waiting lines, dreary surroundings, etc., often characterize our society's "welfare medicine". This unfortunate and unintended result clearly is related to the way in which Medicaid is administered (for example, by welfare agencies) and the extent to which it is funded. However, provider response to Medicaid patients – the unwillingness to locate in their areas, the greater willingness to exploit them for profit, etc. – has also played a role. This raises the question of whether there might be factors other than a lack of funds that influence the nature and quality of health care the poor receive. If so, providing funds (vouchers) might not be a total solution.

### *A. Correlates of Poverty that Affect Consumer Behavior*

It is the thesis of this section that there indeed are reasons why care for the poor has differed from care for the better off, and that these reasons will not dissipate with the provision of vouchers. The reasons inhere in certain correlates of poverty, which include a lack of education, a somewhat different attitude toward and emphasis on preventive care, a more crisis-oriented attitude towards illness, and a greater difficulty in asserting rights against impartial bureaucracies or persons of higher status. The impact of these factors on consumer effectiveness will be discussed in detail shortly. First, however, a few caveats. These correlates of poverty obviously do not apply to all the poor – they are generalizations, no more. Nor should their recognition displace other important factors explaining the poor's lower usage of some sorts of care, such as the still existing economic barriers to access, and the additional barriers imposed by crowded conditions, long waits, and impersonal treatment at many of their health care facilities ([13], pp. 350–351). Moreover, this discussion is not intended to "blame the victim", i.e., to suggest that disparities in health status of rich and poor are the poor's own fault. It is simply intended to show that the social background from which any health reform must spring is more complex than free market proponents are willing to recognize. The insights of psychology and sociology must be taken into account, and they suggest that it is an unfortunate oversimplification to say that the only difference between rich and poor is the number of dollars they each possess.

It is a truism that the poor are, as a group, less well educated than the better off. But anyone knows when they are sick, one might argue, so why should education matter. Inadequate education, however, may decrease ability to discern differences between plans and to assess their relative merits, and it may fundamentally influence views about what sorts of care are necessary. Sociological studies show that the poor have significantly less faith in the value of preventive care than the upper classes ([13], p. 359), and, as one might expect, national health surveys show that a much smaller percentage of their treatment goes to such care ([25], p. 7). Hence the poor may underemphasize prevention in their health care package, or may mistakenly choose plans that cover services the young and healthy may want but exclude ones the poor or chronically ill might need.

The poor may likewise fail to take advantage of those preventive services that are available. In regard to prenatal care, for example, some women who do not seek such care simply believe it is unnecessary if a pregnant woman feels well ([34], p. 2). This is highly correlated with education: over half the women who do not receive prenatal care are ones with less than a high school education ([34], p. 72). Services such as amniocentesis are predictably provided far more frequently to well-off white women than to poor and non-white ones ([27], p. 89). In a consumer oriented system, where financial incentives favor decreased utilization, the poor may fail to receive potentially beneficial services simply because they are not aware that such services are useful or available.

Even when the poor are aware of the benefits of certain health services, they are less likely, as a group, to assert their desires. Some analysts suggest that this is due to lowered self-confidence or self-esteem ([46], p. 143). Whatever the cause, members of lower classes are repeatedly described as being markedly subordinate in relations with medical personnel ([46], p. 146). Being more docile, they make fewer demands on the medical system. They are also less likely to understand the system and its organization, and hence are less able to have a significant impact upon it. These disparities in assertiveness between the poor and other consumers, especially in a system that relies heavily on consumer demands, may lead both to underutilization and to a lower quality of care: i.e., if the poor do not complain, why should a cost-conscious provider spend money to encourage utilization or improve medical care or associated amenities?

There are also differences between the poor and the wealthier in patterns of seeking care which could affect the health status of the poor. Because the poor are less apt to seek care outside their neighborhoods ([23], p. 83), they are often a "captive audience" for inner-city providers. Captive consumers, of course, are not very influential ones. Also, perhaps because lives of poverty are often more crisis-filled, the poor are more likely to postpone seeking medical care until their condition becomes severe or even incapacitating ([46], p. 142–144; [12], p. 140). A failure to attach much significance to certain symptoms may likewise exacerbate their tendency to seek care later in the course of an illness ([25], p. 13). Whatever the precise psychological, sociological, or economic cause, studies show that paying patients have cancers diagnosed at earlier stages than non-paying ones, and that welfare patients with appendicitis are twice as likely as insured patients to be admitted at an advanced stage of the disease ([36], p. 74). Enthoven's emphasis upon the elective nature of much of health care thus is far less applicable to the poor's more crisis-oriented mode of seeking care.

The poor have likewise grown accustomed to second class care – or, at least, care in second class surroundings. Because of this, they may be less vocal in protesting such conditions if they crop up at their new, profit-oriented providers. They may likewise be intimidated by the sense that medical personnel look down upon them because of poverty, race, or behavior that differs from the middle class.[3] Because medical personnel may in fact harbor such feelings, or at least may feel less kinship with impoverished or minority clientele, they may be tempted to increase profits at such persons' expense. Moreover, even if race and economic status have no effect, the poor are more likely to suffer from chronic health conditions ([36], p. 63). Such conditions, unfortunately, are often less attractive to medical personnel, being labor intensive, not curable by high technology, and "less interesting" than certain other conditions. Profits can probably be enhanced by discouraging people with such chronic conditions from becoming customers by, for example, failing to cover treatment for their conditions. This would be perfectly rational provider behavior, since markets "are designed to encourage firms to segment, 'cream' and 'dump' to find the most profitable niches while ignoring areas of low profitability" ([48], p. 1571).

All this suggests that the poor may be less effective in voicing their demands than the better off. That the differences between rich and poor are not solely economic, is illustrated by the fact that in any health care system the middle classes tend to make better use of the available health services ([26], p. 103). Even creating an egalitarian system will not completely solve this problem, as studies of the British system show, since social resources such as information, self-confidence, and articulateness may be as important as financial resources ([26], p. 103). But acting as if the difference is only economic, and hence simply plugging the poor into a market system, may make this disparity worse, because this type system relies heavily on the sort of consumer advocacy at which the poor do not excel. The poor fare less well than others in markets in general, even when they are given purchasing power ([49], p. 213). In light of these various social and psychological correlates of poverty, and the financial incentives for providers to capitalize upon them, it is hard to be confident that the poor would fare all that well in a competitive health care market. It is even harder to believe that the disabled, mentally ill, mentally retarded, impoverished elderly, and chronically ill would effectively induce profit-oriented providers to meet their often labor intensive and costly medical needs. Indeed, the further a group is from the idealized cost-conscious, educated, and assertive consumer, the less confidence one can have that a competitive market system will serve its needs.

### *B. Special Health Needs of the Poor*

Along with the above correlates of poverty is the problem that the poor may have certain special health needs that: (a) differ from those of the middle class; and/or (b) go unrecognized by the poor themselves. For example, children in urban ghettoes are twenty-five times more likely to have high lead levels in their blood than affluent white children ([22], p. 393). Children of the poor see doctors substantially less often than children of the affluent ([13], p. 349). A provider serving the poor in a competitive system may not find it profitable to encourage increased utilization among poor children and/or to provide special screening.

Certain screening programs for the adult poor are similar. For example, blacks are four times as likely as whites to suffer kidney disease requiring dialysis or transplantation ([36], p. 70). This is, at least in part, because they are much more likely to have unrecognized and untreated hypertension ([36], p. 70). Because hypertension is a silent, asymptomatic disease, screening for it again requires encouraging preventive care among the poor (and educating them about its benefits) as well as performing the screening. Again, since this is not the sort of behavior for which profit-oriented providers are known, there is a danger that the poor's special health needs may not be met. This is doubly true for diseases such as tuberculosis, because they seldom affect the middle class and thus require focusing particularly on the poor.

For the society as a whole, prevention is probably cost-effective. The estimates vary, but one estimate is that a moderately effective prevention program could save 400,000 lives, six million person-years of life, and more than six per cent of total health spending in the U.S. ([22], p. 393). Unfortunately, however, generalized screening, especially when it requires educating consumers as to their needs for the service, may not appear profitable for an individual provider. And even if it is in fact profitable, a market system is ordinarily a consumer-initiated one, where the provider waits passively until treatment for illness is sought. This model itself underemphasizes prevention, and is most likely to harm those people unaware of their health needs. For example, the mentally ill, to whom providers are unlikely to reach out, may suffer along with many of the elderly, who may believe their symptoms are simply part of the aging process. With regard to the elderly, it has been noted that nonreporting of symptoms "is an especially dangerous phenomenon when coupled with the passive American organizational structure of health care delivery, which lacks prevention-oriented or early-detection efforts" ([41], p. 830). Clearly, a voucher system will do nothing about this.

Probably the area where prevention and outreach is most necessary is that of prenatal care. As is well known, the United States lags far behind other industrialized nations in decreasing infant mortality. The 1980 U.S. mortality rate of 13.6 per 1,000 births is a striking contrast to Sweden's low of 7.8 ([40], p. 72).[5] For minority infants, the statistics are appalling – they are nearly twice as likely as their white counterparts to die before age one ([39], p. 702). In 1979, the death rate for black infants was 21.8 per 1,000, compared to a death rate of 11.4 for whites ([39], p. 702). If outcome of a health care system is even at all relevant, one test of an efficient and equitable system would be a reduction in infant mortality among the poor.

A health care system cannot do everything, of course. Poverty exists in the United States in a way that is virtually unknown in Western Europe. The sub-standard housing, sanitation, nutrition, etc., that accompany poverty all contribute to our infant mortality rate. However, effective prenatal care can unquestionably help. For example, public health outreach programs that increased prenatal care among Alabama's rural poor yielded a huge drop in infant mortality rates ([36], p. 54–55). Prematurity and low birth weight are leading causes of infant mortality and morbidity, and they are significantly decreased by appropriate prenatal care ([39], pp. 703, 710).[6] Because poor and/or uneducated women may be less aware of the need for prenatal care than the middle class, they need not only access, but also education about needed care and encouragement to seek it.

This sort of outreach would likely be cost-effective on a societal scale. Prematurity and low birth weight cause not only early deaths, but lengthy hospital stays, cerebral palsy, mental retardation, lung disease and other problems. The federal government has estimated that every dollar spent on prenatal care yields $4.00 in savings as a result of reduced hospital and long-term institutional care for handicapped infants ([39], p. 710).[7] Given the costs of neonatal intensive care, good prenatal care would likely also be in the financial interests of individual providers. Providers will undoubtedly cover maternal care, so as to attract young, healthy customers. But whether they will reach out to encourage the poor to obtain care, rather than succumbing to the short-term temptation to decrease utilization, or at least not encourage its increase, is another question. The California prepaid plan that had a large poor clientele resulted in poor services and significant delays, which are especially problematic for maternity patients, who need prompt and continuous access to care ([39], p. 718).[8] This is not to imply that prepaid plans serving the poor necessarily would be the scandal that California's was: some such plans have been very successful. The outreach activities necessary to ensure that access to care translates into utilization are not, however, a typical function of profit-oriented providers, but are more commonly associated with public health agencies. As such, they require a longer and broader view of the health needs of the poor than private providers are likely to have.

One final point in this regard is that another major contributor to low birth weight and prematurity is teenage pregnancy ([39], p. 731). Teenagers need outreach as to contraceptive care if our country's teenage pregnancy rate is to be reduced ([39], p. 731, n. 298). It is not surprising that in this country, with its far higher rate of teenage pregnancies than other Western nations, teenagers are much less likely to get free or low cost contraceptive service ([24], p. 54). This is especially important since the teenage mother is likely to be a victim of poverty. Again, however, whether competing providers would undertake this affirmative responsibility is problematic. A competitive consumer orientation probably will not emphasize the same sorts of programs that a prevention-oriented public health service would. Since these programs are necessary to improve the health status of the poor, the outcome of a competitive system is called into question.

### *C. Middle Class Surrogate Advocates*

Market system proponents have attempted to respond to the suggestion that less effective consumer advocacy by the poor could result in an unacceptably low quality of care for them. Enthoven's response to this potential problem is a trifle vague. Basically, he suggests that so long as vouchers are set at a sufficiently high level that the poor are not concentrated in one plan or facility, they will have access to mainstream medicine. Lowered ability to promote their interests would then be less crucial, since middle class consumers in the same facility would act as advocates. Good quality care, which Enthoven defines as "that which is satisfactory to educated middle-class consumers over the long run", would then ensue ([16], p. 60). According to Enthoven, "For a competitive market system to produce good results, it is not necessary that every consumer be perfectly informed and economically rational. Markets can be policed by a minority of well-informed, cost-conscious consumers" ([16], p. 138).

A problem with relying on advocacy by the middle class, however, is that HMOs in urban ghettoes might have few educated, assertive middle class customers. Havighurst faces this problem more directly. He has suggested requiring that any HMO, before it can accept welfare vouchers, must enroll fifty percent paying customers ([21], pp. 729–732). As "proxies" for the poor, these paying customers could assure quality control. This proposal is important in that it takes seriously the problem that the poor might be somewhat less effective advocates for high quality medicine. However, John Arras, an articulate critic of the free market proposals, argues that it is thoroughly socially unrealistic. Arras asks:

> Since the urban middle class can reasonably be expected to continue frequenting its own neighborhood and suburban health facilities, where will the proxy shoppers for welfare HMO's be found? Will they, one wonders, be bused in from the suburbs? If proxy shoppers are not forthcoming, then inner-city HMOs will presumably fail to qualify for the federal voucher plan and the urban poor will then find themselves in the same position as rural populations, equipped with vouchers that they cannot redeem ([1], p. 36).

Even if a sufficient number of middle-class customers are somehow present in inner-city HMOs, the differences between rich and poor may mean that the interests of these two groups are not identical. As discussed previously, the poor may need outreach and education about services that many middle class persons take for granted, such as screening for hypertension, tuberculosis, lead levels or diabetes, prenatal care, genetic screening, etc. Because the working poor may have less flexibility than higher income workers to miss work so as to keep medical appointments ([25], pp. 8–9), they may need evening or weekend hours to an extent that the middle class does not. Although most of their health needs will be similar, in some areas the needs of rich and poor may diverge to such an extent that the middle class would not effectively represent the poor's interests, even if we were confident that middle class consumers would in fact be customers of inner-city HMOs. Enthoven argues that when designing a system for the ninety-five percent, it should not be seriously distorted to serve the other five percent (here he has migrants, "derelicts", etc., in mind) ([16], p. 132). The poor, however, for whom voucher proponents claim their system will work, are far more than five percent of our population. To improve their health status, it is likely that public health programs are necessary. Unless competing providers can be induced (coerced?) into providing such services, there is indeed reason to fear that the poor will not be effectively served by the marketplace.

### *D. The Charge of Paternalism*

It has been argued that it is paternalistic and even derogatory to suggest that the poor are less effective health care consumers than the affluent. According to one such writer:

> The most important – because most pernicious – objection to vouchers is that poverty disqualifies one as a reasonable and effective chooser. To be poor is to be as a minor, one for whom choices must be made by more competent benefactors. It would not be surprising to see this argument made by flinty Social Darwinists. What is startling is that it is proclaimed by the alleged friends of the poor. Would the poor themselves be pleased to have champions maintaining that the most severe disability of poverty is not lack of money but rather functional deformity? Some ghetto residents might respond that, in order to survive day after day, they have had to be able to make some fairly shrewd decisions ([28], p. 41).

It has similarly been argued that inasmuch as our society allows people to refuse life-saving treatment and choose to smoke, it would be paternalistic and condescending to restrict the health care choices of the poor ([32], p. 89).

It is true that the poor manage to survive in a highly adverse environment. But this proves neither that there are no detrimental influences on their decision-making nor that poverty is nothing more than the relative absence of money. Market proponents tend simply to stipulate equal rationality, rather than looking to the complex of psychological, social, and environmental factors that merit consideration as much as economic ones. This entire section has been devoted to showing why it is dangerous to ignore these other factors. Because of this, rebuffing this charge of paternalism would largely consist of repeating what has already been said. In the interests of mercy to my readers, I will resist that temptation.

I do, however, want to emphasize two points. First, we are not talking about coercion of the poor: forcible prenatal care is not the issue here. We are talking instead about outreach, education, taking special needs into account, and making special efforts to help. It is not paternalistic for social policy to take into account circumstances other than economic ones that influence people's choices. For example, workers are not offered the "free choice" of working eighty hour weeks in factories, not because they are in general incapable of choice, but because there is too great a disparity in status and bargaining power between them and managers. Likewise, prisoners are not given the option of participating in risky medical experiments, because of the recognition that the conditions to which they are subjected overly influence their choices.[9] When social factors contribute to the total picture, there is no reason to focus only on the economic and ignore them altogether.

Second, we have tragic examples from related areas of the dangers of stipulating rationality and ignoring reality. Having long been subjected to involuntary confinement in warehouse-like institutions, the mentally ill began, in the 1960s and 1970s, to receive some societal attention. Much-needed reforms, such as improvements in mental hospitals, tighter standards for civil commitment and retention, etc., were adopted, and the plight of the mentally ill is undoubtedly far better than it was before the mental patients' rights movement. However, the more radical of the reformers suggested that there should be no commitment and no institutions. They claimed the mentally ill were, but for their unfortunate label and history of ill-treatment, just like anyone else.[10] To treat them differently was to be paternalistic and to deny them their right of autonomous choice. As these advocates fixated on the ideal of autonomy, they lost sight of social and psychological reality.

Government officials and mental health care workers, of course, did not in general accept the radical notion that if released from their confinement, even the chronically mentally ill could function in the community just like anyone else. But the social policy of deinstitutionalization was implemented *as if* they accepted it. Causation here is manifold, but a major factor was cost. Officials realized how much cheaper it was simply to release patients and assume that if there was a need for community care facilities, such facilities would spring up ([45], p. 146). To the extent that they did indeed spring up, many were private, for-profit nursing homes, and many have had a scandalous history of preying on their helpless clientele (some "consumers" being more assertive than others, and the chronically mentally ill being in the very low range) ([43], pp. 17–21). Thus a policy evolved that treated a group of people as if they were fully capable of caring for themselves, until the advent of widespread homelessness, with its many mentally ill victims, caused policy makers to pause and consider the wisdom of operating on the basis of this psychological fiction. Unfortunately, much damage had already been done, since the community facilities necessary for a middle course between a complete focus on autonomy (which led to abandonment) and an unremitting paternalism (forcible confinement) had not been created. Small, community-based facilities with outreach programs that respected autonomy but also recognized special needs and impediments to seeking care, could have prevented some of the devastation. But sadly, autonomy and special care were in this country falsely considered mutually exclusive.

The poor, of course, have nothing like the impediments to decision-making that afflict the mentally ill. But questions of capacity and consumer effectiveness are not either/or propositions, with children, the severely mentally ill and the mentally retarded being incapable and everyone else being equally effective. Capacities vary among all groups, with some children, mentally ill and mentally retarded persons being perfectly capable of making their own decisions, and some people who fall into none of these groups needing extra help and require a more "user friendly" health system than others. For a system to work for all, variations in psychology and social reality must be recognized, as must special needs. It is not condescending to insist on this: it is simply to recognize that the human situation is not a variable that can be assumed or stipulated to fit a tidy economic model. Free market systems that take only economic reality into account ignore this at their peril – but more importantly, they ignore it at the peril of every American citizen who, for one reason or another, deviates from that paradigm of the fully autonomous, knowledgeable, assertive, cost-conscious consumer.

## V. CONCLUSION

Having looked at some of the assumptions underlying both pure market and voucher proposals, it is apparent that while the latter contain significant and laudable modifications, they partake of some of the problems of the former. They focus on consumer choice, but ignore the societal impediments to effective choice. As a consequence, the outcome, in terms of health care, is likely to be highly correlated with the ability of an individual or group to press its demands aggressively. Just as the deinstitutionalized mentally ill were given the choice whether to seek health care, but were not provided the sort of outreach necessary to improve their plight, those among us who need extra help as well as freedom of choice may suffer under these voucher proposals. For a health care system to work, the nature and needs of its recipients must be studied, not stipulated. Because voucher proposals fail to do this, they may well fail to serve the health needs of the poor.

*College of Law*
*Ohio State University*
*Columbus, Ohio, U.S.A.*

## NOTES

[1] See also Brown ([5], 176–177), noting that other market proponents likewise do not believe that there should be a minimum, federally defined benefit package, and quoting Alfred Kahn as saying that the market approach should be free "to offer consumers the widest range of choices they are willing to select", statement at the United States Senate Committee on Health, Subcommittee on Finance, Hearing on Proposals to Stimulate Health Care Competition, 96th Congress, Second Session 1980 (p. 192).

[2] While Enthoven's proposal avoids this problem, it has, however, been pointed out that the near poor, whose vouchers do not cover the full cost of a plan, could choose not to supplement their vouchers and hence go without coverage ([1], pp. 35–36).

[3] A large percentage of patients on welfare said they felt discriminated against by their doctors because of being welfare recipients and/or being black ([23], pp. 79, 87).

[4] See generally Caplovitz [8].

[5] Rosenthal relies on statistics from WHO and the United States Department of Health and Human Services, 1980. Other sources give a slightly lower U.S. rate for 1980. See Wegman [49] (U.S. rate of 12.5 per 1,000 live births in 1980).

[6] Among the studies cited by Rosenbaum in support of this is one from California, where outreach maternity services provided to poor pregnant women resulted in low birthweight rates 50% lower than those among comparable infants whose mothers did not participate in the project ([39], p. 710, n. 100).

[7] This data is from the National Center for Health Services Research. Other studies have reached similar conclusions. For example, a 1978 report from the Harvard School of Public Health concluded that for each dollar spent on prenatal care, $3 were saved in hospitalization costs, and a 1983 Virginia state study concluded that the state could save $49.8 million in expenditures for long-term institutional care for the mentally retarded through the provision of better perinatal services ([39], p. 710, note 101).

[8] See generally Chavkin and Treseder [9].

[9] Menzel, who takes the autonomy thesis to its logical conclusion, predictably implies that he would honor the choices of prisoners to participate in risky medical experiments ([32], p. 95). For an analysis of the constraints on prisoners' choices, see generally National Commission [33].

[10] See generally Scheff [44], suggesting that by viewing mental illness as a disease, medical science legitimizes the labeling of nonconformity as mental illness and likewise legitimizes deprivations of rights based on this label. For examples of legal writings heavily influenced by views such as Szasz's, see Plotkin [35].

## BIBLIOGRAPHY

1. Arras, J.: 1981, 'Health Care Vouchers and the Rhetoric of Equity', *Hastings Center Report* **11**, 29–39.
2. Bayer, R., Caplan, A., and Daniels, N.: 1983, *In Search of Equity: Health Needs and the Health Care System*, Plenum, New York.
3. Beauchamp, D.: 1976, 'Public Health as Social Justice', *Inquiry* **13**, 3–14.
4. Brody, B., 1981, 'Health Care for the Haves and Have-Nots: Toward a Just Basis for Distribution', in E. Shelp (ed.), *Justice and Health Care*, D. Reidel Co., Dordrecht, Holland, pp. 151–159.
5. Brown, L.: 1981, 'Competition and Health Cost Containment: Cautions and Conjectures', *Milbank Memorial Fund Quarterly* **59**, 145–189.
6. Buchanan, A.: 1985, *Ethics, Efficiency, and the Market*, Roman and Allenheld, Totowa, New Jersey.
7. Burton, L., Smith, H., and Nichols, A.: 1980, *Public Health and Community Medicine*, 3rd ed., Williams & Wilkins, Baltimore, Maryland.
8. Caplovitz, D.: 1963, *The Poor Pay More: Consumer Practices of Low Income Families*, Free Press, New York.
9. Chavkin, D. and Treseder, A.: 1977, 'California's Prepaid Health Plan Program: Can the Patient be Saved', *Hastings Law Journal* **28**, 685–760.
10. Daniels, N.: 1983, 'Equity of Access to Health Care: Some Conceptual and Ethical Issues', in *President's Commission for the Study of Ethical Problems in Medicine and Biomedical and Behavioral Research, Securing Access to Health Care*, Vol. 2 (Appendix), pp. 23–49.
11. Dickman, R.: 1983, 'Operationalizing Respect for Persons', in R. Bayer, A. Caplan, and N. Daniels (eds.), *In Search of Equity: Health Needs and the Health Care System*, Plenum, New York, pp. 161–182.
12. DiMatteo, M. and Friedman, H.: 1982, *Social Psychology and Medicine*, Oelgeschlager, Gunn & Hain, Cambridge, Massachusetts.
13. Dutton, D.: 1978, 'Explaining the Low Use of Health Services by the Poor: Costs, Attitudes, or Delivery Systems?' *American Sociological Review* **43**, 348–368.
14. Engelhardt, H.: 1981, 'Health Care Allocations: Responses to the Unjust, the Unfortunate, and the Undesirable', in E. Shelp (ed.), *Justice and Health Care*, D. Reidel Co., Dordrecht, Holland, pp. 121–137.
15. Ennis, B.: 1972, *Prisoners of Psychiatry – Mental Patients, Psychiatrists, and*

*the Law.*

16. Enthoven, A.: 1980, *Health Plan: The Only Practical Solution to the Soaring Cost of Medical Care*, Addision-Wesley, Reading, Massachusetts.
17. Fein, R.: 1981, 'Social and Economic Attitudes Shaping American Health Policy, in J. McKinlay (ed.), *Issues in Health Care Policy*, MIT Press, Cambridge, Massachusetts, pp. 29–65.
18. Halper, T.: 1981, 'The Double-Edged Sword: Paternalism as a Policy in the Problems of the Aging', in J.B. McKinlay (ed.), *Issues in Health Policy*, MIT Press. Massachusetts, pp. 199–226.
19. Hansen, C.: 1977, 'Thorny Problems with LRA Cases', *Mental Health Law Project Summary of Activities* **3**, 9–10.
20. Havighurst, C.: 1982, *Deregulating the Health Care Industry: Planning for Competition*, Ballinge Co., Cambridge, Massachusetts.
21. Havighurst, C.: 1971, 'Health Maintenance Organizations and the Market for Health Services', *Law & Contemporary Problems* **35**, 716–795.
22. Himmelstein, D. and Woolhandler, S.: 1984, 'Pitfalls of Private Medicine: Health Care in the U.S.A.', *The Lancet* (August 18), 391–393.
23. Holloman, J.: 1983, 'Access to Health Care', in *President's Commission for the Study of Ethical Problems in Medicine and Biomedical and Behavioral Research, Securing Access to Health Care*, Vol. 2 (Appendix), pp. 79–106.
24. Jones, F. *et al.*: 1985, 'Teenage Pregnancy in Developed Countries: Determinants and Policy Implications', *Family Planning Perspectives* **17**, 53–63.
25. Kane, R., Kasteler, J., and Gray, R.: 1976, *The Health Gap: Medical Services and the Poor*, Springer Publishing Co., New York.
26. Klein, R.: 1984, 'The Politics of Ideology versus the Reality of Politics: The Case of Britain's National Health Service in the 1980's', *Milbank Memorial Fund Quarterly* **62**, 82–109.
27. Lappe, M.: 1981, 'Justice and Prenatal Life', in E. Shelp (ed.), *Justice and Health Care*, D. Reidel Publishing Company, Dordrecht, Holland, pp. 83–94.
28. Lomasky, L.: 1981, 'The Small but Crucial Role of Health Care Vouchers', *Hastings Center Report* **11**, 40–42.
29. Mechanic, D.: 1976, 'Rationing Health Care: Public Policy and the Medical Marketplace', *Hastings Center Report* **6**, 34–37.
30. Mechanic, D.: 1976, *The Growth of Bureaucratic Medicine: An Inquiry Into the Dynamics of Patient Behavior and the Organization of Medical Care*, Wiley-Interscience, New York.
31. Mechanic, D.: 1981, 'Some Dilemmas in Health Care Policy'., *Milbank Memorial Fund Quarterly* **59**, 1–15.
32. Menzel, P.: 1983, *Medical Costs, Moral Choices*, Yale University Press, New Haven, Connecticut.
33. National Commission for the Protection of Human Subjects of Biomedical and Behavioral Research: 1976, *Research Involving Prisoners*, U.S. Government Printing Office, Washington, D.C.
34. Poland, M.: 1984, 'How Women Feel About Prenatal Care', unpublished manuscript.
35. Plotkin, R.: 1977, 'Limiting the Therapeutic Orgy: Mental Patients' Right to

Refuse Treatment', *Northwestern University Law Review* **72**, 641.
36. President's Commission for the Study of Ethical Problems in Medicine and Biomedical and Behavioral Research, *Securing Access to Health Care*, 1983.
37. Rhoden, N.: 1982, 'The Limits of Liberty: Deinstitutionalization, Homelessness and Libertarian Theory', *Emory Law Journal* **31**, 375–440.
38. Robitscher, J.: 1976, 'Moving Patients out of Hospitals – In Whose Interest?', in P. Ahmed and S. Plog (eds.), *State Mental Hospitals: What Happens When They Close?*, Plenum, New York, pp. 141–175.
39. Rosenbaum, S.: 1983, 'The Prevention of Infant Mortality: The Unfulfilled Promise of Federal Health Programs for the Poor', *Clearinghouse Review* **17**, 701–733.
40. Rosenthal, M. and Frederick, D.: 1984, 'Physician Maldistribution in Cross-Cultural Perspectives: United States, United Kingdom, and Sweden', *Inquiry* **21**, 60–74.
41. Rowe, J.: 1985, 'Health Care of the Elderly', *New England Journal of Medicine* **312**, 827–835.
42. Rushefsky, M.: 1981, 'A Critique of Market Reform in Health Care: The "Consumer-Choice Health Plan"', *Journal of Health Politics, Policy, and Law* **5**, 720–741.
43. Santiestevan, H.: 1977, *Out of Their Beds and Into the Streets*, AFSCME, Washington, D.C.
44. Scheff, T.: 1975, 'Schizophrenia as Ideology', in T. Scheff (ed.), *Labeling Madness*, Prentice-Hall, Engelwood Cliffs, New Jersey.
45. Schull, A.: 1977, *Decarceration: Community Treatment and the Deviant: A Radical View*, Prentice-Hall, Engelwood Cliffs, New Jersey.
46. Strauss, A.: 1976, 'Medical Organization, Medical Care, and Lower Income Groups', in R. Kane, J. Kasteler, and R. Gray (eds.), *The Health Gap: Medical Services and the Poor*, Springer Publishing Company, New York, pp. 126–173.
47. Szasz, T.: 1974, *The Myth of Mental Illness*, Harper & Row, New York.
48. Thurow, L.: 1984, 'Learning to Say No', *New England Journal of Medicine* **311**, 1569–1572.
49. Vladeck, B.: 1981, 'The Market vs. Regulation: The Case for Regulation', *Milbank Memorial Fund Quarterly* **59**, 209–223.
50. Wegman, M.E.: 1982, 'Annual Summary of Vital Statistics – 1981', *Pediatrics* **70**, 835–843.

HANS-MARTIN SASS

# MY RIGHT TO CARE FOR MY HEALTH – AND WHAT ABOUT THE NEEDY AND THE ELDERLY ?

...Yet is their strength, labor
and sorrow, for it is soon cut
off and we fly away.
(*Psalm 90*)

It has become very popular to apply the highly inflated *language of rights* in addressing issues of personal health status and of priorities in allocation of taxpayer's resources. Such an intellectual and political fashion fad, however, seems to have little or no basis in our everyday and common sense approach to health and health care [14]. Whenever I personally think about health, apart from those intellectual debates and public policy discussions, it is not the term "right", nor the term "obligation", nor "resource" that comes to mind first. I feel I have a *responsibility* to take care of my health. It is not only an asset; it is an end in itself, inasmuch as its absence or deterioration is painful, harmful, fatal. It is also a commodity, a precondition, which I use to achieve other goals in life: satisfaction, social integration, love, reputation, power, contribution to the common good, and working hard. I am using health as a means; I might trade it for other values, assets, or valuables, according to personal choice or milieu. Protecting or improving or trading health, its quantitative length as well as its qualitative standard, is a result of living, of voluntarily or involuntarily making decisions, choosing, managing risks. Health is a risk factor in life; I therefore deal with it appropriately when using the arts and sciences of risk analysis, risk assessment, and risk management. It seems to me that the *risk language* is more appropriate to deal with issues of personal health care as well as of health policy than the rights language [13, 15].

## I

(1) We are witnessing *inflationary demands for particular rights in developing welfare states*: the right to choose and the right to be compensated for choosing wrongly, the right to be unequal to the other and the right to redistribution against the lottery of nature and milieu, the right to be provided with a job and the right to not be forced into labor, the right to procreate and to abort, the right to learn and the right not to submit to educational stress or introjected values, the right to drink, to smoke, to develop recreational addictions and preferences and the right to medical treatment of lifestyle-related dysfunctions of lung, liver, kidney, and for



*T.J. Bole III and W.B. Bondeson (eds.), Rights to Health Care*, 243–255.

treatment in clinics for addicts [5]. Claiming to have these and similar rights expresses a logical confusion as well as a moral deficiency. It is a logical shortcoming to demand from someone not to interfere with my free choice and thereafter to request his payment for the side effects or outcomes of my choice, which was not his or her choice in the first place [13]. Such a behavior is not only paternalizing the other by making him a partner in the losses in my business of managing personal risks; it is also immoral to keep benefits private while socializing costs related to those benefits. Our forefathers fought kings and slaughtered dictators to free themselves from those parasites and their bureaucratic edifices of heteronomous care and preferential justice. Revolutions have been made, wars fought, and the buildings of injustice reduced to utter ashes, rightly so, in the name of the individual right for self-care and self-determination. There is no biological right to be equal or healthy, but there is a politically achieved and morally and culturally defended right to civil liberty and free choice within reasonable constitutional and legal limits, including the right to care for my health or to trade it for values or valuables I might appreciate more. If we indeed would claim a right to health protected, or even provided, by magisterial authorities, government would have to regulate standards of housing, food stuffs, eating habits, recreational activities, quantity and quality of labor, each and every risk area of life and lifestyle [5, 13]. Necessarily, therefore, advocates of specialized civil rights to be claimed from government, have to turn into advocates for authoritarian state rule. Values have consequences; if we seriously commit ourselves to basic civil rights, the exercise and management of rights to choose freely within as wide as possible limits has consequences in regard to outcome.

The right to have as equal and as supportive as possible opportunities will be undercut by permanently redistributing opportunity ranges, thus making the entire issue of "equality of opportunity range" a joke logically and a deception morally.

(2) While intellectual discussion and egalitarian political combat focuses on the "rights" issue, actual health policy concentrates on *"equality"*. The United States President's Commission ([9], Vol. I, p. 4) concludes "that society has an ethical obligation to ensure equitable access to health care for all". Putting the principle of societal obligation in primary place, the Commission in a subsequent thesis adds that societal obligations should be "balanced by individual obligations". It is argued that "the origins of health needs are too complex, and manifestations too

acute and too severe, to permit care regularly denied on grounds that individuals are solely responsible for their death". Such a statement indirectly includes a political judgment on other forms of need, love need, job need, orientational or religious need. Is our need for love, or for transcendental or religious orientation, less complex? Is its nonfulfillment less acute or less severe on the individual? Why governmental intervention in and management of health care, not of love care, or transcendental care? And is the complexity of an issue, which includes its rather complex cultural and moral assessment, a good reason for handling it over to governmental care. Recall the good reasons of our forefathers to topple the systems of heteronomous care. Care issues that are complex, acute, and severe, are best addressed first by the individual, and only thereafter by solidarity, beneficence, or insurance networks, and, as a last resort, by direct governmental intervention.

Also, it would be wise to look into other forces taking care of health or improving health status. Addressing the World Health Organization's goal for the year 2000 of "health for all", Christopher Wood ([16], p. 103) holds that such a goal eventually might be achieved, but that as far as the developing countries are concerned, "predominately disease orientated medical services need re-directing towards more health promotion and disease prevention ... [and] assistance given to agriculture and water development programmes". This may, in the long term, "contribute more to health than aid given to some of the less relevant medical programmes". To transfer such a broad approach to the complex issue of health care (not sickness care) in highly developed countries vexed by the high social costs of public medical service programs, one would have to strengthen personal responsibility, individual health risk competence and management. John Knowles, once President of the Rockefeller Foundation, put it this way:

> The cost of sloth, gluttony, alcohol intemperance, reckless driving, sexual frenzy, and smoking is a national and not an individual responsibility. This is justified as individual freedom – but one man's health is another man's shackle in taxes and insurance premiums. I believe the idea of a 'right' to health should be replaced by the idea of an individual moral obligation to preserve one's own health – a public duty if you will. The individual then has the 'right' to expect help with information, accessible services of good quality and minimal financial barriers (quoted in [10], p. 16).

"Equitable access" to health care in developing countries calls for an improvement in sanitary and nutritive conditions, together with health information; in developed countries it calls for educating health responsibility and encouraging personal health risk management. In both cases it is the improvements of preconditions for *self-help*: give a man a fish and you feed him for a day; teach him to fish and you feed him for life. In complex matters like health, equity and equality is primarily provided through the improvement of the opportunity range via education and responsibility, and only subsequently and additionally through direct and acute intervention. Such an *opportunity principle* better meets the demands of justice than does any *intervention principle*, because it does not patronize lifestyles by a regulatory heteronomy that as a matter of course restricts personal autonomy and free choice. Also, it does not redistribute unfairly by socializing the costs while keeping the benefits, if any, of improper personal health management private. Herder-Dorneich ([7], p. 277) identifies "spiraling claims" (*Anspruchsspiralen*), in which the level of claim continuously outgrows the level of supply, as what is typical of overdeveloped welfare states. Using the opportunity principle rather than the intervention principle would cut the unbearable moral costs of permanently redistributing; it would transform public care of the sick into a private-public mix of health care. We are discussing the *moral overload*, not the economic overload, of welfare states, even though it is obvious that total societal obligation for care would finally lead to a financial, and not just a moral, collapse of society.

(3) Because neither a principle of rights nor a principle of equality carries us far enough in discussing the issues of health care, we might want to *subdivide the issue* into (a) *basic needs*, (b) *supplementary needs*, and (c) *additional needs*. Basic health care needs would include: certain educational, nutritive, and sanitary standards, availability of low-cost/high-benefit medical intervention, e.g., availability of antibiotics for pneumonia, surgery for appendicitis, plaster cast for broken legs, and access to pain reducers or pain killers for diseases the treatment of which is either marginally beneficial or more than marginally costly. Supplementary needs are those for costly chronic or lifestyle-related health care, e.g., organ transplants, hemodialysis, and lung cancer treatment for smokers. Additional needs for health care may be grounded not in individual, rather in societal benefits, e.g., healthy and hard working populations, less troublesome and dependent upon governmental guidance.

If we differentiate between these three forms of health care, it will be easier to identify areas which indeed would need governmental involvement and social solidarity as we have in other areas of public risk management ([13], pp. 24–27). It is a part of our concept of democratic, peaceable societies and educated citizens that basic needs are directly (via public schools, safety regulations, police) or indirectly (via private schools, licensing, self-regulation and insurance, law) risk managed by government: educational and information services (direct or indirect) for health risk management have to be a part of it, as does setting standards for housing, food, pain care, and low-cost/high-benefit medical intervention. In pluralistic societies, rich in values and valuables, it is questionable and debatable how much the government by way of regulation, and the taxpayer by way of payment, should get involved in regulating and assuming costs and benefits in more controversial areas of medical care, such as redistributive benefits for lifestyle-related diseases and high-cost/marginal-benefit interventions. These are areas which are highly controversial politically as well as extremely costly economically, and which therefore cannot be resolved unless we strictly regulate and standardize diagnosis-related treatment, lifestyle, mandatory check-ups, penalties for non-compliance, etc. The degree to which we want public health policy to be directly involved in managing these health risks is equal to or exceeds the degree to which we will have to reduce personal responsibility and threaten both individual competence and the plurality of lifestyles in a free society. There may be additional demands for public health policy that favors certain individuals being given preferential medical treatment, or underwriting the costs of catastrophic illness, or providing permanent care for the elderly, or even proposing socialized medicine as part of what is necessary for state survival in international competition. Fortunately, such demands do not constitute major considerations in Western health-care policy discussions. Prolonged debate on public involvement in supplementary health needs might lead, however, more and more to such demands. If it does, we would diagnose such a situation as a catastrophic illness of a nation, caused by its own faults, and to be cured only by surgically cutting out the cancerous growth of heteronomous care that undermines the domain of self-care.

To illustrate this, we can plot the benefits of health care to the individual on a horizontal axis, and the amount of public involvement on a vertical (see Figure 1). We can then draw a slowly rising curve of improving health benefits that keeps moral, political, and economic costs of governmental intervention relatively low and personal responsibility relatively high, and even increases personal risk management outcome (A-B).

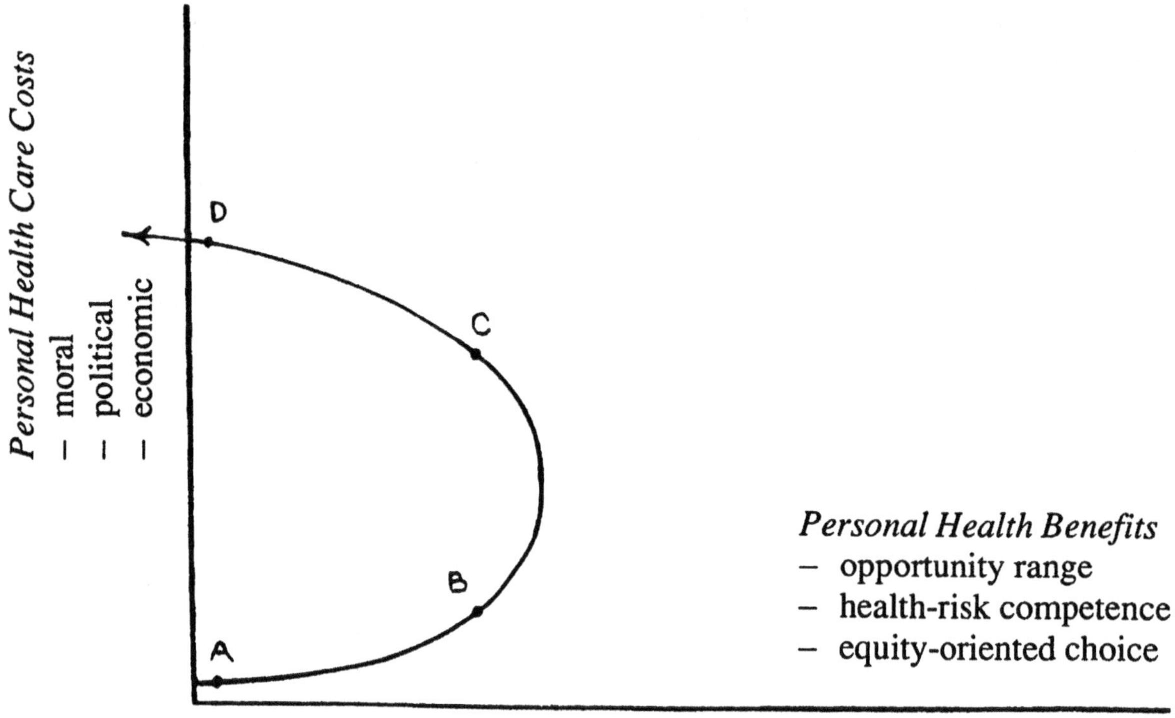


Figure 1. Benefits of public health care.

*Impact on right to health*

A-B: Sanitary and nutritive improvements
Health education
Low-cost/high-benefit catastrophic intervention
Pain care

B-C: High-cost/low-benefit medical intervention
Unsolvable public debate on "equity"
Increase in lifestyle-related illness
Decrease in personal risk management

C-D: Preferential treatment for persons of high public worth
Standardization of lifestyle and medical intervention
No-treatment for the "undeserving" or "worthless"
Rapid decrease in personal and professional ethics

In a second part (B-C) of the curve, the quantity of governmental involvement would increase rapidly with only marginally improving health care, while personal responsibility and willingness for and competence in managing personal risks would decrease, thus making the loss in personal risk management competence the salient thrust, even though the system might still be able to be financed for a certain period of time before spiraling claims and growing technical capabilities eventually brings it to economic collapse. But it will have collapsed morally much sooner, because of cheating and the moral double-standards involved in using the autonomy principle on the one side to justify beneficial individual actions, and using the principle of the right to equal treatment on the other side to justify public health care and intervention costs.

## II

I could label my approach to welfare and health care issues the *responsibility approach.* Such an approach seems to be morally more consistent with the concepts of human dignity, educated choice, and autonomy of the individual. Because the rights approach and the equality-equity approach both suffer argumentative shortcomings and moral imbalances, they weaken personal initiative and competence in managing health risks. The responsibility approach can profit from progress in medical intervention technologies, from the effective deployment of information regarding both preventive and actual health care, and from a utilization of available intangible and tangible resources that is rendered efficient by free exchange. The responsibility approach most likely would work best in a competitive system of vouchers and insurance, but it can also be made a part of socialized medical intervention care also, if done pragmatically. The limits of the responsibility approach would lie in the areas where people have little responsibility, or limited capacity for responsibility, or a lack of opportunity for responsibility. This would be the case for some poor, some elderly, and the incompetent. Even though the rights approach and the equality approach are not very helpful in either one of these special areas of health care and welfare, it is a good test of the responsibility principle to address these issues.

(1) As Nancy Rhoden [12] provides strong arguments for, health care for the poor is not just an economic issue, it is to a great extent also an issue of attitudes, e.g., a lack of risk awareness, poor knowledge regarding nutrition and unhealthy lifestyles, an inability to care for one's self, a crisis-orientated approach to health as well as to other issues in life. All of this makes the poor the most dependent and most captive consumers of either direct or indirect public medical care, which always is understood to be crisis care or emergency care. The most promising approach would

be to fight the roots of poverty: milieu, lack of education, absence of will to fight and to work, risk management competence, good example, social rewards; and all the substandard housing, substandard nutrition, crime, teenage pregnancy, ghetto mentality, social discontent, etc. The poor are not just in need of better health care, they are in need of all sorts of care. We have enough data demonstrating that the simple handout, cost-abolishing, intervention-type welfare approach makes the poor even poorer, i.e., more dependent, less self-reliable, less equipped to overcome poverty, thus caught in a *circulus vitiosus* that only provokes more handout programs. *Rewards for responsibility* would be the best approach to free welfare recipients from welfare dependency. Those rewards have to be of some appreciable value to the recipient: rewards for buying nutritious food instead of "junk" food; rewards for joining education, vocational and health information groups; rewards for achieving good results in such programs; rewards for staying in a job over a certain period of time; rewards for not becoming pregnant before having the social or economic means to give the newborn a decent upbringing; rewards for obtaining free routine medical check-ups; rewards for living according to recommendations based on health check-ups, etc. The *reward for responsibility* approach would help us to get rid of the entire poverty syndrome, which is based on lack of opportunity, lack of will, milieu, ghetto attitudes; it will break the ghetto mentality and the frustration circle. Other reasons for substandard economic capabilities will remain, caused by simple incompetence, technological change, family problems, and personal or social crises. We cannot fight all these causes, and we might not want to interfere with personal choices in a paternalistic manner; nor do we want the government to compel uniformity.

It might also be financially beneficial to channel the vast majority of welfare money through programs rewarding responsibility. Exploiting innovative programs of both charitable organizations and profit-orientated groups might decrease poverty more efficiently than bureaucratically administering the programs. This would call for a revival of the traditional approaches to care for the needy according to the principle of *subsidiarity*, which dictates that social problems should be resolved as close to the grassroots as possible. Where uniform governmental welfare programs have, as an unfortunate side effect, undermined the usual channels of subsidiarity, they might out of sheerly economic grounds promote the use of these channels, including profit-orientated ones; the results will be less and less poverty, more and more incentive and rewards, naturally better health, and less medical costs due to crisis intervention.

(2) The *opportunity range* principle of Norman Daniels ([2]; [9], vol. 2, pp. 265–291) comes quite close to my concept of responsibility and health risk management competence. In as much as the *elderly* are concerned, however, he seems to put too much emphasis on redistributive justice rather than concentrating efforts to broaden and to open opportunities for the non-mature under the age of 21 or 25 years. Establishing opportunities that are as fair as possible for the young generation would include favorable social and family environments, education, skills, and early medical intervention where inequalities and unfair disadvantages can be corrected, e.g., a disorder which can be treated or avoided by special nutrition, hearing or eye disorders which can be corrected by surgery, even taking children out of extremely unfavorable social and family systems which would deprive them of any reasonable opportunity to ever break the circle of dependency, illiteracy, crime, and dissatisfaction. I agree with Daniels that surgical correction of eyesight disorders for the young, e.g., have a much higher moral benefit-cost relationship than hemodialysis for the elderly ([9], vol. 2). If we follow the responsibility approach, everything should be done as early as possible and as effectively as possible, to increase the range of opportunity by the development of educational and social skills, as well as by medical well care. This would ease the problem of age-related forms of care for the elderly. As getting old is already a result of either good fate or successful personal risk management, getting even older should rest on the same principles. The elderly during a relatively long lifespan already got their preferential treatment in the lottery of nature: good genetic origin, good and supportive milieu, low risk exposure; or they have acquired higher forms of risk competence than others as a result of having been exposed to, and fated to, overcome more risks and challenges in life. Such happy conditions call first of all for the appreciation of having survived so many possible risks. Such an appreciation leads to understanding old age as a privilege, not a right. Also, the "stage of life opportunity" for the elderly is higher as a result of good luck, better risk management, and/or more prudent lifestyle. Therefore, protecting age-relative "normal" opportunity ranges should require of society less heteronomous medical intervention in and care for the mature and matured. Health considerations will have played or not played a high priority during a long life, with its value preferences and its allocations of financial resources and insurance priorities. It would

be unfair to redistribute costs for choices made in early- and mid-life by taking the risk out of value-orientated and health-related decisions. Old age is a very natural stage in life; and nothing is more natural and human than to die. To grow old with dignity and to face death and old age with intelligence and prudence is a matter of culture, not of medicine, and definitely not of public intervention to cover medical costs. Health care for the elderly is less of a moral and political challenge than caring for the young and for their opportunities. We have very strange and unjustifiable priorities in public morals and public policy: we subsidize the marginal prolongation of old life by extremely costly intensive care, but we tolerate the irresponsibility of teenagers who give birth to children who can only enter the same circle of despair, discontent, irresponsibility and illiteracy which provokes their existence in the first place. We appreciate individual care traffic, having a proven probability of causing 1 in 5000 deaths every year, but we ban nuclear energy, which has a potential of causing perhaps 1 death in 100,000 (*cf.* [13], pp. 16–24).

(3) Some of the elderly become incompetent. This, also, is natural, and it would be responsible to arrange for such an event while one is competent and of means, by determining as much as possible value choices of that condition providing means or setting guidelines for proxy decisions. Treating the *old and incompetent* in predominantly medical settings in fact is based on life-care proxy decisions [11]; it should be called life care, not health care. Because of the difficulties in determining and defining equity – i.e., what is good for them – for the incompetent, we set general medically governed uniformity rules for a general standard good. What is the value of our heteronomous arguments when we prescribe risk-free and fun-free and responsibility-free, uniform, medically averaged life? We certainly cannot apply the rights principle or the autonomy principle or the equity principle directly; but how then do we want to redefine them adequately for the incompetent? We cannot. The responsibility principle can be applied, however, in taking care of the lives and health of the incompetent: We would want to give them as much responsibility and risk-taking opportunity as possible. If they can learn, we want them to learn out of experience; if they cannot learn or have lost their capacity to learn, we want them to have rewards which they can enjoy, and which can be defined as Engelhardt defines beneficence: "Do to others their good" ([4], p. 87). It is said that some of us never have had the privilege of being competent and that some of us might become very incompetent. Care and responsibility, risk management, and the

willingness and the competence to act, are prerequisites not only for being human persons, but also for forming peaceable societies, a part of which is taking care of health individually and collectively. By including the incompetent, who are not able to make educated choices or who never were able, in our public health care, by making it life care for them, we do so, often very unconsciously, for two reasons: *beneficence* and *solidarity.* If we overemphasize the principles of rights and equality, we tend to forget not only about the responsibility principle but also about the major principles which traditionally have guided our attitude towards the needs of those fellow humans who were in need, in pain, or in despair: beneficence and solidarity ([4], pp. 74–87). Doing good for no other reason than for doing good, not as a means, i.e., helping someone out of beneficence, was traditionally regarded as a great virtue. Let us encourage and re-appreciate beneficence on the personal level, and also apply it in public policy, wherever people cannot claim rights, autonomy or equality. Would it not be more honest to call mercy mercy, and beneficence beneficence, rather than talking about rights an equality inappropriately? It is the foundation of self-respect of educated persons, the ethos of the peacable society, and the dignity of human solidarity in being fragile, often mistaken, exposed to pain, suffering, and loss, which lets us include those who are in need, who are mistaken, and who are in pain, into our welfare networks; they are our image, they are us. If we are equal, we are equally mortal, mistaken, fallible; this does not create any rights to be claimed, but it is the basis for *solidarity* even beyond the boundaries of the responsibility principle.

## III

The last argument leads us back from the critique of the overplayed "right" principle and the "equality" rhetoric to the 2500-year-old wisdom of Micah: "He hath shown thee, O man, what is good; and what doth the Lord require of thee, but to do justly, and to love mercy, and to walk humbly with thy God?" (*Micah* 6:8). To *do justly* is the basis and the goal of the peaceable society formed by the free choices of free persons under just rules and laws. Justice is balance according to Aristotle. In health issues, a just balance means broadening the opportunities via education and other preconditions for good and prudent health-care management, and solid action in making the best use of available technology, economic resources, and responsibility ranges. A just balance would forbid permanently redistributing health assets at public cost or forcing people into mandatory health maintenance programs. Even if we do good in justly balancing, certain things we will never be able to manage fully: pain, despair, distrust, incompetence. If we *love mercy*, we might be able to comfort people beyond the legal and administrative limits of even the most effective networks. Beneficence does not have to reason or argue, it does not need a majority vote to come into existence. As our life is short and as our nature is fragile, we want to *walk humbly*. Walking humbly leads to the re-appreciation of mercy, of solidarity, of beneficence beyond the claims of law, regulation, and right. Applying the language of rights to crucial and complex issues like health is not only not fruitful; it is counterproductive insofar as it conceals the real issues at stake.

*Ruhr Universität Bochum, Institut für Philosophie,*
*Bochum, Germany*

*and*

*Georgetown University, Kennedy Institute of Ethics,*
*Washington, D.C., U.S.A.*

## BIBLIOGRAPHY

1. Buchanan, A.: 1985, *Ethics, Efficiency and the Market*, Rowman and Allanheld, Totowa, New Jersey.
2. Daniels, N.: 1991, 'Equal Opportunity and Health Care Rights for the Elderly', in this volume, 201–212.
3. Enthoven, A.: 1980, *Health Plan: The Only Practical Solution to the Soaring Cost of Medical Care*, Addison-Wesley, Reading, Massachusetts.
4. Engelhardt, H.T. Jr.: 1986, *The Foundations of Bioethics*, Oxford University Press, New York.
5. Fried, C.: 1976, 'Equality and Health Care', *Hastings Center Report* 6, 29–34.
6. Havighurst, C.: 1982, *Deregulating the Health Care Industry*, Ballinger Company, Cambridge, Massachusetts.
7. Herder-Dorneich, P.: 1983, *Ordnungstheorie des Sozialstaates*, J.C.B. Mohr, Tübingen.
8. Moskop, J.C.: 1983, 'Rawlsian Justice and Human Right to Health Care', *The Journal of Philosophy and Medicine* **8**, 329–338.
9. President's Commission for the Study of Ethical Problems in Medicine and Biomedical and Behavioral Research: 1983, *Securing Access to Health Care*, U.S. Government Printing Office, Washington, D.C. (Vol. 1: Report; Vol. 2:

Sociocultural and Philosophical Studies; Vol. 3: Empirical, Legal and Conceptual Studies).
10. Reiser, S.J.: 1983, 'Responsibility for Personal Health: A Historic Perspective', *The Journal of Philosophy and Medicine* **8**, 7–17.
11. Rhoden, N.K.: 1980, 'The Right to Refuse Psychotropic Drugs', *Harvard Civil Liberties Review* **15**, 363–413.
12. Rhoden, N.K.: 1991, 'Free Markets, Consumer Choice and the Poor', in this volume, 213–241.
13. Sass, H.M.: 1985, *Verantwortung unter Risiko*, Köllen Druck und Verlag, Alfter-Ödekoven.
14. Sass, H.M.: 1983, 'Justice, Beneficence or Common Sense: The President's Commission's Report on Access to Health Care', *The Journal of Philosophy and Medicine* **8**, 381–388.
15. Sass, H.M.: 1988, 'National Health Care System: Concurring Conflicts', in R.M. Massey and H.M. Sass (eds.), *Health Care Systems: Moral Conflicts in European and American Public Policy*, Reidel, Dordrecht, pp. 15–36.
16. Wood, C.H.: 1986, 'The Role of Education in Development: With Special Reference to Health?', *The Royal Society of Arts Journal* No. 5354, 96–106.

SECTION V

# HEALTH CARE AS A COMMODITY

DAVID FRIEDMAN

# SHOULD MEDICINE BE A COMMODITY? AN ECONOMIST'S PERSPECTIVE

## INTRODUCTION

In our society, probably in any society, goods are produced and allocated in several different ways. One very common pattern is production for and sale on the market. Butter, computers, and plumbers' services, to take three examples out of a multitude, are produced by individuals and firms, acting in what they perceive to be their own interest, and sold, usually for money, to those who wish to consume them. While this is a common way in which goods are produced and allocated in our society, it is not the only way, nor is it even clear that it represents a larger part of the total economy than alternative ways.

What are the alternatives? The two most important are household production – the way in which children are reared, homes cleaned, clothes washed, and most meals cooked – and political production. While household production represents a substantial fraction of the economy, and perhaps even of total medical services (parents serving as nurses for their sick children, grown children taking care of aging parents, and the like), its role is not at the moment a subject of much controversy and will be ignored in this essay. The two alternatives I will be concerned with are production and allocation on the market and production and allocation by government. The essential question I will try to answer is whether one form of production should be preferred, and if so which.

The meaning of market provision of medical services is fairly clear, although the form may vary; individual physicians, group practices, private hospitals, payment by individuals or by insurance companies, are all possible market arrangements. The meaning of governmental provision is a little less clear. For the purposes of this essay I include three different sorts of government intervention under the general category of government provision and allocation – production, payment, and regulation. An obvious example of governmental production of medical services is a state hospital, or, on a larger scale, the British National Health System. Government payment would include systems



*T.J. Bole III and W.B. Bondeson (eds.), Rights to Health Care*, 259–305.

such as medicare, where the service is produced on the market but paid for, in whole or in part, by government. Regulation includes such widely accepted activities as licensing of physicians and control by the FDA over the introduction of new drugs.

While it is convenient to speak of government as if it were an independent being, it is also highly misleading. As should become clear in Part II of this essay, I regard government not as some outside force accomplishing its own good or evil objectives, but as a shorthand description for a mechanism – a set of rules – under which individuals interact in order to achieve individual objectives. From this standpoint, calling the alternatives "market" and "government" is itself somewhat misleading, since government can be viewed as merely a second and different market, a political market in which exchanges occur and decisions are made under a different set of rules than in the economic market. The question then is whether it is better for medical services to be produced entirely in the private market – what I mean by medicine being a commodity – or entirely in the public market, or by some combination of the two.

In trying to answer this question I will start, in Part I, by discussing the relation between economic analysis and normative conclusions. Doing so will involve an unavoidably lengthy discussion of what economists mean by efficiency and what, if anything, it has to do with conclusions about what should happen. I will go on, in Part II, to discuss what economics can tell us about the relative efficiency of the economic and political market. In Part III I discuss, first, some general criticisms of the economic approach, and then several arguments that suggest that medical care is, in some fundamental way, a different sort of good, so that the ordinary economic arguments used to discuss how butter or automobiles can best be produced are irrelevant to the production of medical care. In Part IV I discuss arguments that treat medical care as an ordinary good but claim that it, like some other goods, has special characteristics that justify producing it partly or entirely through the political market. In Part V I will attempt to summarize my conclusions.

## I. APOLOGIA PRO VIA SUA – A PHILOSOPHICAL INTRODUCTION

The purpose of this essay is to discuss, from the viewpoint of an economist, the arguments against use – or at least, against exclusive use – of the market in health care. This section deals with the relation between the philosophical question of what is desirable and the economic question of what is efficient. Part II will be concerned with the economic and philosophical question of whether we have any basis for judging whether the efficiency of particular market arrangements is likely to be increased or decreased by governmental interventions.

Economic efficiency is not a self-evident goal; most people who clearly understand what it means would agree that there are some circumstances in which an inefficient outcome is preferable to an efficient one. Arguments based on efficiency therefore depend on further moral arguments. I will spend much of this paper – almost all of Parts II and IV – discussing the case for and against the market in terms of economic efficiency; the purpose of this part of the essay is to explain why.

Before doing so it seems only fair to describe, at least briefly, my own position on moral philosophy, in order to help the reader make proper allowances for the biases of the author. This is particularly important since, for reasons shortly to be explained, my views may not be obvious from the arguments I will be using.

The position in moral philosophy that I find least unsatisfactory is that there exist natural rights, that they can be described in terms of entitlements, and that to be entitled to something is not the same thing as to deserve it. Seen from this position, entitlements may not be complete statement of what I ought to do, but they are a complete statement of my claims against others and theirs against me – what each of us may properly compel the other to do. My position is in this regard similar to that defended by Robert Nozick in *Anarchy, State and Utopia.*

The rules of original entitlement and transfer that I find plausible correspond fairly closely to the laws of a pure free market society. In the course of this article I will argue that those rules have other attractive features – that the same institutions I consider morally attractive also approximate a utilitarian optimum more nearly than any alternative I know. One should perhaps be suspicious of such a convenient coincidence. One explanation of it is that I am, consciously or unconsciously, distorting and selecting economic arguments in order to justify as efficient institutions I support for other reasons. A second is that I am, consciously or unconsciously, distorting my moral philosophy to justify institutions that I support because they are efficient. A third – and the most interesting – is that the "coincidence" reflects some underlying connection between natural rights and utilitarian arguments.

*What is Efficiency Anyway?*

The conventional definition of economic efficiency, due to Pareto, is that a Pareto-improvement is a change that benefits someone and injures no one and a situation is efficient if it cannot be Pareto-improved. The rule for judgment that this seems to suggest – reject all inefficient outcomes – is less value-free than it at first appears. I therefore prefer to use a slightly different approach, due to Marshall. I define an improvement as a change such that the total benefit to the gainers, measured by the sum of the numbers of dollars each would, if necessary, pay for the benefit, is larger than the total loss to the losers, similarly measured. I call this a Marshall improvement. I define a situation as efficient if it cannot be Marshall-improved.

While my definition of an improvement is not equivalent to the conventional definition, my definition of efficient is; a situation that is Pareto efficient is also Marshall efficient, and vice versa.[1] To see why, we must look a little more carefully at the word "can" in the definition of efficient; what does it mean to say that a situation "can be improved?"

A determinist would argue that nothing can be except what is; in this sense whatever occurs must be efficient.[2] More generally, how strong or weak a requirement efficiency is depends on how wide a range of alternatives we think of as possible. In practice, economists have defined efficiency using "can be improved" to mean "could be improved by a central planner who had complete economic knowledge and complete control over individual behavior." This represents a sort of outer bound on the outcomes that could actually be produced by varying economic arrangements. It is hard to see how any economic institutions could produce outcomes that could not be produced by such a "bureaucrat god" but easy to imagine that the bureaucrat god might be able to produce outcomes that could not be produced by any real world economic institutions – since we do not have gods available to run our economy. So if an arrangement is efficient there is no institutional change that can improve it; if it is inefficient there *may* be one.

This very broad definition of "can be improved" is the reason why the best available institutions may be inefficient. It is also the reason why Pareto-efficient and Marshall-efficient are equivalent. The proof goes as follows:

Any Pareto improvement is a Marshall improvement; since there is at least one gainer and no losers, gains must be larger than losses. Hence a situation that can be Pareto improved can also be Marshall improved. Hence a situation that is Marshall efficient (cannot be Marshall improved) is also Pareto efficient.

Suppose there were a situation that was Pareto efficient (could not be Pareto improved) but not Marshall efficient. There would then be a possible Marshall improvement – a change that would benefit the gainers by more, measured in dollars, than it would injure the losers. A bureaucrat god could make that change and simultaneously transfer from gainers to losers a sum larger than the losses and less than the gains, taxing each gainer an amount less than his gain and giving each loser an amount greater than his loss. The combination of the Marshall improvement plus the transfer would be a Pareto improvement. But a situation that can be Pareto improved is not Pareto efficient – which contradicts the original assumption. Hence any situation that is Pareto efficient is also Marshall efficient. Hence the two definitions of efficiency are equivalent (but see note 1).

If the two definitions of efficiency are equivalent, why have I bothered to introduce and use the unconventional one? Because the conventional definition, as it is commonly used, is a way of making interpersonal utility comparisons while pretending not to; Marshall's approach makes the same comparisons but is honest about what it is doing.

What justification can there be for making interpersonal utility comparisons in a way that, by comparing gains and losses as measured in dollars, implicitly assumes that the utility of a dollar is the same to everyone? Marshall's answer was that for most economic questions it does not much matter how you weight utilities. Most issues involve large and diverse groups of gainers and losers; unless there is some systematic tendency for one group to have a higher marginal utility of income than another, the differences among individuals average out, so if the utility gain to the gainers is larger than the loss to the losers when measured in dollars, it is probably also larger measured in utiles. Utility measured in dollars is observable, since we can observe how much people are willing to pay to achieve their objectives; utility measured in utiles is not. We use the former as a proxy for the latter. Seen in this way, judgments of efficiency are really approximate judgments about utility. "$O_1$ is more efficient than $O_2$" means "going from $O_2$ to $O_1$ is a Marshall improvement" means "utility is (probably) higher in $O_1$ than in $O_2$". I will make a careful distinction between efficiency and utility only in those special cases where there is a reason to expect the approximation to be a bad one.

*Why Bother With Efficiency?*

My justification for discussing the case for or against the market in medical care in terms of economic efficiency rests on three propositions. The first is that more is known about economics than about moral philosophy, so we are more likely to reach true conclusions and be able to convince others of them through the former than through the latter. The second is that most real world moral philosophies overlap considerably, with utilitarian considerations an important part of the overlap. While only utilitarians will claim that utility is all that matters, and even utilitarians will not claim that efficiency as defined by economists is all that matters, most people believe that utility, and efficiency as an approximation thereof, are among the things that matter. The third proposition is that the difference in efficiency among different institutions is large – large enough to affect the conclusions, with regard to those institutions, of many who are not utilitarians. I suspect, although I cannot prove, that market arrangements are so much superior to any workable alternative that most people, including most non-utilitarians, would prefer their consequences to those of any alternative [9]. If I am right, then political disagreement is fundamentally a disagreement about the economic question of what consequences different institutions produce, not the ethical question of what consequences we prefer. This is in part an economic opinion about the efficiency of the market, and in part an economic opinion about the range of workable alternatives.

*In Sum*

I believe that utilitarian comparisons among the outcomes of different institutions are neither irrelevant nor conclusive; I view the economic concept of efficiency as the most practical way of making such comparisons. I will therefore try to use economics to analyze the arguments against market health care, mostly in terms of economic efficiency.

Economics is not limited to arguments about economic efficiency; indeed, some economists would argue that such "welfare economics" is too much involved with issues of value to be a real part of the science. From my viewpoint, efficiency is simply one useful way of summing up positive results – results about the consequences of particular institutions. It provides a convenient compromise between the questions people ask economists and the questions economists are equipped to answer. It is not, however, the only thing economics can say about the outcome of institutions, hence not the only contribution that economics can make to disputes as to what institutions are best. I shall, to take one example, feel free to use economic arguments not only to discuss whether government interference with the market has a cost in decreased efficiency, but also whether it produces any benefit in increased equality.

## II. IS THE BEST THE ENEMY OF THE GOOD ?

### *Efficiency Proofs*

Economic efficiency is a strong requirement for the outcome of any real world system of institutions, since an outcome is efficient only if it could not be improved by a bureaucrat god – a benevolent despot with perfect information and unlimited power over individual actions. While it may be seen as an upper bound on how well an economic system can work, one might think that using that bound to judge real systems is as appropriate as judging race cars by their ability to achieve *their* upper bound – the speed of light.

Surprisingly enough, it is possible to prove that a market system that meets certain assumptions is efficient in this strong sense. There are several sets of assumptions that will do; two of them may be roughly stated as follows:

I. *Perfect Knowledge*: Individual producers know the cost of all alternative ways of producing and the market price for what they produce; individual consumers know the price and the value to them of goods they consume.

II. *Private Property*: Property rights are defined and costlessly enforced and can be costlessly transferred for all scarce goods.

*Either*

III. *No Transaction Costs*: Any transaction mutually advantageous to the parties involved can be arranged at no cost.

*Or*

IV. *Perfect Competition*: Every producer and consumer is so small a part of the market that the quantity he produces or consumes does not affect the market.

V. *Private Goods*: Every producer can control the use of the goods he produces.

These assumptions are chosen to eliminate market failures traditionally associated with the terms public goods, externalities, imperfect competition, and imperfect or asymmetric information. Assumption III substitutes for assumptions IV and V because of the Coase Theorem; under conditions of zero transaction costs all of the inefficiencies associated with imperfect competition, externalities, and public goods can be eliminated by appropriate bargains among the affected parties [5].

Of course, no real world economy satisfies either assumptions I, II, and III or I, II, IV, and V exactly. This fact is frequently used to advocate government intervention in the market, on the grounds that it can improve the inefficient outcome due to one or another sort of market failure. The difficulty with such arguments is that there is no adequate theory of government behavior that implies that government would choose to do the right things – that its intervention would make things better rather than worse.

It is one thing to show that there is something government *could* do that would improve on the outcome of the unregulated market; it is an entirely different, and much more difficult, thing to show that what government *would* do, given the power, would improve on that outcome. That would require a theory of governmental behavior comparable in power and precision to the theory of market behavior from which the original efficiency theorem, and the inefficiencies due to failures of its assumptions, were derived. No widely accepted theory of that sort exists, and much of the large and growing literature that attempts to produce such a theory seems to suggest that government intervention is more likely to worsen than to improve market outcomes [7, 14, 17, 18].

Even if one drops the traditional argument that every deviation of the market outcome from perfect efficiency justifies a countervailing intervention by the state, one can still argue that since real markets do not meet the requirements of the efficiency theorem we can expect them to be inefficient and do not know by how much, that since there is no efficiency theorem for the political system we have no way of knowing how inefficient it is, hence we have no way of guessing whether political intervention in the market system will lead to more or less efficiency. A similar argument on a grander scale suggests that since we have an efficiency theorem for capitalist economies, whose assumptions real capitalist economies do not meet, and no efficiency theorem at all for socialist economies (or other alternative systems), we again have no basis for predicting which is superior. By defining a perfect outcome that we cannot achieve, we seem to have made it impossible to choose among imperfect outcomes. The best appears to be the enemy of the good.

This argument is, I believe, wrong for at least three reasons. The first is that, although we do not have an economic theory of the political process as well worked out and broadly accepted as the theory of private markets, we do have enough of such a theory to have some idea of where and why the political market is likely to produce less efficient outcomes than the private market. Second, if the situation really were that we had no theoretical basis at all for predicting how efficient the alternatives to the market would be, we would have some reason to expect the market to be (imperfectly) efficient, no reason to expect its alternatives to be, and thus a (rebuttable) presumption that the market is more efficient than its alternatives. Third, whatever the theoretical situation may be, there exists a large and growing body of empirical studies of the effects of government regulation, most of which cast serious doubt on the idea that intervention in the market is likely to lead to improvement.

### *Public Choice – The Economics of the Political Market*

Conventional economics provides us with an analysis of the economic market, including some understanding of why and to what extent it fails to be efficient. In order to compare the outcomes to be expected from the economic market with those to be expected from the political market, we need some way of analyzing the latter. There exists a body of economics called public choice theory that attempts to analyze the political system by using the same approach with which ordinary economics analyzes the private market. While no single version of public choice theory is as well worked out and widely accepted as is the conventional price theory used to analyze ordinary private markets, public choice theory as it now exists provides economists with some limited ability to predict where and why political alternatives to the market will or will not be efficient. We do not have an adequate theory, but we have something more than no theory at all.

Public choice theory is simply economics applied to a market with peculiar property rights. Just as in the economic analysis of an ordinary market, individuals are assumed to pursue rationally their separate objectives; just as in that analysis, one may first make and later drop simplifying assumptions such as perfect information or zero transaction costs. But the property rights on the public market include the right of individuals to vote for representatives, of representatives acting through the appropriate procedures to make laws, of various government officials to enforce the laws, of judges to interpret them, and so on.

Ordinary economics is greatly simplified if we treat firms as if they were imaginary individuals trying to maximize their profits; in this way we reduce General Motors from several hundred thousand individuals to one. There is some cost to the simplification, since it ignores the conflicts of interest within the firm among managers, employees, and stockholders. So far no alternative simplification seems to work as well, so economists continue to analyze an economy of profit maximizing firms, except when the particular problem being considered hinges on intra-firm interactions – as it does, most obviously, in the theory of the firm.

One of the ways in which different public choice theories differ is in what they take to be the equivalent of the firm on the political market, and what it is assumed to maximize. Downs, one of the founders of public choice, took his unit to be the political party [6]. Niskanen took it to be the individual government bureau [15]. Other analyses take interest groups or politicians as the "firms", or eliminate firms entirely and consider individual voters. For the purposes of the following brief sketch, I will consider the firms on the political market to be elected politicians (or, equivalently, the political organizations of which they are a part), and limit my discussion to the market for legislation. This is a simplification of the political market, and only one of several possible simplifications, but it provides a convenient way of sketching the theory.

Consider, then, the market for legislation. Individuals perceive that they will be benefitted or harmed by various laws. They therefore offer payments to politicians for supporting some laws and opposing others. The payments may take the form of promises to vote, of cash payments to be used to finance future election campaigns, or of (concealed) contributions to the politician's income.[3] The politician is seeking to maximize his long run income (plus non-pecuniary benefits, one of which may be "national welfare"), subject to the constraint that he can only sell his support for as long as he can keep getting reelected.

Is the outcome of this market efficient? That depends on additional assumptions. If all transaction costs are zero, the answer is yes. As long as there is any potential change in legislation – any law to be passed or repealed – that confers net benefits, there is some possible agreement that will produce the change while benefitting all those whose cooperation is required. The Coase theorem applies to political as well as private markets.

Even with transaction costs, we would still expect efficient outcomes as long as the amount individuals or groups are willing to pay for any piece of legislation is proportional to the benefit they receive from it, with the constant of proportionality the same for all. In that case, any law whose passage conferred net benefits would be profitable to pass – more would be offered to support it (by those benefitted) than to oppose it (by those injured).

Alternatively, we might expect an efficient outcome if we assume that all voters are perfectly informed. In that case campaign contributions are useless, since no voter can be persuaded to vote against his own interest, and we have civics class democracy, with the candidate who best represents the public interest getting elected.[4]

The important question, however, is not whether the political market works under conditions of zero transaction costs and perfect information; under those assumptions, the private market is also perfectly efficient. The interesting question is how badly each system breaks down when the assumptions are relaxed.

Consider a political market with realistic transaction costs. A legislator proposes a bill that inefficiently transfers income from one interest group to another; it imposes costs of $10 each on a thousand indviduals and gives benefits of $500 each to ten individuals. What will be bid for and against the law?

The total cost to the losers is $10,000, but the amount they will be willing to offer to a politician to oppose the law is very much less than that. Why? Because of the public good problem. Any individual who contributes to a campaign fund to defeat the bill is providing a public good for all the members of the group. The same arguments that imply underproduction of public goods by the market apply here. Just as in that case, the larger the public the lower the fraction of the value of the good that can be raised to pay for it.

The benefit provided to the winners is also a public good, but it goes to a much smaller public – ten individuals instead of a thousand. A small public can more easily organize, through conditional contracts ("I will contribute if and only if you do") to fund a public good. Even though the benefit to the small group is smaller than the cost to the large one, the amount the small group is able to offer politicians to support the bill will be more than the amount the large group will offer to oppose it.

The effect is reinforced by a second consideration – information costs. I now drop the assumption of perfect information; instead, I assume that information about the effect of legislation on any individual can be obtained, but only at some cost in time and money. For the individual who suspects that the bill may injure him by $10, it is not worth paying much to obtain the information. Not only is his possible loss small, but the effect on the probability that the bill will pass of any actions he would consider taking is also small. The member of the dispersed interest chooses (rationally) to be worse informed than the member of the concentrated interest.

If we abandon the particular example and consider the extreme case of an American presidential election, the argument becomes even clearer. Suppose the average election is won by 2.5 million votes. We may simplify the analysis by replacing the actual probability distribution for the outcome by a uniform probability distribution from –5 million to +5 million. The probability that the election will be a tie, hence that one additional vote can decide it, is then one in ten million. So the return to an individual who figures out who is the right candidate and votes accordingly, instead of voting at random, is an increase by one chance in ten million of the probability that the right candidate will be elected. Unless the voter has an extraordinarily high value for electing the right candidate, it is not worth paying very much in order to increase the probability of that outcome by one in ten million, so he does not pay the information costs necessary to decide for whom he should vote. This is rational ignorance. It is rational to be ignorant if the cost of information is greater than its value.

So far I have discussed only one characteristic of a group – its size. It is useful to think of the terms "concentrated" and "dispersed" as proxies for the set of characteristics that determine how easily a group can fund a public good for that group; the number of individuals in the group is only one of those characteristics.

Consider, for example, a tariff on automobiles. It benefits hundreds of thousands of people – stockholders in auto companies, auto workers, property owners in Detroit, and so forth. But General Motors, Ford, Chrysler, American Motors, and the UAW are organizations that already exist to serve the interests of large parts of that large group of people. For many purposes one can consider all of the stockholders and most of the workers as "being" five individuals – a group small enough to organize effectively. The beneficiaries of auto tariffs are a much more concentrated interest than a mere count of their numbers would suggest. That may explain why such tariffs exist, even though they are inefficient – the costs they impose on consumers of automobiles and American producers of export goods (both dispersed interests) are larger than the benefits to the producers of automobiles.

The reason the public good problem leads to inefficiency on ordinary private markets is that the amount a group can raise to buy a public good benefitting that group is less than the total value of the good to the members of the group, hence some public goods that are worth more than they cost to produce fail to get produced, which is inefficient. The problem on public markets is that both costs and benefits are only fractionally represented on the market, due to the public good problem – and if the weights are different, as they almost always will be, laws that impose net costs may be passed and laws that impose net benefits may not be. This again is inefficient.

To compare the efficiency of the two systems – the private and the political market – we must specify the degree of publicness of the good and the nature of the publics involved. If, for example, we consider the private production of a pure public good for which the public is the entire population of the country, the political alternative may be an attractive one. The private market will produce the good only if total benefit times the weighting factor relevant to that public (the percent of the benefit that can be raised to pay for the good) is larger than the cost of producing the good. The public market will produce the good only if total benefit times the weighting factor (the same as before) is larger than total cost times a weighting factor representing the fraction of the tax money saved by not producing the good that taxpayers (also a public with a public good problem) can raise to oppose its production. Since the weighting of cost as well as of benefit is (much) less than one, the chance that weighted benefit is greater than weighted cost (so that the good will be produced politically) is greater than the chance that weighted benefit is greater than unweighted cost (so that it will be produced privately). The public good is more likely to be produced publicly than privately.

This is not entirely a good thing. If the weighting factor on cost is smaller than on benefit – if taxpayers are a more dispersed interest than beneficiaries, as they would be if the good benefitted only a small group – then the good may be produced even if cost is greater than benefit. This too is inefficient, although in the opposite direction – producing a good that should not be produced rather than not producing one that should.

In many cases, although perhaps not in the case of defense, the size of the public and the degree of publicness are quite different on the private and political markets. Consider immunization against contagious disease. Immunization is only partly a public good; if I have myself immunized I receive a private benefit (I am safe from the disease) and confer a benefit on others (they cannot catch it from me). If the public and private parts happen to be of equal value, I will choose to get immunized as long as the total benefit is at least twice the total cost. Private immunization is then only "half public".[5]

Suppose we decide to provide immunization publicly instead of privately. This does not mean that we decide to provide it if and only if it is worth providing – systems that automatically generate right answers are not one of the options available to us. It means rather that we have a system in which what immunizations are given to whom – and who pays for them – are decided politically. The decision is determined by a tug of war between two interests – those benefitted and those paying – both large groups whose political efforts are restricted by the public good problem. The groups would have to be very well balanced indeed in order to generate a more efficient outcome than "produce if benefits are at least twice costs" – the outcome of the private system.

This article is an essay on the role of the market in medical care, not a textbook on public choice theory. I have tried to sketch enough of the theory to show the reader why the efficiency of the public market generally breaks down much faster as we weaken our simplifying assumptions (perfect information and zero transaction costs) than does the efficiency of the private market. On the private market, an individual who pays for information gets the benefit of the information; if I spend time and energy deciding what car to buy, my conclusion will determine what car I get. On the public market, gathering information typically involves producing a public good for a large public; if I gather information on which politician to vote for, the result is a miniscule increase in the probability that I (and everyone else) will get that politician and the legislation he supports. Hence we may expect people to be much better informed about their private than their public decisions – a conclusion that seems consistent with casual observation. Imperfect information, while a problem for both markets, is a much more serious problem for the public market.

Similarly with the set of problems associated with public goods and externalities. The typical good produced on the private market is a mostly private good – an office building that provides $50,000,000 worth of benefit to the individuals renting space in it, plus a $1,000,000 benefit spread among all the inhabitants of the city whose skyline it ornaments. If expensive buildings strike you as a bad example, substitute my front yard (mowed) or your neighbor's phonograph (a negative externality if you have sufficiently different tastes in music – or in when you wish to listen to it). Exceptional cases are pure or almost pure public goods (scientific research of a non-patentable nature, the invention of the supermarket) or goods for which external costs are a large fraction of all costs. In the normal case we have something close to an efficient outcome – the building is constructed as long as benefit is at least 1.02 times cost. Inefficient outcomes are unlikely to occur, and if they do occur the inefficiency is small – the failure to produce something that is worth slightly more than it would cost to produce, or the production of something worth slightly less than it costs to produce. Only in the exceptional cases is inefficiency likely to be a serious problem.

On the public market, on the other hand, public goods with large publics and goods whose costs or benefits are mostly external are the rule rather than the exception. When I support a bill to benefit me at your expense, all of the cost of what I am getting is external – imposed on you and ignored in my calculations. When a politician tries to pass a good law, he is attempting to produce a public good with a very large public. He would almost certainly be better off, from his own standpoint, using the same tax money to buy the votes (or bribes) of some concentrated interest instead. Even if the politician has a public spirited desire to pass good laws, he will find that unless he is very rich he cannot afford to – just as a capitalist cannot afford to give his customers what he thinks they should have instead of what they are willing to pay for. Unless the politician has large political assets that his competitors lack, his attempt to do good will result in his defeat by someone who follows a more nearly vote maximizing policy.

The conclusion of this analysis seems to be that the public market is preferable to the private one, at most, only in cases of extreme market failure. As long as the private market is anywhere close to efficient, it is probably preferable to the political alternative.

How good are our reasons for believing that the analysis I have sketched accurately describes the real world? There is some evidence, but it is not conclusive. The observed outcomes of the political system – what industries get tariffs, how professions come to be regulated, and the like – appear to fit the patterns suggested by the analysis, but neither the evidence nor the theory is sufficiently good to make us as confident about the economic theory of government as we are about the economic theory of ordinary markets [20].

*Evidence*

This brings me to my third reason for believing that the market is typically more efficient than its alternatives. There now exists a substantial and growing body of studies of the effects of government regulation. The conclusions suggested by many of these studies are that regulation rarely achieves its stated purpose, that it frequently imposes serious costs, and that it frequently serves as an indirect way of transferring income from one group to another.

Consider, for example, Linneman's study of the effects of minimum wage legislation [14]. The minimum wage is often defended as a way of helping low income workers by raising their wages. Linneman obtained, for a large sample of individuals, data on employment, wages, and a variety of characteristics (such as age, education, and race) that might be expected to affect employment. Using data from before and after a large increase in the minimum wage rate, he estimated the effect of the increase on the wage rate and employment of the individuals in his sample. He concluded that the effect had been to lower the earnings of the subminimum population – the workers who would have made less than the new minimum wage – and increase the earnings of high wage and union workers. In effect, the increase priced unskilled labor out of the market, increasing the demand for skilled labor.

Consider, as a second example, Peltzman's study of the effects of the Kefauver amendments to the Pure Food and Drug Act [17]. The stated purpose of the amendments was to force drug companies to demonstrate the effectiveness of "New Chemical Entities" (NCEs) before putting them on the market; the argument was that, because of imperfect information, consumers paid unnecessarily to buy, and drug companies to produce, new drugs that were no better than the old drugs. Peltzman showed that the effect of the amendment was to reduce the rate of introduction of NCEs by more than half without any detectable improvement in their average quality; he estimated that the effect was equivalent to imposing a ten percent sales tax on all consumption of drugs.

The point of these examples is not merely to show that the best laid plans of mice, men, and legislators gang aft agley. It is rather to suggest that insofar as we have evidence on the effect of intervention, the evidence is that the intervention imposes net costs, and does so without achieving the objectives – sometimes involving efficiency and sometimes not – for which it purports to exist. It is further to suggest that the observed pattern is more consistent with rational behavior aimed at objectives other than those announced than it is with the other obvious explanation – repeated error. In the case of the miniumum wage, for instance, it is worth noting that the congressmen supporting it have generally represented high wage, not low wage, states and industries.

In summary, then, the conclusion of this section of the paper is that both economic theory and empirical studies provide a basis for preferring market to political arrangements except, perhaps, in cases of extreme market failure. The conclusion does not have the rigor of a mathematical theorem – but neither, so far as I know, does any conclusion about the real world, even in the most precise of sciences. In addition to the uncertainty associated with any attempt to apply a theoretical construct to a real-world situation, the conclusion has the additional uncertainty associated with a body of theory – public choice – that is still developing and many of whose elements are still matters of dispute.

## III: THE CASE AGAINST ECONOMICS

In Part I of this paper, I attempted to deal with general philosophical arguments concerning the appropriateness of economic efficiency as a criterion for judging alternative arrangements; in Part II I tried to show how one might judge the relative efficiency of the market and its alternatives. In Part IV I will try to apply that analysis to the question of how medical care should be produced and distributed.

Before doing so, there are several potential objections to this approach that I would like to try to deal with. Some are objections to the economic approach in general, some objections to its application to medical care. There is little point in my subjecting you to a lengthy economic analysis of the efficiency of providing medical care in various ways, without first making some attempt to persuade you that economic analysis is a legitimate way of learning things and that its conclusions are relevant to questions of the provision of medical care.

I will start by discussing three objections to the economic approach that are, I think, widely held. They are, first, that economic analysis depends on the unrealistic assumption of individual rationality, second, that economic efficiency fails to take account of the fact that money, which it uses as its measure of value, is of different value to different people, and third, that it ignores the infinite value of life relative to money or material goods, and the fact that medical care, like food, is necessary for life. I will go on to discuss the claim that medical care has some sort of special moral status, making it a "priority good" or implying that people have a "right to medical care".

### *Rationality*

The central assumption of economics is rationality – that people have objectives and tend to choose the correct way of achieving them. While the assumption can be modified to deal with information costs, individuals are still assumed to make the correct decision, in an uncertain environment, about how much information to buy.

The use of the term "rationality" to describe this central economic assumption is somewhat deceptive, since it suggests that people find the correct way to achieve their objectives by rational analysis – using formal logic to deduce conclusions from assumptions, analyzing evidence, and so forth. No such assumption about how people find the correct means to achieve their ends is necessary.

One can imagine a variety of other explanations for rational behavior. To take a trivial example, most of our objectives require that we eat occasionally, so as not to die of hunger (exception – if my objective is to be fertilizer). Whether or not people have deduced this fact by logical analysis, those who do not choose to eat are not around to have their behavior analyzed by economists. More generally, evolution may produce people (and other animals) who act rationally without knowing why. The same result may be produced by a process of trial and error. If you walk to work every day, you may by experiment find the shortest route, even if you do not know enough geometry to calculate it. "Rationality" does not mean a particular way of thinking but a tendency to get the right answer, and it may be the result of many things other than thinking.

Half of the assumption is that people tend to reach correct conclusions; the other half is that people have objectives. In order to do much with economics, one must strengthen this part of the assumption somewhat by assuming that people have reasonably simple objectives. The reason for this additional assumption is that, if one has no idea at all about what people's objectives are, it is impossible to make any prediction about what they will do. Any behavior, however peculiar, can be explained by assuming that that behavior was itself the person's objective (why did I stand on my head on the table while holding a burning thousand dollar bill between my toes? Because I *wanted* to stand on my head on the table while holding a burning thousand dollar bill between my toes).

This element in economic theory may be partly responsible for the idea that economists assume that "all anyone is interested in is money". Put in that way, the assertion is wrong; economists usually assume that people desire money only as a means to other objectives. What is true is that, although economics can, in principle, take account of the full richness of human objectives, it is necessary for many practical purposes to assume away all save the most obvious – the consumption of goods, leisure, security, and the like.

Economics is based on the assumption that people have reasonably simple objectives and choose the correct means to achieve them. Both halves of the assumption are false; people sometimes have very complicated objectives and they sometimes make mistakes. Why then is the assumption useful?

Suppose we know someone's objective, and also know that half the time he correctly figures out how to achieve it and half the time he acts at random. Since there is usually only one right way of doing things (or perhaps a few) but very many wrong ways, the rational behavior can be predicted but the irrational behavior cannot. If we predict his behavior on the assumption that he is always rational, we will be right half the time; if we assume he is irrational, we will almost never be right, since we still have to guess which irrational thing he will do. We are better off assuming he is rational and recognizing that we will sometimes be wrong. To put the argument more generally, the tendency to be rational is the consistent and hence predictable element in human behavior. The only alternative to assuming rationality (other than giving up and concluding that human behavior cannot be understood and predicted) would be a theory of irrational behavior, a theory that told us not only that someone would not always do the rational thing but also which particular irrational thing he would do. So far as I know, no satisfactory theory of that sort exists. As I argued in Part II of the essay, it takes a theory to beat a theory; until some better alternative is found, rationality is the best we have.

One possible response to this is that although rationality may give us the best available theory, a theory built on so weak a foundation is not very useful; perhaps we are better off depending on "common sense" answers instead.

This sounds plausible, but when it comes to analyzing a market – a complicated interacting system – "common sense" turns out in practice to mean a poorly thought out, inconsistent, and untested theory. If analyzing such a system were easy, economics would be easy too. Any good economist can provide a collection of horrible examples from casual conversation and his daily newspaper.

*Economics and the Poor*

A second set of objections to the market and the economic approach is based on the claim that neither gives proper consideration to the implications of income inequality. This involves two different arguments, one of which I in part agree with. The first is that the market – and the criterion of efficiency according to which its outcome is often judged – measures individual values in dollars, not in intensity of feeling; since some people have fewer dollars than others, their desires receive less attention, even if they are as strong as, or stronger than, the desires of the more fortunate. The second argument is that economic analysis, and the economic defense of market outcomes, is put in terms of choice, but that choice is irrelevant to the poor; how can one say that an individual chooses not to have that which he cannot afford?

The first argument can be restated by going back to my original defense of the efficiency criterion as a proxy for the utilitarian objective of maximizing total welfare. I conceded there that the measure of improvement used in defining efficiency (explicitly in the Marshallian approach, implicitly in the more conventional Paretian approach) compares costs and benefits to different individuals measured by how much money those individuals would spend, if necessary, to get the benefit or avoid the cost. This is equivalent to doing an interpersonal utility comparison on the assumption that the marginal utility of a dollar is the same for everyone.

I defended that way of defining efficiency, from a utilitarian standpoint, ([12], Chapter 15) by arguing that the particular rule used to weight utilities did not matter very much. Most of the decisions an economist is interested in affect large and heterogeneous groups of people. Unless there is some reason to expect that a particular weighting rule favors the losers more than the gainers, or vice versa, differences among individuals can be expected to average out when we consider the effect on the whole group.

There are some cases in which we do have good reason to expect that the choice of weighting rule will differently affect the different groups. Most of us believe that people with high incomes have, on average, a higher marginal utility of income than people with low incomes. If so, there will be a systematic discrepancy between efficiency and utility when we are considering decisions that have differential impact on high and low income people. One obvious case is the decision of whether or not to provide free medical care (or housing, or food, or money) to the poor at the expense of the rest of us. If we evaluate the transfer in terms of efficiency, we conclude that what the poor get is worth at most its cost to them (if it were worth more, they would have bought it) while the cost to us of paying for it is at least its cost (more if there is some cost to collecting taxes), so the transfer provides benefits that are at most equal to the cost and almost certainly less. If we evaluate it in terms of utility, we conclude that even if a hundred dollars worth of medical care is worth only ninety dollars to the poor recipient, that may well represent more utility than the hundred and ten dollars that it costs the not-poor taxpayer.

This is the traditional utilitarian argument for redistribution. Three things are worth noting about it. The first is that the assumption that marginal utility of income is correlated inversely with income, while plausible, is not necessarily true; under some circumstances one would expect the opposite. The second is that the argument has no special relevance to medical costs, except to the extent that differing medical bills may be an important cause of inequalities in the marginal utility of income among individuals; it is a general argument for redistribution from the rich to the poor. The third is that while the argument suggests that we should prefer outcomes that are biased towards the poor, everything else being equal, it does not tell us that we should favor political over market processes. That conclusion depends on some further argument to show that the outcome of political processes is likely to modify the market outcome towards equality and not away from it – that the poor can be expected to do better on the political market than on the economic market.

I begin with the first point. The standard argument for redistribution begins with the claim that, for a given individual, the marginal utility of income declines as income increases. This seems plausible in terms of introspection, observation (of behavior under uncertainty) and theory, although one can construct counterexamples. The next step is to claim that, absent information about differences in individual utility functions, we must treat each individual utility function as a random draw from the same population, so declining marginal utility of income applies not only to the same individual with different incomes but (on average) to different individuals with different incomes. It follows that the poor have, on average, a higher marginal utility for income than the rich.

This argument assumes that there is no causal connection running from utility function to income. That is plausible enough if income is determined exogenously – by inheritance, or lottery, or some other chance mechanism. It is quite implausible if differing income is the result of differing effort. An individual who greatly values the things that money buys will be more willing than others to give up other goods, such as leisure, in order to get income, so he will, on average, end up with a higher income. If, instead of assuming that different individuals have the same utility function but different incomes, we assumed identical opportunities but different utility functions, we would expect income and marginal utility of income to be *positively* correlated. So our opinion about the sign of the relationship will depend very much on our opinion about the origin of observed inequalities in income.

This brings me to my second point. The cost of medical services represents a large, and to a considerable extent random, subtraction from income. Even if we believed that people with low incomes are, on average, poor because they do not value money, rather than that they value money because they are poor, we should still make an exception for those who are poor because of medical expenses. In that case, at least, the traditional argument seems to hold rigorously. The income difference is the result of a random exogenous force, so differences in utility functions should be unrelated to differences in income net of medical expenses, so those whose consumption of other goods is low because of their medical expenses should have a high marginal utility of income. It follows that the utilitarian case for transferring income to those who are impoverished by high medical expenses is stronger than the conventional utilitarian case for transferring income to the poor.[6]

Is this a conclusive argument, within the utilitarian framework, for political intervention to alter the outcome of market provision of medical services? That depends on the costs of such intervention – if large enough, they might outweigh the gains. This argument is usually put in terms of the costs resulting from the distortion of incentives introduced by both tax and subsidy.

While this is a legitimate argument for limiting the size of income transfers to the poor, whether for medical or other purposes, it gives us no reason to expect that the optimal transfer would be zero. Elsewhere I have made a different argument that does suggest that conclusion – essentially that giving government the power to transfer income sets off an expensive free-for-all, with individuals expending considerable efforts to ensure that they will be the beneficiaries of transfers instead of the ones who pay for them [9].

There is, however, another important problem with political redistribution. It is usually assumed without discussion that having government intervene to redistribute in favor of the poor, while leaving most of the rest of the system unaffected, is one of the available alternatives. This is the same sort of assumption made by those who argue that wherever the market outcome is inefficient, government should step in; in both cases, government is treated as a *deus ex machina*, introduced to produce whatever outcome we have decided is desirable. But in discussing equality, just as in discussing efficiency, it is not enough to show that there is something government *could* do that would improve on the outcome of the market. The question is whether what government *will* do will be an improvement [8].

Public choice theory does not give any clear answer as to whether government, given the power to redistribute, is likely to increase or decrease inequality, to distribute to the poor or from them. On the one hand, votes are more evenly distributed than income, which should tend to make the political market more egalitarian than the private market. On the other hand, many of the characteristics that give groups and individuals political influence are closely related to income. Education reduces information costs, labor skills differentiate their possessors into (relatively) concentrated interest groups, stockholders have their interest represented by well organized firms and skilled (hence highly paid) workers by well organized unions, and so forth. So far as theory is concerned, it is difficult to predict whether the political system is more likely to transfer money down the income ladder or up.

The evidence is also unclear. There are a variety of well publicized programs to help the poor at the expense of the not poor – food stamps, welfare, and the like. There are also a considerable number of government programs to help the not poor at (on average) the expense of the poor – state subsidies to higher education being a notable example. Social Security, by a large margin the biggest income transfer program in our society, has ambiguous effects; a poor individual who works and collects for the same number of years as a rich individual gets a somewhat better deal, but poorer individuals, on average, start work earlier and die earlier, hence pay for more years and collect for fewer – which may more than balance the progressive element in the payment and benefit schedules [21].

The greatest weakness in the (utilitarian) argument for government intervention, in the market for medical care or for anything else, is that it is an argument for an outcome. It simply assumes that government intervention will produce that outcome. If the assumption is correct, if government intervention results in redistribution down the income ladder with relatively small costs, then the argument is a correct one. If government intervention results in large costs, as I have argued elsewhere, or if the direction of the redistribution is ambiguous or even perverse, so that there is no gain in equality to balance the loss in efficiency, then the argument is wrong.

So far I have been discussing the traditional utilitarian argument for government intervention to redistribute to the poor. There is a second, and to my mind less defensible, set of arguments sometimes employed against the market – especially in medical services and other "necessities". This is the claim that poor people cannot really be said to choose poor medical care, or poor nutrition, or whatever, since they cannot afford anything else. Hence, it is argued, such goods should be provided to the poor, whether or not they are able or willing to pay for them.

Insofar as this is simply a repeat of the argument for redistribution from differing marginal utility of income, I have already discussed it. What is disturbing about it is that the argument seems, in practice, to be used to justify programs to provide "necessities" to the poor, whether or not such programs are actually redistributive – and often enough they are not[7]. Such programs, in effect, force the poor to spend their own money on what someone else has decided they need. It seems odd to argue that the poor family cannot afford to pay for adequate medical care (because there will not be enough money left for food?) and should therefore be compelled to pay for adequate medical care (and starve?). Yet this is what a program of governmentally funded medical care implies, unless it also redistributes. And if one is going to redistribute, there seems no obvious reason why the subsidy to the poor should take the form of medical care. Why not give money and let the poor family decide for itself what it is most important to spend it on? The claim that the rich choose among their wants while the poor "must" buy their "necessities" may be effective rhetoric, but what it actually says is that choice is less relevant to large values than to small ones, which seems peculiar.

### *The Value of Life*

One popular and persuasive criticism of both market provision of health care and the economic analysis thereof is the claim that both embody a fundamental error – the failure to realize that life is infinitely more valuable than money. Medical care, it is argued, is not merely a want but a need. Without medical care we may – some certainly will – die, and without life all other values are meaningless. Hence neither willingness nor ability to pay is relevant to who should or should not get medical care; the only proper criterion is who needs it.

This argument embodies several errors. The first, and least important, is the confusion between money as a thing of value and money as a measure of value. While money is convenient in defining economic efficiency, it is not essential – any tradable commodity will do. The real comparison is not between life and money but between life and other things people value – leisure, consumption goods, education for their children, housing, *et multa caetera*.

But the question still remains: Is not life, and hence medical care as (sometimes) a necessity for life, infinitely more important than other values?

Not if we accept the behavior of ourselves and others as evidence about our values. Anyone who both smokes and believes the conclusions of the surgeon general's report is deliberately trading life – an increased probability of dying of heart disease or lung cancer – for the pleasure of smoking. Anyone who spends time and money on anything that does not increase his life expectancy while there remains some unexploited opportunity for an expenditure that does – an extra visit to the doctor, a marginal improvement in nutrition, a slightly safer car – is demonstrating that life, while it may be valuable, is not infinitely valuable in comparison to other things.

Perhaps we do trade life for lesser values but should not. I know no way to prove what we should do, but the sort of life that would be implied by that prescription does not seem very attractive. Not only does the assumption that life is infinitely valuable imply that we should take no avoidable risks – no sky diving, no skiing, no skin diving – it also implies a society wholly devoted to achieving a single goal. It is all very well to say that we need food, or air, or housing, or medical care, but how much do we need? In the case of medical care, at least, the level of expenditure per capita at which additional care produces additional returns in life expectancy is surely far above our per capita income. If so, a society that treated life as infinitely valuable would have no resources left to spend on anything else – including the things that, for many of us, make life worth living.

Another argument that could be made for the infinite value of life is based on the economist's principle of revealed preference. Few of us would agree to be hung tomorrow in exchange for a payment of a million dollars, or even ten billion. Does not that imply that our value for life is very high and perhaps infinite?

No. It proves that money is of no use to a corpse. What is special about the situation being considered is not the high value of what is being bought but the low value of what is being paid with.

Consider the difference between an offer to buy all of someone's life immediately and offers to buy half of the remaining years, or payments for some specified probability of death. In both cases, many more people would accept than with the first offer – and those who would sell all their life for some price would sell part of it for a much lower price. This is not merely speculation; studies of wage differentials in hazardous professions generate value of life estimates well below a million dollars. The reason for the almost discontinuous change as we move from offering to buy all of a life to offering to buy a fraction of it is not the reduction by a factor of several in the amount of life we are buying but the increase, by a much larger factor, in the amount of life available in which to enjoy the money [11].

While I have very little intellectual sympathy with the claim that life is infinitely more important than other goods, I have a good deal of emotional sympathy with it – it appeals more to my moral intuition than to the ideas with which I try to make sense out of both my moral and economic beliefs. To show why it does so, and why I nonetheless reject it, I will discuss a case in which the claim seems at first very plausible.

Suppose there is some individual who requires – and does not get – a ten million dollar operation to save his life. Further, suppose that ten million dollars is precisely the sum spent, during a year, by all the people in the U.S. in order to have mint flavor in their toothpaste. Surely, this is a monstrous outcome – a man losing his life in order that others can have the trivial pleasure of mint flavor in their toothpaste.

The conclusion seems unavoidable, but I believe it is wrong. The problem, I think, is a fault in my (and I presume your) moral intuition – our inability to multiply by large numbers. To most of us, a number such as two hundred million has only a vague meaning – we have no intuition for how much the importance of a trivial pleasure is increased when it is multiplied by two hundred million. In contemplating the situation I have described, we end up comparing the value of one life to the value of a trivial pleasure to one person, or perhaps a few. Seen that way, the answer appears obvious.

In trying to test my intuition, I find it useful to remove the other people from the problem and convert numbers into probabilities. Suppose I know that by eliminating mint flavor from my toothpaste I can avoid a one in 200 million chance of my own death. That is, in some sense, an equivalent problem – at least if we specify, in the first situation, that nobody knows whether or not he will be the one in need of medical treatment.

Put this way, the answer is far from obvious; this time it is the tiny probability that my intuition finds it difficult to deal with. In order to help it out, I imagine that I get to make the same choice two hundred times, each time eliminating some minor pleasure. It seems far from clear that giving up two hundred minor pleasures – mint flavored toothpaste, the availability of my favorite flavor of ice cream, lying in bed an extra minute each morning, rereading a favorite book one more time – in order to eliminate a one in a million chance of dying is a good deal.

As a further crutch to my intuition, I try converting one chance in two hundred million of losing all of my life into a certainty of losing one two hundred millionth of my life; while there is no obvious reason why I must be risk neutral with regard to years of life, it seems the natural first approximation. Assuming that I have fifty more years to live, one two hundred millionth of my life is about eight seconds. That is not an obviously exorbitant price for mint flavor in my toothpaste.

*Rights to Life and Similar Claims*

A variety of different arguments are used by philosophers to defend the proposition that medical care has a special moral status and that choices involving it cannot be properly made on the same basis as most other choices. I have neither the space nor the competence to explore all of these claims, but I will briefly discuss three – the attempt to derive a right to medical care from a right to life, the claim that medical care involves objective needs and hence that the correct amount of medical care is not an economic issue, and the claim that medical care has a speical status because of its connection with the ability to make choices in the future.

It is sometimes claimed that individuals have a self-evident right to life, hence to that necessary to life, hence to medical care. One difficulty with this argument is that it proves too much. Medical care is not the only thing whose consumption affects life expectancy. Nutrition, clothing, housing, education – a very large fraction of all desirable things have some effect on how long one lives. If a right to life means a right to anything that increases life expectancy, then we each have a right to most of what everyone else possesses – since many of those goods, like medical care, continue to increase life expectancy at levels of consumption well beyond their current average.

One possible reply would be that there is a fundamental difference between denying someone a kidney machine, hence killing him with certainty, and denying him the additional amenities that would increase his life expectancy by a few days. I find this argument unconvincing. A kidney machine does not, after all, prevent death – it only postpones it. Nothing, so far as we know, prevents death. It is hard to see any profound difference between giving someone an additional five years with certainty and increasing by ten percent his chance of an additional fifty years.

My own view is that to talk of a right to life in this sense is already a mistake, even before it is translated into a right to medical care. A right to have life is by its nature contradictory. The concept of a right to life makes sense as my right to have other people not kill me. It does not make sense as a blank check against the rest of the human race for anything that extends my life.

A second argument for the special status of medical care holds that economic considerations are appropriate for choices that involve differing tastes, but inappropriate for choices where the right answer is a matter of objective fact. Thus, it could be claimed, one may properly decide who gets a piece of land by who is willing to offer more for it, but one should not decide who wins a lawsuit, or which scientific theory is true, by how much each party is willing to offer the judge. In the first case, the question being decided is one of value; in the second, it is one of truth.

The problem with this argument is that decisions about medical care, like most decisions human beings make, involve issues of both fact and value. When I buy a steak at the grocery store, my decision depends both on what I think the characteristics of the steak are and how I value them. Precisely the same is true of medical care. The expert can provide information about the chance that an operation will save the patient's life, or about possible side effects of a drug. But the information does not by itself compel a conclusion. The patient may care or not care about particular side effects, value his life more or less in comparison to monetary or non-monetary costs of the operation, and so on.

A final argument goes roughly as follows. It is said that there is something peculiarly important – essentially human – about the ability to make choices. One consequence of lack of medical care may be a drastic reduction in the number of available alternatives – a cripple cannot become an athlete, to take a particularly sharp example. Hence, it is said, while poor people may not have any general right to be given money, they do have a right to be provided with medical care. It is claimed that this argument justifies medical vouchers – payments to the poor that can only be used for medical expenses.

There are some obvious problems with this. Just as in the first case discussed, the argument proves too much – many inputs other than medical care affect our future choices. Further, the argument for vouchers appears to contain a simple error of logic – the proposal to increase choice in fact reduces it.

To see why, consider, not the issue of whether poor people with medical needs should be given money, but the question of whether, if they are given money, they should be compelled to spend it on medical care. The claim is that they should. The argument is that our objective is to increase the range of choice available to the recipient, that spending money on an operation does that while spending it on a vacation does not.

There is only one unambiguous sense in which choices can be increased – if the new set of alternatives includes the old, plus at least one additional alternative. In this sense, removing the requirement that the money be spent on medical care obviously increases choice. The individual who must spend the money on medical care is free to choose among alternatives A-M – all the things he can accomplish if healthy. The individual who can choose how to spend the money can also choose A-M, by spending the money on medical care, but has additional choices N-Z. It is hard to see how the former can be said to have a wider range of choice than the latter. Of course, after spending the money, on either a vacation or medical care, the individual has reduced his choices – but that is the usual consequence of choosing.

In ending this part of the essay, I should perhaps apologize to the philosophers among my readers for trespassing on their domain. I will refrain from doing so, on the grounds that many of the arguments propounded by philosophers trespass on territory that properly belongs to economics. Instead, I invite the philosophers to correct my errors in their field, as I attempt to correct theirs in mine.

## IV. TECHNICAL ARGUMENTS

So far, I have discussed – and tried to establish – the relevance of economics and of economic efficiency to choice in general and to choices associated with medical care in particular. In this part of the essay I will consider a variety of economic arguments that might be used to demonstrate the desirability of governmental involvement in the production and allocation of medical care, discussing the different ways in which market imperfections can be expected to occur in connection with the market for medical care and the possible interventions in the market that might improve the market outcome. In all cases "imperfection" means "failure to achieve an efficient outcome" and "improve" means "make a change that produces net benefits" in the sense described in Part II.

In each case, the discussion contains four parts. The first is a description of the reason for market failure. The second is a discussion of the (imperfect) market solutions – the ways in which the market, while failing to achieve the efficient outcome, approaches it more closely than a casual consideration of the problem might suggest. The third step is to discuss the ideal governmental intervention, the efficient solution that would be imposed by an all wise and all powerful benevolent despot – a bureaucrat god. The fourth step is to discuss the consequences of actual governmental involvement, and their attractiveness relative to the outcome of the market.

The first and third steps of this process, taken alone, constitute the traditional theory of regulation – the theory that regulation consists of government stepping in to correct market failures. The fourth step is what distinguishes the modern theory of regulation. Instead of asking what an ideal government could do, we ask what a government, possessing the authority to intervene in certain ways, will do.

The problems discussed will be grouped according to the source of the market failure: imperfect information, asymmetric information, imperfect competition, and externalities/public good problems.

### *Imperfect Information*

The assumption of perfect information seems most appropriate for goods that are purchased repeatedly and whose characteristics are easily observed by the user. If I buy packages of fruit from a store, I discover whether the fruit at the bottom is worse than the fruit at the top as soon as I open the package, and I discover whether the fruit tastes as good as it looks as soon as I eat it. Since each individual purchase represents a tiny fraction of my lifetime expenditure on fruit, the cost of shopping around to determine the quality of fruit offered by different stores is small compared to the cost and value associated with consumption of fruit, and can for most purposes be ignored.

Some medical purchases seem to fit this pattern – cold medicines, for example, are used repeatedly, providing the customer an opportunity to determine which ones do noticeably better than others at relieving his symptoms. But the value of a cold medicine depends only in part on how well it relieves symptoms; other relevant considerations are its effect on how fast you recover from the cold and its side effects, if any. Measuring these effects is difficult both because your health depends on many different factors and because side effects may be very long term. So the information on the quality of medicines that we get by consuming them is very imperfect. The same applies to information on the quality of medical services that are consumed regularly – such as the services of a pediatrician patronized by a large family.

For many medical services the situation is far worse. Few of us break our bones often enough to form a competent opinion of the skills of those who set them; still fewer are so unfortunate as to acquire a large sample of the services of oncologists or brain surgeons. The same is true for our experience with medicines used for the rarer and more serious ailments. In these cases we may be poorly informed. If so, our willingness to pay for drugs or services reflects only very approximately their real value to us, hence providers of goods and services may find it in their interest to provide them even when their real benefit is less than their cost, or not to provide them even when it is greater. In either case we have an inefficient outcome.

There are a number of market solutions to problems of this sort. One is to shift voluntarily the decision, and the associated costs and benefits, to some organization better informed than the individual consumer. By buying health insurance, for example, and allowing the insurance company to provide expert advice on what doctors I should patronize or what drugs I should use, I transfer the decision to a firm that has better information and more expertise than I do.

This raises some additional problems associated with insurance, which will be discussed below. It also raises the problem of how to decide which insurance company to trust. One solution is to purchase life and health insurance in the same package – since the seller then has a strong incentive to keep me alive, his incentives are at least roughly the same as my own. Another is to rely on published reports of performance of insurance companies. While this again imposes information problems on the customer, they are less severe than the original problems – there are fewer insurance companies than doctors, so it is easier to get some measure of their relative performance.

Another solution is for some expert body to certify the quality of drugs or physicians; a familiar example in another field is the Underwriter's Laboratory. In the case of prescription drugs, and to some degree of non-prescription drugs, the customer hires the physician to give him expert advice – and the physician hires the people who produce medical journals. In the case of physicians, group practice provides one form of certification and the acceptance of a physician by a hospital provides another. In a market without governmental licensing of physicians, other forms of certification, by medical associations, insurance companies, or the like, would presumably arise, as they have in other professions.

Another solution is a guarantee. If customers believe that they are ignorant about the effects of drugs and that the drug companies are not, they should strongly prefer drugs produced by companies that assume liability for unexpected side effects – and be willing to pay more for the drugs sold by such companies. Under those circumstances, a company's refusal to guarantee its product is tantamount to an admission that it knows something the customers do not. Similarly, if patients believe that physicians vary widely in quality, they may choose to patronize those who voluntarily make themselves legally responsible for their errors.

There are two sides to the question of guarantees – or more generally, of liability. Given that the court system is costly and imperfect, the cost to a physician of being liable for his mistakes – or what a court decides are mistakes – may be larger than the value to his patients of having him so liable. Under a market system the rule is freedom of contract. The legal system establishes a default rule – *caveat emptor* or *caveat venditor* or something in between – and the physician is free to transfer the liability to himself by offering a guarantee or away from himself by requiring patients to waive some or all of their rights to sue in exchange for his treating them. Under such a system the market differential between "guaranteed" and "non-guaranteed" physicians, or between the services of the same physician with or without the guarantee, reflect the cost to the physician of being liable – the cost of additional malpractice insurance, lawyer's fees, damage payments, and the like. The customer is free to decide whether the additional security is worth the additional price. Under our present system liability rules are determined by the courts and waivers are unenforceable. The customer ends up paying for the malpractice insurance whether or not he thinks it is worth the price.

So far I have discussed private solutions to problems of imperfect information. What about governmental solutions? The obvious one is for the government to generate information, leaving the customer free to decide for himself how to make use of it. A familiar example is the labeling of cigarettes; the customer is informed that the government believes they cause cancer, and is free, if he wishes, to reject the conclusion – or to decide that he is willing to pay a price in increased cancer risk in exchange for the pleasure of smoking.

As long as we are dealing with rational – albeit imperfectly informed – consumers, it is better for the government to leave the consumer free to utilize the information it provides as he wishes. Although the government may have superior information about the side effects of a drug or the consequences of smoking, the consumer has superior information about his own values – how much he enjoys smoking, how important side effects are to him, how much he is willing to pay for a superior product or treatment.

In the case of physicians, this is an argument for certification and against licensing. An individual who, knowing that the government believes a practitioner to be insufficiently skilled, still prefers to go to him – perhaps because he is the only one available, is less expensive, or speaks the same language as the customer – is then free to do so. This has the further advantage of allowing the customer to ignore the government's opinion if he concludes that it is probably wrong. If his experience with government generated information is more extensive than his experience with physicians, he may be able to evaluate the former even if not the latter.

If we look not at what the government should do to deal with problems of imperfect information but at what it does do, we find that certification is an uncommon solution for either physicians or drugs. Physicians are licensed, and unlicensed physicians are forbidden to practice. Similarly, although there is some control over the labelling of drugs, the main effect of government intervention is to keep drugs off the market until the FDA has approved them.

One explanation is that present policies assume customers who are not only poorly informed but irrational as well – even when the government tells them what they should do, they refuse to do it. An alternative explanation is that the chief objective of at least some regulation is the welfare not of the patient but of the doctor. Whether or not medical licensing improves the quality of medical care, it surely holds down the number of physicians and so holds up their salaries. There is extensive evidence that the American Medical Association has used medical licensing for this purpose, in some cases supporting requirements, such as U.S. citizenship, that made more sense as ways of restricting entry to the profession than as ways of maintaining quality. And in many cases of licensing – not only of physicians but of barbers, manicurists, egg graders, yacht salesmen, tree surgeons, potato growers, and a host of others – it is clear that the political pressure for licensing came not from the customers but from the profession.

This raises a general issue frequently ignored by those who wish to substitute governmental for private decisions. Even if the government is better informed than the individual, the individual has one great advantage in making decisions about his own welfare – he can be trusted to have his welfare as one of his principal objectives. That is less true of anyone else, including the set of interacting persons called government.

In the case of FDA regulation of drugs, the analysis of what happens and why is less clear than in the case of medical licensing; while it is possible to interpret FDA policy as the enforcement of a cartel agreement on behalf of the producers of existing drugs, it may also be interpreted as a response to public pressure produced by widespread publicity on the hazards of new drugs, in particular due to the Thalidomide case.

Even if the only interest group affecting the legislation is the general public acting on free information (because of rational ignorance, it rarely pays the general public to base its political decisions on anything else), the results may still be undesirable, since free information is often not very accurate. If the FDA licenses a drug that turns out to have disastrous side effects, the result is a front page story and the end of the career of whoever made the decision. If it refuses to license a useful drug, the result is to keep a cure rate from rising – say from 92% to 93%. The total cost may be very large, but it is not very visible, so the FDA may have a strong incentive to be overcautious, possibly with lethal effects.

This point was illustrated some years ago when the FDA put out a press release confessing to mass murder. That was not, I should add, the way the press release was phrased, nor the way in which it was reported. The release announced that the FDA had approved the use of timolol, a beta-blocker, to prevent recurrences of heart attacks; it was estimated that its use would save between seven and ten thousand lives a year.[8] Since beta-blockers were already widely used outside of the U.S., the FDA was, in effect, confessing that by preventing their use for over a decade it had killed about a hundred thousand people – a sizable cost, even when compared to the benefit produced by the FDA's decision to keep Thalidomide off the market.

Peltzman's study of the effect of the Kefauver amendments concluded that they had imposed large net costs. So far as I know, nobody has yet done a comparable study attempting to estimate the net effect, in either dollars or lives, of FDA regulations restricting the introduction of potentially dangerous drugs. The case of beta-blockers suggests that it is not obvious whether there has been a net gain; absent a complete study, it is hard to say much more than that.

### *Asymmetric Information*

Asymmetric information involves a more subtle kind of problem, and different solutions, than imperfect information. It involves situations in which one party to a transaction has information that the other lacks, and there is no (convincing) way to share the information with the other party. A standard example is the used care market [1]. Owners of used cars have information about their quality, based on experience, that potential buyers cannot reproduce. The seller can tell the buyer that a particular car is a cream puff rather than a lemon – but since the seller has an incentive to say that whether or not it is true, the buyer has no reason to believe it.

While this is unfortunate for the buyer who ends up with a lemon, why does it lead to an inefficient outcome? The problem is that a car may fail to be sold even though it is worth more to a potential buyer than it is to its present owner. The buyer, who cannot distinguish good cars from bad, makes an offer based on the average quality of used cars being sold. The seller accepts or rejects the offer based on the actual quality of the particular car, which he knows. In the case of a lemon, the buyer is paying for a better car than the owner is selling, so he is likely to make an offer the owner will accept. In the case of a cream puff the reverse is true; the buyer's offer assumes the car to be average, the seller knows it is better than average, so he may be unwilling to sell at any price but the buyer is willing to offer. The average car sold is then worse than the average care offered for sale, since below average cars are more likely to be sold than above average ones.

That fact further depresses the amount the buyer is willing to offer – the relevant average for him is the average of all cars sold, not of all cars offered for sale. The acceptance of his offer will be evidence that the car is below average. In extreme cases, this process may proceed so far that only one car is sold – the worst car on the market, sold at a price correctly reflecting its quality. In more realistic cases, there is a partial market failure – some of the better quality cars fail to be sold even though they are worth more to the potential buyer than to the seller, because the potential buyer's offer does not take account of their actual quality.

In the context of health insurance, the same problem is called adverse selection. Individuals know more about their own health, past and future, than insurance companies can learn. An individual's knowledge of how dangerously he drives, how well he takes care of himself, what medical problems he has been having that he has not yet reported, how willing he is, if he does get sick, to sit in the hospital for as long as possible at the insurance company's expense, and the like, is relevant to how much it costs to insure him and is inaccessible to the insurance company. With regard to such information the company must treat each customer as an average over the population of individuals buying insurance, and set its rates accordingly. This makes insurance more attractive for the customer who knows he is a bad risk – more precisely, a worse risk than the insurance company thinks he is – and less attractive for the customer who knows he is a better risk than the insurance company thinks. So the people choosing to buy insurance represent a biased sample of the population as a whole – a sample biased towards the bad risks. The insurance company will allow for this in setting its rates. That makes insurance even less attractive to the good risks, reducing even further the number of good risks who buy it.

One market solution is a group policy. If an insurance company insures all the employees of a firm together, the sample of insured individuals is only slightly biased towards bad risks, since the existence of the insurance is only a minor factor in determining who chooses to work for that company. This is an imperfect solution, since it means that some people get insured who would not want insurance at a price that correctly reflects their medical circumstances – individuals whose risk aversion is not sufficient to make up for the administrative cost of providing the insurance. It is also imperfect because it applies only to members of suitable groups.

The obvious governmental solution is a group policy for the entire population – national health insurance. This eliminates one of the imperfections of the private solution – its limited applicability. If provided by a bureaucrat god perfectly informed about the appropriate level of coverage and all the associated administrative details, and suitably tailored to the requirements of different customers, it would be more efficient than the private alternative.

If we consider the governmental solution under more realistic assumptions, the case for it becomes far weaker. National health insurance affects many people other than patients – and some, such as physicians and hospitals, are more concentrated and better organized. The arguments of Part II of this paper suggests that the public good problems associated with providing "the right amount of health insurance in the right way" may be much greater than the advantages gained by eliminating adverse selection, especially given that a good deal of adverse selection can be eliminated privately by group plans.

In addition to adverse selection, insurance in general and health insurance in particular involve a second efficiency problem. While it is due to externalities rather than to asymmetric information, this seems the most convenient place to discuss it. The problem is called moral hazard.

Consider an individual deciding whether to stay an extra day in the hospital. The cost of doing so is $200. The value to him, in terms of a slight reduction in the chance of a relapse, is $50. If he is paying his own bills, he goes home. If the insurance company is paying more than 3/4 of the cost, he stays. Since he is buying something whose cost is more than its value, the outcome is inefficient. Put differently, his decision imposes an external cost on the insurance company; since he ignores that cost in his decision he may make an inefficient decision. The same problem occurs whenever the individual either makes decisions affecting benefits he receives and the insurance company pays for, or makes decisions affecting costs he pays and benefits that go, at least in part, to the insurance company. An example of the latter would be an expense – a medical exam not covered by the insurance, or an improvement in nutrition – designed to reduce future medical problems.

There are several ways in which the costs of moral hazard can be reduced. One is *coinsurance* – if the insured is responsible for part of the bill, he has at least some incentive to keep it down. Another is for the insurance company to make some of the decisions – to pay for only the number of days in hospital that its doctor recommends, for example.

Moral hazard applies to government health insurance just as it does to private health insurance – indeed, it applies to any government program that transfers some of the cost (or benefit) of individual decisions from the individual concerned to others. The advantage of the private system is that insurance will occur only if the gain due to risk sharing (or other advantages) at least balances the cost imposed by moral hazard – otherwise the insurance company will find that there is no price it can charge at which it can both cover its costs and sell insurance. There seems to be no comparable constraint on the political alternatives. In this case the market outcome is as good as the best possible political outcome, and better than any outcome that we would expect the political system to produce.

### *Imperfect Competition*

One of the assumptions that is usually used in proving the efficiency of the market outcome is perfect competition – every consumer and producer is assumed to be a small enough part of the market so that the amount he consumes or produces does not affect the price. To see why this matters, consider the situation of a producer who knows that the higher his rate of output the lower the market price will be. If he produces at a rate of 101 units a month instead of 100 units a month, his revenue will rise by the price of one unit and fall by 100 units times the amount of the resulting price drop. His marginal revenue – the change in revenue due to one more unit of output – will be the sum of the two changes. Since his objective is to maximize profit – revenue minus cost – he will produce up to the point where marginal revenue equals marginal cost. But it is price, not marginal revenue, that measures the value to a consumer of one more unit. Units for which the cost of production is more than marginal cost but less than price will not get produced; this is inefficient, since it means that there are units not produced that would be worth more than it would cost to produce them. Similar problems arise if the assumption of perfect competition is violated for consumers.

Much of the medical market – most obviously the services of general practitioners in large cities – has the characteristics necessary for an almost perfectly competitive market. The main hindrances to competition on that part of the market are the result of government interference – the prohibition on advertising the price of medical services and the restriction on entry to the profession, both enforced by state regulation of who can practice medicine. The same applies to the retailing of medicine; advertising of the prices of prescription drugs has frequently been illegal.

There is an important point here that I have made before and will make again. It is not sufficient for the supporters of intervention in the market to say "we oppose those particular interventions – what we are in favor of are only the interventions that increase efficiency." Unless they have some solid theoretical or empirical grounds for claiming that they know how to achieve that objective, we must take existing regulation as evidence of what government does – and will do – given the opportunity. Since regulation has often consisted of the use of government power to drive up the price of some good or service for the benefit of the producer, we must treat that fact not as a repeated accident but as evidence. That is particularly true if the evidence fits such theory as we have – and it does. Producers are typically a more concentrated interest than consumers.

What about imperfect competition resulting from economies of scale in the medical industry? Examples are drug research, hospitals, and physicians in sparsely populated areas. In such cases there exist policies that a bureaucrat god could follow that would produce improvements. In order to follow the optimal policies, however, the real world regulator must know the cost curves of the regulated firms – how much it costs to produce any level of output, where level includes quality as well as quantity – and the demand curves of consumers. This is difficult in any industry, and there is a good deal of evidence that regulation of monopolies does not and perhaps cannot force them to charge efficient prices [23]. It is particularly difficult in an industry such as medicine, where quality variables are important and hard to measure. Regulation has the additional disadvantage of providing the industry with a tool that may be used to maintain a monopoly position that would otherwise be eliminated by technological change; a classic example is the regulation of canal traffic and trucking by the ICC in order to defend the railroads from competition.

My own conclusion, considering both the theoretical arguments and historical experience, is that regulation of monopolies may never be desirable, and certainly is not in cases that fall substantially short of complete and very long-lived monopoly. The question is one on which there is a long literature, and many economists would take a less extreme position.

*Externalities / Public Goods*

The final category of market imperfection that I will discuss is the category of externalities and public goods. I combine them because they are, to a considerable extent, two ways of looking at the same problem.

A public good is a good that, if produced, will be available to all the members of a preexisting group: the producer cannot control who gets it. A pure public good is a public good for which consumption by one individual does not interfere with consumption by another. The fact that a good happpens to be produced by government does not make it a public good – postal service, for example, is a private good that happens to be produced by government – nor does the fact that something is a public good mean that it cannot be produced privately. Radio and television broadcasts are pure public goods, yet both are produced privately. The fact that something is a public good does, however, pose a problem for the producer who wants to be paid for what he produces – if he cannot control who gets it, how can he make consumers pay him? This in turn implies a problem from the standpoint of economic efficiency. If a good is worth more to the consumers than it costs to produce, it should be produced; but it will be produced only if the amount the producer can get paid for producing it is at least as great as his cost of production. Since the amount that can be raised to pay for a public good is generally much less than its value to the consumers, there will be public goods worth producing that do not get produced – which is inefficient.

One solution to this problem is to have the good produced by government. This solution, as I pointed out in Part II, raises a second public good problem. Production of good law – or bad law for that matter – is a public good from the standpoint of the group benefitted by the law, so getting the government to do things requires that someone solve a public good problem.

There are also a number of private solutions to the public good problem. One is charity. Another is for the producer to organize a contract among those who will benefit from the public good by which each agrees to contribute only if the others do. Each member knows that either his signature has no effect (if someone else refuses to sign) or it gives him the package "good minus payment" (if everyone else signs). As long as the specified payment is less than the value of the good to the consumer, he signs and the good gets produced.

This solution runs into problems in a world of imperfect information and non-zero transaction costs, especially if the public is large. It pays each individual to pretend the good is of little value to him, or to claim that even though he values it he is too stubborn to agree to pay for it. If he succeeds in convincing the entrepreneur drawing up the contract, his name will be omitted from the list and he will get the public good without having to pay for it. Such bargaining problems imply that for any save a very small public, the amount that can be raised to pay for a public good is a small fraction of its total value, and that the fraction gets smaller as the public gets larger. So methods of this sort can be used only for small publics or for goods whose cost is small compared to their value.

There are other ways of producing public goods, including the ingenious solution employed by the broadcast media, but a discussion of all of them would exceed the bounds of this chapter. The important results are that public goods can be produced privately, that the outcome of private production is not in general efficient (some goods that are worth producing do not get produced), and that the problems tend to increase with the size of the group.

An externality is a cost (or benefit) that one individual's actions impose on another. A public good can be described as a positive externality, and the prevention of a negative externality is a public good. As a rule, the term "public good" gets used to describe situations where the purpose of the action is producing the good (radio broadcast) and the term "externality" to describe situations where the action has a separate purpose and the injury or benefit is a side effect. The line between the two terms is a blurred one, and the decisions of whether to describe a particular problem in terms of externalities or public goods is to some degree arbitrary.

The reason that externalities lead to inefficient outcomes is straightforward. An individual producer produces a good if and only if the benefit he receives (by selling or consuming it) is at least as great as the cost he pays. As long as he pays all of the associated costs and receives all the associated benefits, this is equivalent to the efficient rule – "produce if and only if total benefits are at least equal to total costs". In the case of good produced without externalities on a competitive market, the condition is satisfied; the price he receives equals the marginal value of the good to the consumer, the prices he pays for his inputs equal their marginal cost of production, so cost and revenue on his balance sheets equal the "total social cost" and "total social benefit" produced by his actions.

With externalities, this is no longer true. Since the producer pays only part of the cost (or, in the case of positive externalities, receives only part of the benefit), the revenue and cost figures that determine his decision of what and how much to produce no longer equal the corresponding total values and total costs. He may choose to produce a good even though total cost (including the external cost) is greater than total benefit, or choose not to produce even though total benefit (including the external benefit) is greater than total cost. The outcome is no longer in general efficient.

How does all of this apply to the market for medical care? As on most markets, some of the goods impose externalities – in the case of medical care, typically positive ones. If I get inoculated against a contagious disease, that reduces the chance that I will infect you – a positive externality. If my drug company discovers a new family of drugs, that provides information useful to other companies. If I spend money on keeping myself healthy, that benefits all those who care for me and would be made unhappy by my illness; it may even benefit people who have never met me, but feel a glow of pleasant pride at knowing that "my country is the healthiest in the world" or "year by year the human race is getting healthier". All of these are properly counted as externalities. Economic theory suggests that the market will underproduce inoculations, drug research, and health because in each case the individual paying the cost receives only part of the benefit. Similarly, if the use of some antibiotics imposes external costs by encouraging the development of resistant strains of bacteria, such antibiotics will be overused, since the individual using them pays only part of the cost.

All of these, however, are what I earlier described as mostly private (or at least, largely private) goods. In each case a large part of the benefit goes to the person who pays for it – I stay healthy because of my inoculation, the drug company makes money off its new drugs, and my own health probably gives more pleasure to me than to even the most altruistic of my friends. If ninety percent of the benefit goes to me, then I will make the wrong decision only in those cases where the cost is more than ninety and less than a hundred percent of the value, so the result, although inefficient, will not be very inefficient.

I argued in Part II that the political market is usually very inefficient, since the political pressures for and against legislation (and other government activities) represent the values to the groups affected, weighted by their widely varying ability to exert political influence in defense of their interests. My conclusion was that the substitution of the political for the private market was justified, if at all, only in cases of extreme market failure on the private market. None of the cases I have described involve such extreme failure. I conclude that, although the outcome of the market could be improved by a bureaucrat god, it is likely to be worsened by real world regulation.

I believe I have now covered all of the obvious causes of market failure on the market for health care. In most cases the failure implies that a sufficiently wise, powerful, and benevolent authority could improve – in terms of economic efficiency – the outcome of the market. In no case is there any clear reason to believe that assigning additional power to government – as government actually exists – would improve the situation; in many there is reason to believe that doing so would make it worse. In several cases existing problems are the direct result of government interference with the market.

## SUMMARY

In the course of this essay, I have attempted to make plausible a thesis many readers will find absurd – that health care should be provided entirely on the private market, just as shoes and potato chips are now provided. Obviously I have not, and cannot, answer all imaginable arguments against doing so; I have tried to answer the ones I find most persuasive.

One argument that I have not yet answered – and find very unpersuasive – is the claim that "health is too important to be left to the market". My response would be that the market is, generally speaking, the best set of institutions we know of for producing and distributing things. The more important a good is, the stronger the argument for having it produced by the market.

Both barbers and physicians are licensed; both professions have for decades used licensing to keep their numbers down and their salaries up. Government regulation of barbers make haircuts more expensive; one result, presumably, is that we have fewer haircuts and longer hair. Government regulation of physicians makes medical care more expensive; one result, presumably, is that we have less medical care and shorter lives. Given the choice of deregulating one profession or the other, I would choose the physicians.

I suggest to those readers who remain entirely unconvinced that they may be making the error of judging a system by the comparison between its outcome and the best outcome that can be described, rather than judging it by a comparison between its outcome and the outcome that would actually be produced by the best alternative system available. If, as seems likely, all possible sets of institutions fall short of producing perfect outcomes, then a policy of comparing observed outcomes to ideal ones will reject any existing system.

It is easy, and satisfying, to pick some unattractive outcome – a poor man, actual or imaginary, turned away from the expensive private hospital that could have cured his disease – and describe it as "intolerable", "unacceptable", or some similar epithet designed to prevent further discussion. This is, however, a game that any number can play. It is equally easy, as I demonstrated earlier in this essay, for the defender of the market to orate about the hundred thousand people who died of heart attacks because the FDA refused to permit American physicians to prescribe beta blockers to American patients. In a large and complicated society, it is likely that any system for producing and allocating medical care – or doing anything else difficult and important – will sometimes produce outcomes that can plausibly be labeled as intolerable.

The question we should ask, and try to answer, is not what outcome would be ideal but what outcome we can expect from each of various alternative sets of institutions, and which, from that limited set of alternatives, we prefer. I have tried to do so. My conclusion is that there is no good reason to expect governmental involvement in the medical market, either the extensive involvement that now exists or the still more extensive involvement that many advocate, to produce desirable results.[9]

*University of Chicago Law School*
*Chicago, Illinois, U.S.A.*

NOTES

[1] While this statement is true under most circumstances, there are some exceptions [13].

[2] This position has sometimes been taken by George Stigler, in criticizing both welfare economics and (other people's versions of) public choice theory.

[3] Stigler has suggested that one reason politicians are so often lawyers is that, since most of the politician's "customers" have legal business to be done, paying the politician's firm above-market rates for its services is a convenient way of concealing payments to the politician for political services. Similar arguments apply to connections between politicians and other widely used firms – banks, for example [22].

[4] This situation may, however, result in majorities organizing to exploit minorities, and produce large organizational costs associated with everyone trying to end up on the winning end of that game. Since I am not presenting a coalition model, I will ignore such problems here.

[5] If the beneficiaries can organize to pay part of the value of the public part of the good, the result is even more favorable; if, for example, the indirect beneficiaries of inoculation each pay half the value of their benefit, I will get inoculated as long as benefit is at least 4/3 times cost.

[6] This argument depends on the medical expenses being uninsurable; otherwise exactly the same utilitarian argument, applied to an individual, will result in his insuring against them. Expenses may be uninsurable because the outcome can be observed before the potential victim can buy insurance. Someone who is born blind cannot insure against blindness.

[7] For example, a study by the Institute of Economic Affairs found that British National Health Insurance provided higher income clients more for their money than lower income clients; the implicit redistribution was up, not down, the income ladder.

[8] The estimate of 7,000–10,000 lives a year was by Arthur Hayes, *HHS News*, Nov. 25, 1981. Prior to 1981 another beta blocker, propranolol, had been approved (and widely used) for other purposes, but not for preventing heart attacks.

[9] This essay originated as a comment on a manuscript version of Buchanan's essay in this volume. My debt to my target, and to the organizers of the conference at which both papers were given, will be apparent to anyone familiar with Buchanan's work.

## BIBLIOGRAPHY

1. Akerlof, G.: 1970, 'The Market for "Lemons"', *The Quarterly Journal of Economics* **70**, 480–500.
2. Becker, G.: 1958, 'Competition and Democracy', *The Journal of Law & Economics* **1**, 105–109.
3. Buchanan, A.: 1985, *Ethics, Efficiency and the Market* Roman & Allenheld, Lanham, Maryland.
4. Buchanan, J.M. and Tullock, G.: 1962, *The Calculus of Consent*, University of Michigan Press, Ann Arbor.
5. Coase, R.: 1960, 'The Problem of Social Cost', *The Journal of Law & Economics* **3**, 1–5.
6. Downs, A.: 1957, *An Economic Theory of Democracy*, Harper & Row, New York 1957.
7. Ecker, R.D. and Hilton, J.W.: 1972, 'The Jitneys', *The Journal of Law & Economics* **15**, 293–325.
8. Friedman, D.: 1989, *The Machinery of Freedom*, 2nd edition, Open Court, Peru, Illinois.
9. Friedman, D.: 1980 'Many, Few, One-Social Harmony and the Shrunken Choice Set', *The American Economic Review* **10**, 225–232.
10. Friedman, D.: 1980 'What does Optimum Population Mean?', in J.L. Simon (ed.), *Research in Population Economics*, Vol. III, Jai Press, Greenwich, Connecticut.
11. Friedman, D.: 1982, 'What is Fair Compensation for Death or Injury?', *International Review of Law and Economics* **2**, 81–94.
12. Friedman, D.: 1990, *Price Theory: An Intermediate Text*, 2nd edition, South Western Publishers, Cincinnati.
13. Friedman, D.: 1988, 'Does Altruism Produce Efficient Outcomes? Marshall vs Kaldor', *Journal of Legal Studies* **17**, 1–13.
14. Linneman, P.: 1982, 'The Economic Impacts of Minimum Wage Laws', *Journal of Political Economy* **90**, 443–469.
15. Niskanen, W.A.: 1971, *Bureaucracy and Representative Government*, Aldine-Atherton, Chicago.
16. Olson, M.: 1965, *The Logic of Collective Action*, Harvard University Press, Cambridge.
17. Peltzman, S.: 1973, 'An Evaluation of Consumer Protection Legislation: 1962 Drug Amendments', *Journal of Political Economy* **81**, 1049–1091.
18. Peltzman, S.: 1975, 'The Effects of Automobile Safety Regulations', *Journal of Political Economy* **83**, 677–725.
19. Peltzman, S.: 1976, 'Toward a More General Theory of Regulation', *The Journal of Law & Economics* **19**, 211–240.
20. Posner, R.: 1974, 'Theories of Economic Regulation', *Bell Journal of Economics* **5**, 335–358.
21. Stigler, G.: 1970, 'Director's Law of Public Income Redistribution', *The Journal of Law & Economics* **13**, 1–10.

22. Stigler, G.: 1975, *The Citizen and the State*, University of Chicago Press, Chicago.
23. Stigler, G. and Friedland, C.: 1962, 'What Can Regulators Regulate?', *The Journal of Law & Economics* **5**, 1–16.

KLAUS HARTMANN

# THE PROFIT MOTIVE IN KANT AND HEGEL

## I. INTRODUCTION

The topic addressed in this paper may seem academic and far-fetched; its relevance though becomes apparent once we focus on the changes taking place on the medical scene today. Hitherto, medical service has been subject to several ethical considerations. One is the view that physicians have a duty to help their patients (in the sense of being ready to provide service and to provide competent service).[1] Another ethical consideration is the presumed duty on the part of third parties to ensure medical standards and the readiness on the part of physicians to provide service – such third parties normally being medical associations enjoining an ethos among their members, or the state enjoining a certain behaviour on the part of physicians, either directly or, more likely, indirectly via an authorization of the medical associations as ethos keepers. A final ethical consideration pinpoints just allocation of, and access to, treatment.

The changes alluded to concern a development in the course of which medical service will more and more become a commodity on the market. Hospitals will turn more and more commercial, insurance companies will employ physicians to provide service, while competition between hospitals, insurance companies, and physicians serving the various insurance companies will take care of the issues of standards and readiness to serve, at the price, though, of preferential treatment of well-to-do patients. The problem thus is whether the profit-motive on the part of doctors, insurance companies, and hospitals – and/or the market principle – can make the said ethical considerations redundant, or whether, to the extent that the effect of the profit and market principles equals that of ethically enjoined and publicly financed or subsidized service – however unlikely that may be in the matter of just allocation –, ethics can even endorse such a change-over to complete commercialization. Clearly, a philosophical issue is involved, and so one might wonder what philosophers have to say in the matter. Now it seems to me that attention to Kant and Hegel, two major philosophers dealing with ethical questions, may be particularly rewarding. In reviewing them, I will not



*T.J. Bole III and W.B. Bondeson (eds.), Rights to Health Care*, 307–325.

limit myself to an expository statement in terms of the history of philosophy but shall try to keep the ulterior modern problems in mind.

## II. WHAT NOT TO EXPECT

The current development in the medical profession and the medical service, at least in the United States is, of course, not fully covered by a phrase like 'ascendancy of the profit motive'. But in any case, certain implications of commercial medicine will have to be addressed. The physician employed by a commercial hospital and/or private insurance company may be under pressure to work faster or harder in order to maintain himself in his position or help maintain his hospital or insurance company in the market, so that high-powered working schedules might be the order of the day. The self-employed physician or the public hospital may be less subject to stress of this kind. Also, the market principle will only gradually afford transparency as to which agency provides excellent and which less excellent service, while in the meantime fatal consequences may have ensued. Once the running-in lag is a matter of the past, dependability may be satisfactory, or doubts may remain, as they do in the economy at large.

But however that may be: although such questions playing over into micro- and macro-economics are subject to ethical reflection, we cannot expect our two philosophers to have spoken out on them (with a possible qualification in the case of Hegel). So all that seems possible here is to address the issues at hand on the level of abstraction set by the theories concerned.

## III. KANT

### *A. Preliminary Orientation*

Trying to present an analysis of what emerges from Kant for medical ethics, we may first concentrate on the second formula of the categorical imperative: "So act as to treat humanity, whether in thine own person or in that of any other, in every case as an end withal, never as means only ([12], p. 46). If persons are a self-purpose it would seem that whoever could help a person maintain personhood, by saving his or her life, e.g., ought to do so. Now since physicians are singularly qualified to do this, they seem to be called upon to render such service.

The phrase "never as means only" would suggest that another person may in some way be used as a means (let us say, for profit-making), but not exclusively so. Thus Kant would seem to endorse the physician's duty to come to the rescue of another person while a side-effect could be that he makes money that way.[2] But such a side-effect would have to pass for a motive spoiling the otherwise moral action; the action would not be one done 'from duty'. The physician's action would have no 'moral value'.

> I omit here all actions which are already recognized as inconsistent with duty, although they may be useful for this or that purpose, for with these the question whether they are done *from duty* cannot arise at all, since they even conflict with it. I also set aside those actions which really conform to duty, but to which men have *no* direct *inclination* ... For in this case we can readily distinguish whether the action which agress with duty is done *from duty* or from a selfish view. It is much harder to make this distinction when the action accords with duty, and the subject has besides a *direct* inclination to it ([12], p. 15).
>
> But I maintain that in such a case an action of this kind, however proper, however amiable it may be, has nevertheless no true moral worth, but is on a level with other inclinations.[3]

It does not follow, though, that the physician's action is morally or, more correctly, ethically indifferent: although not in Kant's sense of 'moral value', accruing to an action done from duty, its ethical effect would still be something appreciated and valued; it seems better that the physician, from whatever motive, including the profit motive, treat his patient. That this is so also for Kant, emerges from one of his illustrations. Even if a shopkeeper acts from selfish motives, honesty may be the beneficial outcome:

> For example, it is always a matter of duty that a dealer should not overcharge an inexperienced purchaser; and wherever there is much commerce the prudent tradesman does not overcharge, but keeps a fixed price for everyone, so that a child buys of him as well as any other. Men are thus *honestly* served...[4].

The ethical effect of a justice constraint, however heeded for selfish rather than moral reasons in the Kantian sense, is certainly ethically positive. Kant fails to theorize the moral valuation of ethical effects, but there can be no doubt that he admitted the ethical value of certain effects. On our analysis, the physician coming to the rescue with a mixed motivation, including profit, seems to be justified; the effect achieved, or aimed at, must be appreciated.

### *B. The Division of Duties*

So far we have culled from Kant very selectively and, perhaps, misleadingly. We have rested our case on the assumption that the second formula of the categorical imperative, enjoining respect for persons, covers actions necessary to preserve a person's life. But when we inspect the illustrations Kant gives for his various formulas of the categorical imperative ([12], pp. 38–41, 45f), we note two things. First, that his examples of actions *vis-à-vis* another are abstention from making a false promise and beneficence (promotion of another's welfare), and not coming to his or her rescue. Nor do the first and third formulas, expressing the universalizability rule, seem to prompt the injunction to rescue another person's life because he or she is a self-purpose. Kant's injunction against suicide, condemning an action *vis-à-vis* oneself, finds no counterpart in an injunction to ward off the death of another person. We note, second, that in his illustrations, Kant introduces – contrary to our simplistic interpretation of the second formula as enjoining the rescue of another person – the distinction of 'perfect' and 'imperfect' duties, or 'narrow' and 'wide' duties.[5] The examples come in pairs: perfect duties over against myself (prohibition of suicide) or over against another (prohibition of false promises), on the one hand, and imperfect duties over against myself (promotion of one's own talent) or over against another (promotion of the other's welfare, beneficence) ([12], pp. 39–41, 46f.; [11], pp. 84f., 92ff., 110, 120). Thus we have to distinguish between meritorious or, as John Findlay has it, 'hortatory' duties whose non-fulfilment is morally of zero value,[6] and necessary or, as Findlay says, 'minatory' ones.[7] Accordingly, the constraint in the former case is much weaker. (It may not be unfair to say that Kant has not provided a principle for the minatory/hortatory division in his ethics, except in terms of distinctions like necessary vs. optional, or rule-governed vs. purpose-oriented.)

### *C. Respect for Persons and Beneficence*

One difficulty in the present context is that on the above interpretation (in our preliminary orientation), saving a person's life would be a perfect duty, although not a minatory or negative one (as the ban on suicide in fact is). Qua positive, rescuing a person's life should be beneficence only; however, qua necessary, directly resulting from the respect for persons, it should count as something non-optional, indeed as something categorical (optional beneficence notwithstanding).

This impasse is reflected in a recent critique of Kant's position. In *The Foundations of Bioethics*, Engelhardt takes Kant to task for conflating freedom as a value with freedom as a side-constraint – or freedom as a practical reason for acting on a person's behalf and freedom as a practical reason for respecting a person – suggesting that this move commits Kant to a "*particular* ethic":

> One is doing more than elaborating and justifying the fabric of morality itself by valuing freedom. One can consistently treat all persons as ends in themselves while affirming as a moral maxim that individuals may freely decide when to cease to be free by choosing suicide or a term in the French Foreign Legion. Respect of freedom as the necessary condition for the very possibility of mutual respect and of a language of blame and praise is not dependent on any particular value, or ranking of goods, but requires only an interest in resolving issues without recourse to force. When one has distinguished between freedom as a condition for morality, and freedom as a value, one loses the basis for duties to oneself as well. ([12], p. 69).

Conceivably, though, the argument from particularity, ranking freedom as a particular good, is not compelling, if only because Engelhardt's counterrepresentations can be seen to involve Kant's valuation of freedom as a premiss: if morality is committed to mutual respect, to resolving issues without recourse to force, the reason for this must be a Kantian valuation of freedom. The difference between Kant and Engelhardt is due to Engelhardt's readiness to include the negation of freedom in a given case among the things to be respected, whereas Kant maintains the need for consistency between principial (rational) freedom and empirical (existential) freedom (which may want to negate principial freedom in the case of a suicide wish).[8]

From the above it would appear that – in order to return to our ulterior context – a physician is obliged to respect the other in a positive manner by taking him and, by implication, his life as a self-purpose morally commanding his service where the survival or the meaningful exercise of such self-purpose is at stake.[9] Failure to act in such a way would be morally evil (a mixed motivation notwithstanding). Saving another's life, although not a negative or minatory injunction, is a strict injunction.

The trouble is, though, that to say this is to go against the grain of Kant's theory. If strict or necessary duties are negative ones (as *The Doctrine of Virtue* has it), or minatory ones (as Findlay has it), then the positive commitment to another person's life is at loggerheads with the architectonic of the theory. We could put it this way: there is no symmetry between the self-regarding duty not to commit suicide and the other-regarding duty to rescue others. Instead of the latter, Kant, as we noted, lists abstention from false promises as *the* necessary (narrow, perfect) other-regarding duty. There is even a comparative ranking of duties such that truth-telling is held to be superior to saving a person's life by telling a lie.[10]

The difficulty with the division and ranking of duties between self-regarding and other-regarding, as well as between perfect and imperfect duties – a division on which so much hinges for an assessment of how much a person's life counts in Kantian ethics – may be related to the two types of categorical imperatives, the type of the first and third formulas and the type of the second formula. As the first and third formulas of the categorical imperative suggest, Kant is predominantly concerned with moral reflection in terms of universality. These formulas are categorical in making universalizability the criterion for the approval of my maxim: whoever qualifies for personhood has a claim on me to be treated as a rational will. Could I then want to tell a lie and so contradict my rational will? Could I then want to deceive a person and so deny him or her the respect due to theirs being a rational will? The second formula, on the other hand, is categorical in its insistence on the value inherent in persons in a sense in which also their existence is valued. This tension between truth-oriented and existence-oriented versions of the categorical imperative may help explain that Kant has conflicting options for what can count as highest categorical duties: the saving of a person's life or the abstention from euthanasia in the case of a suicide wish, on the one hand, and truth-telling on the other. Much in Kant would lend itself to a view which, like Engelhardt's, tends to reduce the second formula to a side-constraint. Kant may be closer to Engelhardt than Engelhardt thinks, or at least less clear than we tried to make him appear initially.

### *D. The Division of Duties Re-examined*

Another difficulty is not so much bound up with internal inconsistencies of Kant's ethics but prompted by its medical application. Although above we tried to extend Kantian examples to cover medical issues, we cannot have failed to notice that medical examples and applications are singularly absent from Kant's text. This lacuna leads to critical questioning as to the applicability of Kant's ethics to medicine. In our analysis so far, we discussed a categorical duty to save persons' lives (or prevent suicide or euthanasia) – this against Engelhardt – and the meritorious duty of beneficence. Can this division of duties accommodate the full range of ethical issues raised by medicine? We think not. It is easy to see that medicine is not predominantly concerned with welfare in the sense of positive improvement of bodily incarnate persons or with cultivating their bodily potentialities (e.g., in the case of a flabby office sitter egged on to a sportsman-like physique). Rather, medicine is mainly engaged in curative measures, lessening or removing pain, redressing disadvantages impairing a person's life.[11] Now it would not seem acceptable to say of such activities that failure to engage in them is 'morally zero' as we would have to say if we followed Kant's definition of meritorious duties (albeit that Kant has a rider that we must not make such failure our policy). So besides a (controversial) duty to save life and a meritorious duty of beneficence we seem to need a third type of duty urging whoever is able or competent to offer his curative help to do so.[12] The negative of such a duty, failure to render assistance, should count as morally blameworthy. The rest is a terminological matter: whether to split up beneficence into two ethically distinct compartments, depending on the assessment of omissions, or whether to find a name fitting the said third type of duty. One could call such duties duties of assistance or of solidarity.

It is obvious that on our view, subsumption of medical action under categories of duty involves problems in view of a continuity of phenomena, or of gradualism: is the prolonging of life not some kind of rescue? is removal of pain not something quite close? etc. But this difficulty has to be countenanced; the fact of the matter is that there is such a continuum.

Our proposal will, if it is convincing, have considerable influence on one's assessment of Kant's ethics, generally, and on such assessment in connection with medicine; it will also impinge on Engelhardt's rationalization of medical ethics, or bioethics. But we need not go into this here.

### *E. The Impact of the Profit Motive*

We seem to have engaged in a digression not essential to the topic of the profit motive. Or can it be shown that the profit motive and, historically, its ascendancy make a difference to the duty scene, and that Kant's views on duty, to the extent that they pass critical examination, make a difference to the acceptability of the profit motive in medicine? I feel the case can be made.

That the profit motive does not exclude positive ethical effects, even in Kant, has already been noted. The more difficult problem is how its acceptance as a motive – as one among others like duty, or as a sufficient motive? – affects the claims to be made on physicians (in the sense of 'appealing' to their duty) and the assessment of their actions and omissions in terms of the various duty categories. The profit motive as elemental (whether sufficient or concurrent) to a physician's service would bid him respond to requests for service. (In fact, the term 'service', previously used, lapses in favor of 'tender of service'.) Now requests and tenders are neutral over against the respect of persons on a strong reading, beneficence on a weak reading, and duties of assistance or solidarity. The physician will enlist customers, some of whom want to have their life saved, or prolonged, or alleviated, put back in shape, or terminated by way of euthanasia. The profit motive on the part of the physician will thus tendentially (but for extant ethos rulings) cut across, level down, and indeed ignore, the distinctions between categorical and meritorious duties and (if this be accepted) duties of assistance or solidarity. The categorical nature of the imperative (positively) to rescue a life and (negatively) to prevent its termination through euthanasia would certainly lose its paramount importance. But also issues of curative assistance arise: will the physician be ready to make a night-call? Will the night-call not be ruled out because it is beyond the patient's means? There seems in fact to exist an affinity between medical tenders determined by the profit motive and a beneficence reading of medical practice in quasi-Kantian terms: when I fail to tender beneficence, the moral value is zero. What for Kant, though, was still a matter of duty – albeit that he, too, left out the solidarity aspect of duty – turns into a matter of societal arrangement, with prices determining the readiness of the physician concerned (unless an ethos ruling reminds him of the duty aspect). To expect no more than beneficence in the form of tenders means to favor a different view than that held by many – Kantians and others – as to what the calling of a physician should be. Engelhardt's side-constraint vs. beneficence view of Kant dovetails nicely; indeed one might say that his *Bioethics* is a philosophical justification of a changed attitude to the profit motive in medicine).

The impact of the profit motive might be briefly pursued to the point of fully commercialized medicine as envisaged above. In a way, this latter development is a consistent follow-up of the ascendancy of the profit motive. What matters now is how the individual addressee of ethics, so far in the center of our attention, is affected when the market principles give rise to medical companies covering the medical market. Employment by a medical company (for hospitals, for general practitioners, specialists, etc.) would seem to exonerate the individual physician (on pain of disadvantages, though) in the sense that he need not regard himself as the addressee of an ethical imperative as long as others are likely to step in. He is replaceable, and his responsibility is one at one remove. I submit, though, that the patient would want to deal with the person having the responsibility, anticipating his living up to his duty, while under the commercial scheme such expectation gets transformed into a possible liability claim after the fact against a commercial company.[13]

But then again, matters could be turned around in favor of an economic perspective: while duty in Kant could be seen as a non-economic device to ensure performance ([14], pp. 47–50), commercialism and the market might prove a perfectly satisfactory vehicle to ensure performance without regard to duty. Even medical standards might be assured by the market principle through the competition of physicians in the interest of profit: they might have an interest in tendering service under the aegis of a higher-ranking insurance or medical company. Only ethical questions concerning allocation or distributive justice would remain unanswered.

Kant would have to say to all that that, morally, the said exoneration does not relieve the physician of his duty. But then again, as a staunch defender of liberalism in political and economic matters, Kant falls back on a market position himself. He endorses, it is true, some measure of public support for the indigent:

> The general Will of the people has united itself into a society in order to maintain itself continually, and for this purpose it has subjected itself to the internal authority of the state in order to support those members of the society who are not able to support themselves ([13], p. 93).

In this connection, Kant also thinks of "widows' homes, hospitals, and so on" ([13], p. 93). But otherwise, leaving destitution aside, he believes in private initiative:

> For a being endowed with freedom is not satisfied with the pleasure of life's comforts which fall to his lot by the act of another (in this case the government); what matters, rather is the principle according to which the individual provides such things for himself ([10], p. 157; German text: *Academy Edition*, Vol. VII, 87, note).

Admittedly, the text does not mention medicine specifically, but the message, directed against the welfare state, has implications for medicine as well. So by extension, Kant's view would agree to the market principle and the profit motive in matters of medical coverage. For him, moral duty and societal fact are two separate compartments. The physician may have a duty (categorical, minatory, or hortatory) to treat a patient, but then again he is left to function in terms of the economic rules of the game. Kant, we remember, opts for what subsequently came to be called a 'formal constitutional state' (*formaler Rechtsstaat*), leaving societal arrangements largely, except for the abovementioned qualifications, to contract law. That the physician will have to put up with a split consciousness did not bother him unduly. Nor would Kant have objected to preferential treatment according to the financial resources of patients. This seems evident from his somewhat unsavory remarks on social strata:

> The woodcutter ...; the smith in India ...; the private tutor, in contrast to the schoolteacher; the sharecropper, in contrast to the farmer; and the like – all are mere underlings of the commonwealth, because they must be under the orders or protection of other individuals. Consequently, they do not possess any civil independence.[14]

But we cannot expect Kant to endorse all this ethically. He may have concluded that where morality does not ensure performance, contract law will, if with less perfection. It is hard to prognosticate in retrospect whether the ascendancy of the profit motive in medicine, and commercialized medicine in general, would have made him pause to think of some other adjustment of ethics and social well-being.

## IV. HEGEL

### *A. Preliminary Orientation*

Hegel's case is very different. The central shift of perspective may be seen in the fact that the concept of duty loses its paramount status in ethical theory. Duty is indeed discussed: its place is a predicament in which the individual wonders what he or she ought to do. But precisely as long as the individual is concerned in isolation, it is hard to determine what he or she should do. (Hegel would not mind endorsing minatory duties such as not to kill oneself or someone else, but that would not take that individual very far as concerns his or her moral orientation.) As long as duty is considered in the context of an individual over against another individual – in first-person language: how should I relate to myself and to a given other in order to make my will a good will – duty remains abstract; the Kantian criterion of universalizability and non-contradiction will not generate, or incline towards, any content. Hegel says,

> From this point of view, no immanent doctrine of duties is possible; of course, material may be brought in from outside and particular duties may be arrived at accordingly, but if the definition of duty is taken to be the absence of contradiction, formal correspondence with itself – which is nothing but abstract indeterminacy stabilized – then no transition is possible to the specification of particular duties ... The absence of property contains in itself just as little contradiction as the non-existence of this or that nation, family, etc., or the death of the whole human race.... But if duty is to be willed simply for duty's sake and not for the sake of some content, it is only a formal identity whose nature is to exclude all content and specification.[15]

Once social content is brought in – family and state are cases in point, what in the English translation is in general called 'ethical life' – duties figure to the extent that the individual is considered an addressee of the claims made on him or her by the family or the state (or by society once this is sufficiently developed to include institutions such as the Administration of Justice). Thus the individual is said to have a duty to marry[16], duties in terms of the law generally[17], or a duty to be member of a state.[18] More or less, duties become counterfactuals having given way to relations of membership in a concrete ethical formation such as family or state, or society, for that matter; contexts in which conformal behavior is presupposed for the functioning of the formation in question. In his early *Propaedeutik*, Hegel even speaks of 'family duties' *(Familienpflichten)* and 'political duties' (*Staatspflichten*) ([9], secs. 49–58), implying the assumption that, contrary to fact, it might still be an open question what an individual ought to do. Interestingly, he does not mention any societal duties, the idea being that in society, which is modeled on economic competition, it is not a matter of duty to play one's role; one's function in society comes naturally, out of self-interest, just as universality, too, comes naturally ([6], sec. 187).

Clearly, the counterfactual can always resurface as a fact when an individual reflects on what to do, or experiences 'collisions' of duties in which an overriding self-willed attitude, or various types of loyalties or memberships, may loom large. Thus Antigone may be said to have considered it her duty to bury her brother against the city interest represented by Creon.

*B. A Kantian Analogue?*

We normally wish to see persons respected; Kant's second formula of the categorical imperative may serve as a philosophical expression of this desire. Is Hegel dodging the issue, or does his philosophy feature a Kantian analogue? Let us pursue this suggestion. Hegel in fact has plenty of opportunity to speak about persons and about mutual recognition of persons. The concept of person figures prominently in the *Philosophy of Right*: 'person' is a crucial concept in 'Abstract Right' ([6], sec. 35), 'person' is once again crucial in 'Morality ([6], sec. 105), and in the Administration of Justice of Civil Society ([6], secs. 209, 218). It is on these various levels that recognition of persons is being practised. (Another strategic example is recognition between master and slave in the *Phenomenology of Spirit*.[19]) And it is on all these levels that we could construe Kantian analogues. But Hegel's point is that only on comparatively concrete levels (society, state) does recognition constitute a condition of membership in an ethical 'substance'. A merely moral form of recognition and respect does not suffice for it depends on the individuals' good intentions. But Hegel does not prove insensitive to Kantian concerns. When he opts for ethical 'substance', what he adds is reflection on the likelihood of conformal behavior. Or, better still, he realizes that ethical formations can be regarded as coincident with certain ways and levels of recognition, each of which featuring a categorial novelty. To consider respect of persons a duty accordingly appears as a counterfactually abstract version of what is granted, viz., that persons should be respected, in fact are being respected on a given level. It is understood, that a person's life has to be respected, that a person – not so much qua rational will in terms of universalizable maxims but qua subject, or originator, of actions constituting socio-ethical formations – commands every co-member's respect and, in case of danger, can rely on active support.[20] More explicitly, Hegel concentrates on the other's and everybody's well-being ([6], sec. 125), he speaks of a moral person's relation to others as a "positive relation"[21] (thus as one not adequately covered by a minatory duty), and in the context of Civil Society proclaims a "legal recognition" of "property and personality" ([6], sec. 278), but we cannot expect of him an analysis distinguishing between categorical and meritorious duties (or relations) such as we tried to explore in the above section on Kant.

### *C. Ensuring Services and Standards*

The 'staggered' treatment of personhood and of the respect due to persons, as we find it in Hegel, calls for further analysis, especially in view of the possible suspicion that the individual in Hegel's scheme is being transcended by 'higher unities'. So we have to pursue the question of how diverse memberships in the social set-up are construed and how diverse levels, such as society and state, relate. Take Civil Society. Resting on division of labor and self-interest on the part of participant members, it is assumed to generate self-stabilizing institutions such as the Administration of Justice, the courts and the police (in a wide sense covering welfare measures, administrative supervision of the economy, and enforcement of standards ([6], secs. 231–242). By such devices, Civil Society claims to protect each of its members; indeed it regards welfare as a right on everybody's part and considers infringements of a member's rights as infringements of itself:

> But the right actually present in the particular requires, first, that accidental hindrances to one aim or another be removed, and undisturbed safety of person and property be attained; and secondly, that the securing of every single person's livelihood and welfare be treated and actualized as a right, i.e., that particular welfare as such be so treated ([6], sec. 230).

> Since property and personality have legal recognition and validity in civil society, wrongdoing now becomes an infringement, not merely of what is subjectively infinite [i.e., the individual], but of the universal thing ...: the action is seen as a danger to society and thereby the magnitude of wrongdoing is increased ([6], sec. 218).

In fact, Hegel gives a normative turn to services society regards as essential to itself and its members. While it may be a question of an individual's self-interest if he wants to offer, say, medical services, society may see fit to come up with requirements in point of standards, assured performance and allocation.

However, for Hegel, civil society proves incapable of ensuring the interests of all its members, unable as it is said to be to cope with pauperism ([6], secs. 243–246). Accordingly he moves on to the state which, as yet another ethical formation, 'supersedes' civil society. On the new level, things should be organized in terms of a novel concept, that of citizens rather than that of entrepreneurs and workers. Granted that the common good is realizable only at state level, we have to ask how the state relates to society, the level of private interest and profit-making. In the state, much as in fully developed society, individuals have claims addressed to the community, so that the state will have to see to it that they are being met. It will let society and its economy run its course and give it free rein except where interventions are deemed necessary:

In times of peace, the particular spheres and functions pursue the path of satisfying their particular aims and minding their own business, and it is in part only by way of the unconscious necessity of the thing that their self-seeking is turned into a contribution to reciprocal support and to the support of the whole ... In part, however, it is by the direct influence of higher authority that they are not only continually brought back to the aims of the whole and restricted accordingly..., but are also constrained to perform direct services for the support of the whole.

In a situation of exigency, however, whether in home or foreign affairs, the organism of which these particular spheres are members fuses into the single concept of sovereignty. The sovereign is entrusted with the salvation of the state at the sacrifice of these particular authorities whose powers are valid at other times ... ([6], sec. 278).

Society has the instrumental value of supporting the whole, and in the given case, intervention will have to assure its functioning in the interest of the whole. But what about the ethical impact the state would want to make, barring emergency measures? Hegel is cryptic or overly abstract:

Particular interests should in fact not be set aside or completely suppressed; instead, they should be put in correspondence with the universal, and thereby both they and the universal are upheld ([6], sec. 261).

...what the state demands from us as a duty is *eo ipso* our right as individuals, since the state is nothing but the articulation of the concept of freedom. The determinations of the individual will are given an objective embodiment through the state and thereby they attain their truth and their actualization for the first time. The state is the one and only prerequisite of the attainment of particular ends and welfare ([6], addition to sec. 261).

### *D. Constraints*

Supposing medical services are a societal matter, what curbs would the state envisage in what kind of circumstances? What changes will have to be effected in the interest of the universal will *vis-à-vis* which members of society have duties? The passage in § 278 would suggest that society is right unless ..., so that the burden of proof, or the decision, that changes or interventions are necessary is on the state. Medicine, so we may supply, serves particular or private interests and will be an ingredient of society or the economy, but subservient to a higher (universal) interest.

This could be an interest in terms of societal justice, respect of the individual person, standards, or ensurance of performance.

There need be no conflict between duty on the part of the individual physician, or ethical expectations with respect to his conduct on our part, and the profit motive, except that we could now envisage, parallel to a Kantian duty to treat, a right to be treated when life is at stake, a collision resurfacing as a conflict between the particular and the universal will. So there may have to be constraints on profit-seeking in the universal interest, i.e., in the interest of the common good. This in turn will be a matter of national policy and mentality; a systematic assessment of, say, Lockean liberalism vs. a common good view does not seem to emerge. So if it could be shown that profit-seeking medicine had the same ethical effect (in terms of categorically obligatory and beneficial treatment) as state-supervised medicine, Hegel would have little reason to change such an arrangement. But his criticism of particularity suggests that he did not consider that likely. Also in the case of medicine, then, Hegel would in all probability have opted for supervision (much the way the police supervises weights and measures), for the injunction of an ethos to curb abuses due to the market, and for a degree of justice in allocation. As far as the case of the indigent, or market failure, is concerned, his point in making society miscarry as long as it is left to its own resources is precisely that the state must come to the rescue. Since, contrary to the linear arrangement of the theory in which society is superseded by the state, society persists in reality, this means political intervention, supervision, and ethos management undertaken with a view to rectifying societal shortcomings. Conceivably, Hegel would opt for a staggered arrangement of allocation and preferential treatment such as Engelhardt envisages in his *The Foundations of Bioethics*. ([2], Ch. 8, esp. pp. 343f.).

## CONCLUSION

A comparison of the two positions discussed may be almost redundant, so obvious are their differences. But it might still clarify certain points. We drew attention to the fact that, in Hegel, duty loses the paramount status it possesses for Kant in favor of contextual relations in which the individual is embedded. A concomitant feature of this view, which we have not had occasion to stress but which highlights the difference between Kant's deontological and Hegel's teleological positions, is that the Kantian alternative of 'from duty' and 'from inclination' or 'from a selfish motive' is rejected. As in Aristotle, an act may be good, if it is conducive to a good, irrespective of whether the actor performed it from duty or not. The good life will hopefully be a fusion of aspiration to the good and its realization. As in Aristotle, it is important that the good be real, not just intended by a good will. Membership in an ethical universal is the way the reality of the good – which unrealized appears as a duty – supplants the ought by the is.

Kant's deontological position forces him to limit his ethics to the adjudicating reflection of the individual, with no concern for the consequences of his act and, accordingly, with no concern for the social context. Thus this context is theorized in a very different idiom, that of consequentialist liberalism. It is in virtue of this cleavage between ethics and the social context that Kant would seem to be permissive in questions concerning the latter.

Hegel, on the other hand, offers a uniform theory which culminates in what for him is the most affirmative ethical unity. It is its reality that counts and thus my and everybody's membership in it, not a certain maxim. There is, therefore, on the one hand, much leeway for arrangements below that highest level (private and societal arrangements) to the extent that they are not dysfunctional with respect to the whole. Thus economic solutions can – in a spirit of subsidiarity – very largely do duty for what is desired or in demand. On the other hand, Hegel's holistic view will require him to attend to dysfunctional phenomena, to justice, and to assuring that in fact every citizen can affirm the overall arrangement. So, as for our special concern with medicine, Hegel's position sets the stage for intervention, standards, and ethos management. His position resembles what, in the German legal tradition, is called a 'material constitutional state' (*materialer Rechtsstaat*), a position further developed, e.g., by Lorenz von Stein, a jurist and philosopher who gave special emphasis to the state's administrative functions.[22] In any case, we cannot extract from Hegel the thesis that fully liberalist arrangements should be endorsed. In Kant, they could, except for the moral mortgage, which for the moral person will undo all the liberty accorded to him or her in the economic sector.

Neither of the two philosophers under review seems to bear out an unqualified legitimacy of the profit motive in medicine, a view which recent discussion in the United States tends to favor. But perhaps the various inspirations contained in the two theories can help clarify the ethical issue at hand. This much seems clear, namely, that a teleological view with built-in subsidiarity, such as Hegel's, can more easily – and by that I mean: within a uniform theory, without a chasm between ethics and economics or ethics and the law – accept the profit motive in the medical profession, or in the professions generally, albeit with certain limitations in terms of the common good.

*Tübingen University*
*Tübingen, Germany*

## NOTES

[1] The Hippocratic oath contains, as has been pointed out to me by Professor H.T. Engelhardt, no statement of such a self-assumed duty to provide service. It is mainly concerned with what a physician promises to do or not to do if he decides to treat a patient.

[2] We should add reference to the physician's own person: the self-reflexive duty on his part not only not to commit suicide but to rescue himself when he has tried to do so does not, of course, involve any prospect of making money that way.

[3] ([12], p. 16). Kant leaves room for direct (positive) inclination which may accommodate the sentiment school of ethics as well as for a 'selfish view' which may be construed negatively as the attempt to avoid disadvantages.

[4] ([12], p. 15). The illustrated case is treated from an ethical angle, as a matter of prudence. One might, however, view it also as an illustration of what legality requires.

[5] ([12], p. 39). Kant confuses juridical duties as narrow duties ([11], pp. 43, 49, 73 and passim) with ethical duties which may be narrow if no consideration of purpose is involved (as opposed to duties of virtue) ([11], p. 41). That the distinction between narrow and wide duties is intra-ethical is also apparent ([12], p. 39 note 9): "I understand by a perfect duty one that admits of no exception in favor of inclination, and then I have not merely external but also internal duties" (external perfect duties being juridical or legal duties).

[6] "Imperfect duties, accordingly, are only *duties of virtue*. To fulfill them is merit (*meritum* = +a); but to transgress them is not so much *guilt* (*demeritum* = –a) as rather mere *lack of* moral *worth* (= 0), unless the agent makes it his principle not to submit to these duties" ([3], p. 339).

[7] 'Necessary' occurs in ([12], p. 46) and alibi; 'negative' is stressed in ([11], p. 81). Speaking of duties to oneself, Kant says: "The first of these are *limiting* (negative duties); the second, *widening* (positive duties to oneself). The negative duties *forbid* to act contrary to the *end* of his nature...." In view of the confusion of juridical narrow and ethical narrow duties, we find in *The Doctrine of Virtue* no statement of negative duties *vis-à-vis* others. Cf. [3], p. 339.

[8] Engelhardt would of course argue that there may be cases where the rational exercise of freedom has to be weighed against pain or ennui, or where the rational exercise of one's moral agency has become insignificant.

[9] I do not claim that all borderline cases such as Engelhardt may have in mind have been covered by my categorical reading.

[10] When, in his essay, 'On a Supposed Right to Tell Lies from Benevolent Motives' (German text: *Academy Edition*, Vol. VIII, 423–430), Kant plays off an other-regarding duty to save a person's life against truth-telling, he assumes a different stand concerning the division of duties. The representative character of abstention from suicide as the strict self-regarding duty is dethroned in favor of truth-telling which, as opposed to the other-regarding abstention from false promises, appears as a self-regarding duty: "The greatest violation of man's duty to himself merely as a moral being (to humanity in his own person) is the contrary of truthfulness, the lie ..." ([11], p. 92). Two things seem to follow: 1) that the play-off of truth-telling against saving another person's life is in fact a play-off between other-regarding and self-regarding necessary duties, such that the latter prevail; and 2) that a person's duty not to commit suicide is not his or her highest duty. Cf. Kant's discussion of casuistry in this respect ([11], p. 86). So we might have another play-off between truth-telling and saving one's own life.

[11] I realize that there is the problem of preventive medicine. It may be pure beneficence. Cf. also Kant's views on dietetics which, with the help of philosophical reflection, is supposed to prolong life. Kant, I.: *Conflict of the Faculties* (German text: *Academy Edition*, Vol. VII, 95–116).

[12] Such curative help is also often demanded for the benefit of animals.

[13] In conventional malpractice suits, the defendant still is the very person in whom the patient placed his trust, anticipating the physician's commitment to duty; in the new situation, an intermediary is slipped in. To put it in another way: by contrast to ordinary contract law, the issue has existential relevance because life or an impaired life is at stake. There is a weak analogy in the area of real estate: if accommodation is treated as a commodity, the existential concerns of the tenant tend to be overlooked.

[14] Admittedly, the context is that of political rights, but Kant's endorsement of social inferiority has ramifications in the area of medical coverage as well.

[15] ([6], sec. 135, p. 90). It is a pity that Hegel did not in this context consider the second formula of the categorical imperative.

[16] ([6], sec. 162). "Our objectively appointed end and so our ethical duty is to enter the married state".

[17] ([6], sec. 150). "In an *ethical* community, it is easy to say what man must do, what are the duties he has to fulfil in order to be virtuous: he has simply to follow the well-known and explicit rules of his own situation. Rectitude is the general character which may be demanded of him by law and custom".

[18] ([6], secs. 75, 258). "On the other hand this final end [the substantial unity of the state] has supreme right against the individual, whose supreme duty is to be a member of the state" (sec. 258, p. 156).

[19] ([5], pp. 111–119). There are further constructs of recognition in that work, e.g., forgiveness.

[20] There is a problem of is vs. ought here ([6], p. 10): "*What is rational is actual and what is actual is rational*". Hegel's earlier views differed on this point ([7], p. 28 and [8], p. 51). The author tends to speak of a 'normative ontology' to express what Hegel is after.

21 Knox's translation has "a positive bearing of each on the other" ([6], sec. 112, p. 249).

22 For a selection, see ([15], pp. 21–494 and [4], pp. 65–95). Strangely enough, the development in German public law was, to put it simply, that under the Hegelian auspices of a 'material constitutional state' a Kantian emphasis on the dignity of man, answering to the second formula of the categorical imperative, came to be enshrined in the constitution ([1], pp. 117–157).

## REFERENCES


1. Dürig, G.: 1956, 'Der Grundrechtssatz von der Menschenwürde', *Archiv für öffentliches Recht* **81**, 117–157.
2. Engelhardt, H.T. Jr.: 1986, *The Foundations of Bioethics*, Oxford University Press, New York.
3. Findlay, J.N.: 1961, *Values and Intentions*, Allen and Unwin, London.
4. Hartmann, K.: 'Reiner Begriff und tätiges Leben', in R. Schnur (ed.) *Staat und Gesellschaft. Studien über Lorenz von Stein*, Duncker und Humblot, Berlin.
5. Hegel, G.W.F.: 1977, *Hegel's Phenomenology of Spirit*, trans. A.V. Miller, Clarendon Press, Oxford.
6. Hegel, G.W.F.: 1962, *Hegel's Philosophy of Right*, trans. T.M. Knox, Clarendon Press, Oxford.
7. Hegel, G.W.F.: 1983, *Die Philosophie des Rechts. Die Mitschriften Wannenmann und Homeyer*, ed. K.-M. Ilting, Klett-Cotta, Stuttgart.
8. Hegel, G.W.F.: 1983, *Philosophie des Rechts. Die Vorlesung von 1819 in einer Nachschrift*, ed. D. Henrich, Suhrkamp, Frankfurt am Main.
9. Hegel, G.W.F.: 1968, *Philosophische Propaedeutik*, Hegel Studienausgabe, Vol. 3, Fischer Bücherei, Frankfurt am Main.
10. Kant, I.: 1957, *Conflict of the Faculties*, tr. M.J. Gregor, Harper & Row, New York.
11. Kant, I.: 1964, *The Doctrine of Virtue*, tr. M.J. Gregor, University of Pennsylvania Press, Philadelphia.
12. Kant, I.: 1986, *Fundamental Principles of the Metaphysics of Morals*, tr. T.K. Abbott, 18th printing, Macmillan, New York.
13. Kant, I.: 1965, *The Metaphysical Elements of Justice*, tr. J. Ladd, Bobbs-Merrill, Indianapolis.
14. Koslowski, P.: 1985, *Staat und Gesellschaft bei Kant*, Verlag J.C.B. Mohr (Paul Siebeck), Tübingen.
15. Lorenz von Stein: 1972, *Gesellschaft-Staat-Recht*, ed. E. Forsthoff, Propylaeen, Frankfurt.

H. TRISTRAM ENGELHARDT, JR.

# VIRTUE FOR HIRE: SOME REFLECTIONS ON FREE CHOICE AND THE PROFIT MOTIVE IN THE DELIVERY OF HEALTH CARE

## I. FROM ARISTOTLE'S POLIS TO BEYOND HEGEL'S STATE

> Physicians value'd Fame and Wealth
> Above the drooping Patient's Health,...
> Thus every Part was full of Vice,
> Yet the whole Mass a Paradise; ... ([12], p. 20).

A reasonable gloss on de Mandeville is that, by pursuing their own interests, individuals in general, and physicians in particular, contribute to a general societal increase in health and wealth. Self-interest is such an important and powerful motive force that the profit motive requires a central place in any account of societal undertakings, including the development of health care systems. It is, after all, self-interest expressed in the profit motive that brings many to develop capital and produce goods and services of benefit to the commonwealth. On the one hand, this is a matter of principle: the only ways to acquire the services and capital of individuals, other than through coercive force, are to appeal to their altruism, to special interests (e.g., the pursuit of knowledge or the arts), or to their self-interest, of which the profit motive is a general statement. Given the salience and power of self-interest, the profit motive offers one of the major alternatives to the coercive acquisition of capital and services for health care. On the other hand, the profit motive has its "costs". The profit motive by itself will not achieve all common social desiderata. This state of affairs invites detailed analyses of the moral and economic consequences of employing different social structures with different institutional goals for the delivery of health care. But such considerations lead one to ask what one means by these goals, or by happiness or by the good. Because of the salience of non-monetarizable interests, and because of the difficulty of ever knowing what weight to assign to non-monetarizable consequences, such analyses are often inconclusive.

The difficulty lies in providing a canonical ranking of human goals or goods. Depending on the ranking one gives to such important human



*T.J. Bole III and W.B. Bondeson (eds.), Rights to Health Care*, 327–353.

goals as prosperity, liberty, and equality, one will attempt to develop radically different societies with radically different health care systems. For-profit health care is, for instance, likely to contribute to the gross domestic product of a country and to the wealth of a society. Insofar as a society develops corporations that sell their services beyond the country's border, profits will be returned from the provision of health care elsewhere. Whether such is good or bad will depend, *inter alia*, on the relative value one assigns to equality in access to health care services or the equality of opportunity, which is contingent on health status.

Kant's reflections are in part a response to this quandary. If one cannot determine what one ought to do in what priority either by inspecting the design of the world and the character of reality (as had been attempted by traditional natural law theory), or by inspecting the various motives and interests of humans (the nature to which the Scottish philosophers turned in order to establish morality), then one might turn to reason itself. In this vein, Kant attempts to read off from the character of moral reasoning what one ought morally to do. Just as Kant reconstrues objectivity in terms of intersubjectivity, that is, in terms of the necessary conditions for the possibility of experience for any spatio-temporally intuiting discursive knower, Kant reconstrues morality in terms of a person acting morally. But the moral agent under such circumstances is anonymous and need not even be human. As a result, determinations of the will by human inclinations, wants, and sympathies are immoral at worst and morally neutral at best. To act morally, to act in a way that would make an agent worthy of happiness, is to act in accordance with moral relationality itself. As I have argued elsewhere ([5], pp. 68–71), in accord with Hegel's criticism of Kant ([8], 135 *Zusatz*), Kant does not gain content but lays out, to borrow a metaphor he uses in the Prolegomena regarding knowledge, the grammar of morality (*Prolegomena* IV, 322).[1]

I do not deny that Kant attempts to acquire content. But he cannot succeed without importing into his notion of an anonymous law-giving, law-governed agent particular moral values set within a particular hierarchy. Consider Kant's weak defense of beneficence.

> A fourth man, for whom things are going well, sees that others (whom he could help) have to struggle with great hardships, and he asks, "What concern of mine is it? Let each one be as happy as heaven wills, or as he can make himself; I will not take anything from him or even envy him; but to his welfare or to his assistance in time of need I have no desire to contribute." If such a way of thinking were a universal law of nature, certainly the human race could exist ... Now although it is possible that a

universal law of nature according to that maxim could exist, it is nevertheless impossible to will that such a principle should hold everywhere as a law of nature. For a will which resolved this would conflict with itself, since instances can often arise in which he could need [*bedarf*] the love and sympathy of others, and in which he would have robbed himself, by such a law of nature springing from his own will, of all hope of the aid he desires [*wünscht*] ([10], IV, 423, p. 41).

In order to impart content to his moral theory, Kant has to appeal to a contradiction in will (which incorporates a sense of needs[!] and what one may in the future wish), not a contradiction *in sensu stricto.* Similar difficulties would have awaited Kant, had he attempted to develop Hartmann's proposed duty to act to prevent suicide or to act to provide health care. Without importing a particular thin theory of the good or hierarchy of values into the very concept of reason or of the moral agent, one can at best gain an empty formalism in terms of which one knows that one should not break promises but not which promise to make. Or to put the matter in terms of health care, one will know that one should deliver health care when one has contracted to do so, but one will not know with whom and under what circumstances what sort of contracts should be fashioned. Kant's journey into the nature of reason cannot provide the canonical hierarchy of human goals any more than Hume's journey into the nature of feelings and sympathies.

Hartmann's paper is important in adding to the usual contrast between teleological and deontological analyses of the profit-motive a categorial one, a *tertium quid* [7]. Hegel, as Hartmann underscores, is not directly interested in particular duties or obligations. Hegel does not provide a normative ethics. Instead, in the *Philosophy of Right* he seeks a ranking of ethical unities. Each unity, each category, constitutes a particular mediation of being and reason so that each category provides a conceptual perspective on moral structures. If the categorial ranking succeeds, one then understands how the diverse ingredients of the moral life constitute a unity in reason. In particular, Hegal focuses on the tension between the formality of moral rights and duties and the content of moral rights and duties. The more one succeeds in stating them universally, the more abstract they become. The more one succeeds in giving them content, the more parochial they become. Hegal can be seen as criticizing Kant in the first two sections of the *Philosophy of Right*, Abstract Right and Morality [*die Moralität*]. If one remains at these abstract levels, one cannot account for rights or duties.

Because every action explicitly calls for a particular content and a specific end, while duty as an abstraction entails nothing of the kind, the question arises: what is my duty? As an answer nothing is so far available except: (a) to do the right, and (b) to strive after welfare, one's own welfare, and welfare in universal terms, the welfare of others ([7], sec. 134, p. 89).

Again, one can in general know that one ought to honor one's contracts, but one does not know what contracts to make. In general, one knows one ought to be beneficent, but one does not know concretely what beneficence involves. Hegel attempts to solve this problem by appealing to the concept of *Sittlichkeit* ("*Sitten*" can often simply mean "customs"), which is often translated as "the ethical life". Here in terms of a family and a particular community, one comes to understand concretely one's obligations.

After the family, Hegel introduces the concept of civil society, "*die bürgerliche Gesellschaft*". In civil societies one finds for Hegel the workings of the market, what Hegel would term "the external state, the state based on need, the state as the Understanding envisages it" ([8], sec. 183, p. 123). Here I have an important point of contention with Hegel. He needs a notion of a concrete community [*ein sittliches Gemeinwesen*] spanning families and giving them an understanding of the proper rankings of values through at least a thin theory of the good. The state, both as the external state held together by market forces as well as the state as the custodian of constitutional law, must transcend particular communities or moral commitments. The point is this. For the most part, large-scale states have been seen falsely on the model of Aristotle's polis, which, as he describes in the *Nicomachean Ethics*, must include less than a hundred thousand citizens (IX 10, 1170b31–32). In such a polis there can be a commonly held understanding of the goals of society and of the obligations of indviduals. It presupposes a homogeneous population with common antecedents and shared commitments. A common sense of justice and societal obligations is set at jeopardy when these conditions are not fulfilled. Besides, "in an overpopulous state foreigners and metics will readily acquire the rights of citizens, for who will find them out? Clearly, then, the best limit of the population of a state is the largest number which suffices for the purpose of life, and can be taken in at a single view" (*Politics* VII 4, 1326b18–25) ([2], p. 1284). It is ironic that Aristotle pens these reflections looking backward to a form of government that was to become obsolete. Aristotle, after all, was the tutor of Philip's son Alexander.

Hegel recognizes that the state is, as he announces at the beginning of his treatment, the actuality of concrete freedom. This is the case because, on the one hand, "personal individuality and its interest" achieve in the state their "complete development and gain explicit recognition for their right" and, on the other hand, they "know and will the universal" ([8], sec. 260, p. 160). In a number of passages Hegel speaks of the state as if it could transcend the particular concrete moral understandings of particular concrete communities [*sittliche Gemeinwesen*]. One might think of his criticism of the anti-Semites of the time, where he argues that the state not be seen as either Christian or Jewish, but as a neutral structure open to all as citizens (sec. 279, *Zusatz*). One might also think of the very interesting passages where, in contrast with Marx, Hegel argues that the universal class is that of the civil servants because they have the universal ends of the state as their essential activity (sec. 303). Perhaps here the post office might provide an illuminating example. Postmen accept and deliver the mail independently of the sender's or recipient's religion or understanding of distributive justice. The bureaucrats of a particular country can provide services to its citizens in a way that cuts across particular communities, particular poleis.

Without fully acknowledging it, and in substantial conflict with many of his other statements and arguments, Hegel takes a step toward articulating the conceptual standpoint from which one can understand the notion of a secular pluralist state, a state without a particular ideology, a state that successfully spans numerous particular moral communities with their own particular moral commitments. The state in this sense mediates or solves the tension between abstract individual rights and the concrete notions of obligations and duties that characterize the ethical life, *Sittlichkeit*. As a result, though one has an abstract duty of beneficence, and though particular religious or ideological groups may have their own views of beneficence, the state should not allow them to be enforced coercively (except in the case of indentured servitude). Particular moral communities should make do with excommunication and loss of party membership in order to punish those who act from the wrong motives or in the wrong ways, but within their rights as citizens. This recasting of Hegel is important, not only because it allows one, in a Hegelian fashion, to place the significance of Kant (Kant provides the formal structure of moral agency) and Hume (Hume takes account of the concrete passions and sentiments that flesh out actual duties and rights), but because it allows one to talk about societal institutions, which may in many important senses be amoral (i.e., from the standpoint of particular communities with their concrete moral commitments), but which realize a striking generality. Institutions that cut across and encompass various particular communities of moral commitment require a neutrality with regard to those communities, as Hegel foresaw in his account of the universal class, civil servants.

Put in this light, the market with the profit-motive can be seen from the standpoint of the state (so construed) not as a problem, but as offering a special solution to a conceptual difficulty. The market allows each when making his own choices to act in a way that supports the whole. The already-cited passage from Mandeville and much of Adam Smith support this point. But it is not just that the market succeeds in producing wealth and services. The market involves a form of willing, which through particular choices achieves a universal significance: it binds people together freely, respecting a central element of Kantian morality (especially when one reads freedom as a side constraint, not as a value), while providing content to the choices of each of the particular participants. Each individual pursues his or her own concrete individual goals, and the universal is achieved as well. The market incarnates the cunning of reason.[2]

All of this is important because the challenge is to understand the significance of institutions that cut across particular moral communities, whether the institutions are run by governments, publicly traded corporations, or non-profit entities. To understand whether the profit motive should be regarded as a positive or negative factor in evaluating such institutions, or the activities of their employees, depends at least in part on whether one credits this neo-Hegelian suggestion that institutions spanning particular moral communities in a strategically morally neutral sense are necessary elements of understanding a large-scale state. If that is the case, one may no longer assess them, at least directly, in Kantian or utilitarian terms. Rather, such institutions would be seen as allowing individuals from different moral communities and with different concrete understandings of the goals of medicine (and therefore with different schedules for weighing consequences) and of the constraints of distributive justice to address over-arching general concerns with health. One must now see why such morally neutral institutions are morally unavoidable.

## II. A SKEPTICAL INTRODUCTION: JUSTICE, FREEDOM AND PARTICULAR VALUE HIERARCHIES

There is a major difficulty in justifying concrete accounts of the moral life or of justice. As already indicated, any concrete account of what one ought to do, of what contracts one ought to make, of what obligations one possesses, or what is required by distributive justice will presuppose a particular ranking of such major societal goals as liberty, prosperity, and equality. If it is not possible to discover a way to meet this challenge, then the only avenue left will be that of peaceably creating common understandings. The reason is this. If one cannot discover the correct answer to such central moral questions (e.g., what is the correct ordering of such important moral desiderata as liberty, equality, and prosperity?) and if one simply acts with force to impose a particular understanding on the basis of one's own intuitions or inspirations or on the basis of the consensus of the majority, then there is no possibility for individuals who do not share that common understanding to recognize that there is a morally binding common solution they should respect. All that will be open to them will be mediation through force; the moral viewpoint imposed will not have a general justification. However, if they are interested in resolving such disputes other than through a fundamental appeal to force, then they can agree to attempt to resolve such controversies insofar as possible by reason and otherwise by mutual agreement. Those who participate in this endeavor may then, when necessary, defend themselves against and punish those who oppose peaceable negotiation, since these individuals cannot consistently protest in general secular terms when they are met with defensive or punitive force, for they have rejected the eschewal of force as the cornerstone of controversy resolution. Those who appeal to peaceable negotiation can therefore (1) consistently place and handle those who do not participate, (2) sustain a moral fabric that can encompass moral strangers such that they can be regarded as blameworthy or praiseworthy, and (3) fashion common moral endeavors with those who are willing. A grammar for morality is available with a minimum of presuppositions for anyone interested.

The more one is suspicious about the capacities of reason to discover concretely what is morally required, or the more one is skeptical regarding the authority of majorities to enforce particular understandings of proper conduct, the more one will be constrained to recognize that respect of freedom as a side-constraint is the lynchpin of the moral life when moral strangers meet. One will not be valuing respect of freedom, but rather recognizing it as the means for negotiating common moral endeavors and for gaining common moral authority. One will be constrained to recognize rights to privacy not as moral desiderata but as the plausible limits of communal authority when *the* canonical ranking of values cannot be discovered and all cannot be presumed to have consented to a particular undertaking. All who are interested in resolving issues peaceably can be presumed to have consented to defense against murder, rape, and robbery and to the equitable distribution of commonly owned property.[3] In short, the formalism of Kant's ethics and the "Abstract Right" of Hegel can be established as integral to the minimal grammar of a morality shared by strangers. The same, too, can be said regarding Hegel's "Morality" as a general and abstract commitment to beneficence. Concrete understandings of distributive justice and of the other concerns of the moral life will need to be found in particular moral communities. The state then appears as an amoral (from the standpoint of particular moral communities) institution that enforces abstract rights (as well as concrete recognized agreements made on the basis of those rights) and makes possible the continued existence of particular peaceable moral communities.

When one thinks of the state in this fashion, one is forced to ask how it will transact its business of gaining authority in its various macro- or micro-level activities. Plausible answers will be instances of actual agreements, actual conveyals of authority. Those occur in limited democracies, which maintain robust privacy rights and develop open means for determining how to distribute commonly owned resources and their revenues. These also occur through the practice of free and informed consent by which individuals make individual agreements and refuse or accept welfare rights. When one asks how the various complicated deployments of resources will take place within a state, a cardinal answer will be through the market, for it is an ever-present expression of free choices. The market allows individuals who belong to communities with radically different concrete views of proper moral deportment to cooperate with each other to their own benefit and that of society in general. The market, moved by the profit motive, is then not just a practical triumph but a conceptual solution. It provides a moral structure that does not require a particular moral vision.

The more one considers these sorts of issues, the more particular states become less important and the more one looks to an international and (from the point of view of particular moral communities) amoral state encompassing various moral communities that now easily transcend national boundaries. The existence of multi-national corporations, religions, and unions demonstrates how individuals moved by particular goals transcend national boundaries and make contributions to the international common weal, often while only pursuing their own particular interests. The conceptual difficulty of establishing particular canonical hierarchies of values or goals, or of establishing the authority of governments or preponderant majorities, leads one to seek solutions that will be plausible in the absence of such canonical visions and established authority ([5], pp. 17–65). It suggests why one must live with very little content drawn from concrete views of distributive justice and why respect of freedom as a side constraint, not a value, plays such a cardinal role in modern society. One should not see the salience of limited democracy, free and informed consent, and the market as expressions simply of liberal sentiments or of capitalist commitments. They also reflect the realization that these are vehicles through which men and women can collaborate against a background of skepticism regarding the capacities of reason and the scope of public authority.

Given these considerations, one can reexamine the question of the degree to which motives are important in the moral evaluation of the actions of individuals and institutions in the context of health care. From the standpoint of the state, motives should not matter (the state is, after all, not a particular moral community) unless the motive is that of an outright and diabolical rejection of beneficence on principle and in general. The state should also ensure that the abstract rights of citizens (in the case of patients) are respected. Beyond that, it may also attempt to support the realization of certain goods insofar as its constitution and commonly-owned resources allow (e.g., welfare rights that may aid in the acquisition of health care). But in such circumstances the state is at best creating, not discovering, a view of justice. In fact, from the standpoint of the state, the character of welfare programs is not a matter of justice, save with regard to equitable participation in decisions regarding the use of common resources. One is creating a public policy, not discovering a just policy. Also, from the standpoint of the state, there do not appear to be any grounds for being concerned because individuals or institutions will act on a profit motive; after all, the profit motive moves the market, which is the cement of any free society. Still, the state may be concerned regarding the consequences of individual and joint actions and may, insofar as its constitution and resources allow, attempt peaceably to support contrary outcomes.

All of this has been stated in terms of limits. One should realize that, from a general moral perspective, these limits are not desired or endorsed because of particular goals. They express rather the limits of human reason and authority, which should become ever more salient as one reflects on the failures of the Enlightenment project.

## III. PRUDENCE AND CHARITY: BENEFICENCE IN THE FACE OF FUNDAMENTAL LIMITATIONS

The profit motive is not the only one that brings people to invest their resources in health care institutions. Individuals also establish governmentally supported health care systems as forms of insurance against possible future needs. Because of recognized duties of beneficence and feelings of sympathy for the indigent, they invest for charitable reasons as well. Even in the absence of appeals to theories of distributive justice, particular communities as well as large-scale states can act to establish mechanisms to provide health care for the impecunious. Indeed, given the intellectual difficulties that attend the project of establishing concrete duties of beneficence or rights in distributive justice, appeals to such foundations are best abandoned in favor of attempts to account for the proper roles of health care institutions in large-scale states. Examinations of the applicability of theories of justice to actual questions of health care allocation are marked by so much contention and unclarity that practicality would argue (in addition to the theoretical considerations already advanced) for simply asking what level of insurance against losses at the natural and social lotteries members of communities or citizens of states would want to provide themselves.

The insurance metaphor invites citizens of a state to recognize that they are all exposed to the natural lottery (defined as that nexus of causes leading individuals to having greater or lesser natural endowments, including states of physical or mental health), which will determine their needs and desires for health care, and to the social lottery (defined as the nexus of causes leading individuals to having greater or lesser wealth and/or influence), which will determine their ability to acquire health care. Given these uncertainties, prudent individuals will very likely place some funds aside for health care in the event that they should lose at both of these lotteries. But investment in insurance against the natural and social lotteries with respect to providing health care has to be made with an important consideration in mind. Losses at the natural lottery with regard to health are not like losses with regard to food, clothing, or shelter, where individuals usually in principle can be made whole and, indeed, can in principle be made equal in their possessions. Losses at the natural lottery with respect to health often cannot be cured no matter how energetic or costly the medical interventions. A blind quadriplegic epileptic individual cannot usually (if ever) be restored to full health, given any level of exertion. The individual's well-being can be increased. But as further investments are made, returns diminish and full health is never regained. The same is also often the case with regard to postponing death. One comes finally to circumstances where the investment of resources in postponing death brings at best only marginal results and at great costs. The prudent purchaser of an insurance policy against losses at the natural and social lotteries with regard to health care needs will, therefore, purchase insurance, but not a policy that pays a great deal for marginal returns. Again, what will count as "marginal", and what will count as "great cost", will depend on individuals and societies. The definition of "marginal" and "great cost" is often the focus of democratic compromise when the issue at stake is the investment of communal resources. Disagreements among individuals can be solved by dissidents seeking insurance carriers that provide special levels and foci of coverage. Or if worse comes to worst, individuals can self-insure, if they have sufficient disposable resources.

Societies, however, must make decisions in order to frame policy. Those who participate in disputes regarding the creation of a policy may be moved by particular prudential or eleemosynary motives with various religious and ideological roots. However, the policy considered as that produced by a secular state compassing a plurality of moral communities should not be viewed as grounded in any particular theory of beneficence or distributive justice, but as the result of a governmental choice to create a particular response to meet concerns of prudence and/or charity. Such outcomes are cognizable in a general, secular fashion without endorsing any particular theory of justice or beneficence.

These reflections lead unavoidably to endorsing two general tiers of health care services: one provided on the basis of the ability to pay and the other provided on the basis of need from either charitable contributions or a general insurance scheme established by a community. These two tiers are unavoidable as long as (1) all resources are not communally owned, (2) the community does not wish to invest from its funds sufficiently to bring the second tier up to the standard of the first tier, and (3) the community lacks the moral authority to abolish or constrain the private tier so as to realize overriding commitments to particular goals such as equality. Previous skeptical reflections concerning the limits of reason and secular authority make it very implausible that all goods are either commonly or privately owned or that communities have moral rights to such totalitarian control. As a result, in nearly all likely circumstances there should always be two tiers of health care.

One might note that the two tiers cannot be distinguished simply as a for-profit versus a non-profit tier. Health care will often be provided on an ability-to-pay basis by non-profit health care institutions, some of which may in fact be in part supported by philanthropic contributions. In addition, for-profit health care institutions may be paid to provide health care for the indigent because of their perceived efficiency or because of the style in which they provide health care. This is a point to which I will return later.

The heterogeneity of human goals and the limits of human reason and authority will always leave a domain that for-profit health care institutions may with moral right claim as their own. In order to see this more clearly, one must recall how for-profit health care systems come into existence. They do not steal their resources nor impress their employees into service. They peaceably attract capital and employees in a freely constituted joint venture toward the goals of providing an important and desired set of services so as to make profits for shareholders, justify high salaries for management, and allow the payment of salaries to the employees at a level sufficient for the realization of the first two desiderata insofar as these are compatible with labor negotiations and other agreements. As a result, the provision of health care through for-profit institutions is far less problematic morally than licensing schemes that would restrict the freedom of physicians to practice where they would wish, constrain physicians to accept Medicare or Medicaid patients, or draft physicians in order to provide health care for those in need. All such coercive intrusions require establishing that one knows what is correct to do, that one has the authority to impose it, and that the intrusions will generate more benefits than harm. For-profit endeavors need not rely on such presuppositions insofar as they justify their activities in terms of the limits of human reason and authority. They can be agnostic about which social goals are overriding, and rely instead for their authority on the free consent of those who collaborate with them. As with other peaceable consensual activities, for-profit health care institutions have a prima facie right to come into existence, freely solicit capital, hire employees, and sell their services to those with the ability to pay, whether that ability comes from their own resources or from insurance provided to them through some governmental scheme.

To appreciate the heterogeneity of our health care market, it is helpful to note the extent to which American health care is provided through a mixture of private and governmental systems. In 1989 the United States invested $2354 per person in health care expenditures, compared with (on the basis of purchasing power parities, PPP's) $1683 per person in Canada, $1232 per person in the Federal Republic of Germany, $1035 per person in Japan, $836 per person in the United Kingdom, and $371 per person in Greece [15]. 42% of the American health care budget was derived from state, federal, or local taxes (this includes 17% for Medicare, 10% for Medicaid, and 15% for other governmental health care services, of which the Veterans' Administration hospitals consume a significant proportion) [11]. The 3.3% invested in non-Medicaid public assistance programs, state and local public hospitals and other state and local public programs amounts to an investment of approximately $70 per person. Since only about one-sixth of Americans have no insurance, this may be close to an allocation of about $420 per uninsured person. To put these figures in perspective, one would need to add that many of the uninsured are very young and nearly all are under 65. In addition, some are eligible on a bed-availability basis for care in veterans' hospitals. Also, even the uninsured have some access to public health care, and some have sufficient personal resources to pay for much of their health care. Finally, approximately 33% of health care funds came from private health care insurance, and 21% from direct patient payments.

These data underscore the diversity of sources from which funds are acquired to support health care in the United States. This diversity masks the fact that nearly everyone, if not everyone, is concerned about finances. Institutions, profit and non-profit alike, governmental and private, all worry about the bottom line. They do not want their physicians, nurses, or other health-care workers to expend more on the care of patients than they have funds. A good case study in point is that of the National Health Service in the United Kingdom [1]. More than that, employees tend to look at their bottom lines as well. Most physicians, nurses, and allied health care professionals are interested in continuing to earn a living for themselves and their families. Therefore, they will be interested in the financial viability of the institutions in which they work and the ways in which the financial resources of their institutions may be increased, leading to more remuneration for themselves.

All of this should not be surprising. Medicine has traditionally been practiced for profit. Even before the times of ancient Greece, physicians practiced medicine because of the financial rewards and status it brings. It is only the exemplary, dedicated few, the Albert Schweitzers and members of religious orders with a strict vow of poverty, for whom the practice of medicine may have been undertaken with little or no concern for the financial benefits it might generate. However, even religious brothers and sisters, without families to support, may be concerned about the cost to their orders from the provision of health care services. Even their interests may focus on the financial bottom line.

Health care provided through the auspices of national health insurance schemes or through charity comes to be shaped by concerns with financial issues that create analogies with the behavior of managers and employees in health care institutions established to produce profits for their shareholders. Though one might expand the segment of the health care market commanded by non-profit or by governmental health care systems, in having done so one would not have escaped the ever-present concern of institutions to remain financially viable and of managers and employees to earn more money. The latter concern can be cured, if at all, by attempting to increase the number of vocations to religious orders with strict vows of poverty who will treat patients with very little interest in their own financial well-being.

## IV. DUMPING AND SKIMMING: HOW SEEMING VICES ARE OFTEN REALLY VIRTUES

One must now ask why anyone would ever be suspicious of the profit motive [6]. Why would it ever seem wrong to do things for profit? Are there certain activities that must be undertaken without a profit motive? It would appear that individuals have intuitions of an often vaguely articulated character that a central and salient interest in profit changes the character of certain undertakings. Imagine, for example, that one is concerned with saving the souls of children in a particular part of the world, and that one belongs to a religious faith where infant baptism is sufficient to ensure salvation. Imagine also that a for-profit marketing company makes a proposal to the religion's head office to increase (for a fee, naturally) the number of baptisms and souls saved by one hundred percent. It plans to do this through mass-marketing techniques and through hiring members of the religion's clergy to perform the baptisms so that the rituals are completed with proper attention to detail and intention. Would such a proposal be immoral or unacceptable? Of course, the answer will in part depend on the particular scriptures and traditions of the religion. But aside from such restraints, it is unclear to me how one would generate an objection in principle. One might argue that such a commitment to marketing the religion would offend communicants back home, leading to their falling away from the church and being damned, as well as to their contributing less to the support of the religion. One might have a nostalgia for the way things were done by the fratres minores and find the endeavors of Madison Avenue wanting aesthetically in comparison. Such considerations might be decisive for those who are in charge of the religion. However, they do not provide a conclusive moral argument against a for-profit soul-saving corporation's being engaged by the church.

Serious objections, if they can be articulated, will need to show some abuses to which the profit-motive leads, which are sufficient to justify abandoning the profit motive as a cardinal motive force in health care delivery. But even here, one must note that free individuals may choose freely: The mere fact that life is made more unpleasant for some because others choose freely to establish for-profit health care institutions does not mean that one has a moral right to stop them. If one cannot find anyone with whom to dance at the ball, that is unfortunate, but not unfair, unless one can establish a prior claim to the dancing services of others. To constrain the services of others or to limit their freedom of association requires establishing a canonical hierarchy of goals and the authority to impose it, not to mention showing that the imposition will work more benefit than harm. Still, one might be able to demonstrate that for-profit institutions engage in immoral activities because of their interests in profits. One might be able to show that institutions established to return profits to their shareholders transmogrify the character of physician/patient and other health care giver/client relationships in a way that is morally unacceptable.

One of the grounds often advanced for condemning the for-profit motive in health care is that it leads to skimming (where skimming is defined as attempting to attract only those patients who can pay, leaving governmental hospitals to absorb the impecunious) and dumping (where dumping is defined as transferring to governmental hospitals those emergency patients who arrive for care but without the ability to pay). These two practices are criticized because (1) they leave the health care institution of last resort (usually a hospital supported by tax funds) with the obligation of paying for the treatment of the indigent (hence the source of the term skimming, one skims off the economic cream). As a result, the burden on taxpayers is increased. Moreover, (2) if there are differences in the level of amenities or quality of care between governmental hospitals and non-governmental hospitals, inequalities in levels of treatment become manifest. Finally, (3) some individuals may die sooner or suffer harms because they were not treated by the non-governmental hospitals, that is, because of dumping and skimming.

To begin with, it is worth underscoring that it will be in the immediate economic best interest of all health care institutions to skim and dump wherever possible. This holds true not only for non-profit hospitals but for governmental hospitals as well. State, county, and city taxpayer-supported hospitals will be only too glad to harvest patients with some private health insurance or with Medicaid and Medicare coverage, while transferring potentially costly patients where possible to Veterans' Administration hospitals. Hospitals with fixed budgets will want to avoid resource-consumers for whose costs they will not be reimbursed.

First and foremost, one might ask why it is the moral obligation of hospitals to take care of the indigent who cannot pay. Supermarkets are not usually obliged to provide food for the hungry. Rather, they cooperate in providing food on the basis of food stamps which they support through the taxes exacted on corporations. Construction companies are not usually obliged to build houses for the indigent, though they do so on government contract. The reasoning appears to be that the general obligation to address the needs of the indigent for food or shelter falls on the community or the state, not on individual supermarkets or construction firms. One might, given certain views of beneficence, reasonably argue that the owner of a supermarket ought to provide food to a person who would otherwise expire on the spot, though the owner might prudently add that such feeding may have to be conducted discretely, whether in New York or Calcutta, otherwise the purported indigents of the locality might outnumber the customers.

Why is health care different, especially if one grants for the sake of discussion that health care institutions should provide care for individuals when they cannot be safely transferred to public institutions and considers only those who can be transferred safely? In fact, it would appear from a number of perspectives to be morally praiseworthy to skim and dump under such circumstances. This is the case because, in the absence of skimming and dumping, those who happen to be sick and can pay are taxed to pay for those who happen to be sick and cannot pay. To keep within budget as a non-profit hospital or to maintain profits as a for-profit hospital, the charges of all are increased to take account of the costs likely to be incurred when skimming and dumping are forbidden. If at the very least one holds that taxes should be publicly discussed and democratically enacted, and that their costs should be borne by all, then attempts to skim and dump lead to (1) frankness and candor in discussing a special hidden taxing system, and (2) honesty in confronting what is the central issue, the amount of funds that should be allocated in a society to provide care for the indigent.

From the above, it should be clear that it is not for-profit medicine that creates the problems of skimming and dumping. Skimming and dumping exist when some hospitals have fixed budgets and are obliged to provide care for which they are not reimbursed. The problem can be solved either by reimbursing hospitals for the care rendered or by honestly returning to a system of health-care institutions to which the indigent, except in circumstances of true emergency need, are frankly and openly transferred, and which provide that standard of care the community decides should be available to the indigent. If for-profit hospitals in fact exacerbate the phenomena of skimming and dumping, then they aid the community and state to focus more clearly on the issue of what level of care should be accepted as a decent minimum for all citizens.

Since skimming and dumping (i.e., involving only the transfer of those who can be moved without harm) are in themselves not problems at all, and highlight a difficulty independent of the pursuit of the profit motive, why would reasonable individuals ever conclude that these phenomena are the invidious results of profit-seeking in health care? Perhaps the concern about skimming and dumping masks a desire that the inability to pay should never adversely affect an individual's health or likelihood of survival. If that is the issue, the problem is not with for-profit health care, but with a notion that it should not be the case that some individuals should have resources, which, if they were possessed by others, would improve their health or extend their lives. It is surely the case that for-profit health care presupposes at least limited private property rights in services and goods, such that one will at some point be confronted with the unpleasant fact that individuals will have the moral right (i.e., based on the limits of state authority) not to be forthcoming with their resources to save the lives of those in need, even if such denial is seen by many to be grievously immoral. That such must be the case is clear to all who recognize that the community's rights to control private resources and services are not absolute: limits are placed on communal interventions by the limits of reason and by the consequent limits of communal authority.

Freedom is generally upsetting. If individuals are free to dispose of their own goods and services as they wish, the world will be untidy, people will be unequal, and central planning will be in vain. Individuals will act according to their own views of the good life and beneficence. They will pursue their own good and that of those for whom they feel sympathy or acknowledge a bond of duty. The results will not be uniform. Freedom and equality are incompatible. This truism cuts to the quick in health care, where free choices as well as differences in disposable resources and energies lead to differences in health, well-being, and length of life. Individuals with greater resources will generally be healthier and live longer than others. One need not assume that all resources or all services are privately held. However, insofar as any resources are in private hands, and insofar as the services of physicians are not fully owned by their society, a private tier of health care will develop, which tier should make a positive difference for the health status and longevity of individuals who have access to it. For those who endorse equality as an overriding good, this is an unacceptable circumstance.

Moreover, if for-profit, fee-for-service medicine offers benefits of greater health and longer life, those who cannot afford to purchase such services will clamor to have them made available through taxation, leading to an escalation in the amount of funds deployed for health care. If there are diagnostic and therapeutic interventions that can increase the health and extend the life of individuals, and if they are paid through a third-party reimbursement system in a fashion that does not directly impact on the consumer of health-care services, those who can benefit from the services will request them, even when the benefits are marginal at best, although clearly costly. Thus began the ever-upward spiral of health care costs witnessed in the United States since the 1960s. The attempt to purchase equality in health care for the indigent becomes a pursuit without an end, often with little benefit, and often with great costs.

## V. LEARNING TO LIVE WITH DEATH AND FREEDOM

The Catholic moral theological distinction between ordinary and extraordinary care allows health care to be foregone or not provided without moral guilt when such would involve an undue inconvenience because of the financial, psychological, or social costs. This distinction is easier for individuals to accept within that religious tradition than it is for nonbelievers because death is not seen to be the greatest downside risk. That risk is damnation. Moreover, that tradition does not focus on equality in this life, but on our need to repent of our sins, do good works, and accept salvation. It was in the 1960s just as this and similar traditional approaches had less of a claim on the minds of patients and health-care givers that an increased emphasis was placed on equality in health care and resources were invested in entitlement programs in order to erase the obvious differences between the health care provided to the indigent versus to those with the ability to pay. Here, concerns with charity, as well as with prudence (setting aside funds, should one need the new expensive technologies), acted in synergy with interests in equality and solidarity with the poor. Advances in medical technology took place as individuals who consumed its benefits were increasingly insulated from its costs. Traditional charity, which involved a particular individual's choosing to give his own resources to a particular institution, was replaced by taxation schemes that isolated benefactors from direct control over, and evaluation of whether, the funds given were worth the benefits received. The result has been a striking increase in the percentage of the Gross National Product deployed for health care, rising from 5.3% in 1960 to 11.6% in 1983 ([11], p. 14).

The cultural challenge is to accept the fact that there will often be interventions that may possibly extend life or increase health marginally at significant costs, to which many will not have access. For instance, if there is a treatment that costs a dollar per application and will save a life once out of every 10,000 applications, if the disease is not too frequent, most governmental health-care programs will be able to sustain its cost. However, if the cost of the treatment is $100,000, the cost of each life saved will be one billion dollars, at which point the treatment will be too expensive, if there are any significant numbers of people likely to be treated in a year. If there are 2,000,000 in "need" of such care, one would be committing about $200 billion or about one third of the entire United States health-care budget. To give this example content, one might imagine a treatment for AIDS. One would have three options: (1) to attempt to tax sufficiently in order to pay for the treatment for all who might benefit, (2) to forbid the licensing of the treatment until its costs dropped to a level at which all in need could have access, or (3) to allow the rich who would wish to squander $100,000 on a 10,000/1 gamble to do so with the same freedom they now have to purchase a Rolls Royce, but with the added attraction that over time use of the treatment may lead to better and cheaper treatment. Of course, in reality there are many intermediate cases where the diagnostic or therapeutic interventions are less costly, but must be repeated, and perhaps convey only a possibility of extending life for a few years.

It is very probable that the future is full of therapeutic and diagnostic modalities that will offer various margins of benefit at various costs with various likelihoods of offering additional quality years of life. If one acknowledges limits to the amount of money that may be taxed by governments from individuals (a limit set at least by the line between truly private versus communal resources and services), and if one holds that free individuals should not be impeded from peaceably offering to others their services for a fee, then we will have to acknowledge the fact that freedom means that some individuals will purchase a postponement of death not available to others.

This analysis leads one to conclude that the phenomenon of dumping and skimming, rather than the fault of for-profit medicine, reflects a failure of communities and states (1) to confront forthrightly the issue of health care for the indigent or for all through a basic insurance package, and (2) to recognize that a two-tier system, even where there are differences not only in the amenities but in the quality of care, is unavoidable as long as individuals are free and have entitlement to some of their own goods and services. The difficulties are engendered by the failure of many democracies to make responsible choices under the insurance metaphor. They have failed to tie the benefits of health care clearly to its costs, so that instead of seeing governmental policies for the provision of a minimal amount of health care as an investment in services reasonably limited by costs, they have often spoken of individuals having rights to the best health care available. Similarly, when speaking of duties of beneficence or charity, little is made of the many ways in which costs can defeat all duties, even the duty to provide health care. Here one might think of the recent "Baby Doe" regulations, which even in their final form make no reference to costs to parents or to society playing a proper role in determining when a duty to treat is defeated [4]. The difficulties associated with making prudent investments in health care are attributable not to the profit motive, but primarily to the failure to acknowledge the limited character of our lives and resources and the failure to recognize that morally unavoidable inequalities in resources will always give some individuals access to better health care than others.

## VI. SELLING LONGER LIFE AND VIRTUE; OR, IF YOU DON'T BELIEVE THAT MONEY BUYS HAPPINESS, YOU DON'T KNOW WHERE TO SHOP

But can one buy dedication? Will for-profit institutions encourage the kind of commitment to patients that traditionally has been a part of good medicine? Will the very ethos of the market and the explicit commoditization of health care erode the rights of patients and lead to inhumane treatment? The answer is likely to be "yes" to the first and second questions and "no" to the last, at least as long as patients are interested in having their rights respected and in receiving humane care. The market tends to provide what purchasers seek. If patients are interested in having their rights respected and being treated humanely, humane treatment will be a commodity worth retailing. For-profit corporations generally understand that customers want courteous, attentive service that regards their wishes and rights.[4] One might think of an advertisement by Federal Express comparing a stereotypical view of the response of postal clerks vis-à-vis Federal Express employees. Or one might think of comparing sales and service from an American car company, however bad one might think that to be, with the sales and service generally provided by the purveyors of automobiles in Soviet-bloc countries. If the bottom line depends on attracting patients, and if patients are interested in having their rights respected, then the market will respond because bioethics will have cash value.[5]

Lest this analysis seem sacrilegous, one must recall again what the market is. The market affords an opportunity for free men and women to exchange their goods and services peaceably. In the market, in the absence of fraud and in the presence of adequate information, people purchase what they wish, insofar as they have sufficient resources and insofar as others are willing to sell. It must be granted that there are significant difficulties in enabling patients to choose with sufficient information. There is little education in primary and secondary schools regarding what it is to be a patient, though most students will need to confront the challenge of being patients during some periods of their life. There is also less critical information readily available to the general public regarding health-care systems than there is regarding home appliances. But as more of the health-care market comes into the hands of a few large corporations and governmental providers, the easier it will be for data to be collected to supply prospective patients as well as citizens with information concerning which health care systems provide what increase of health or decrease of morbidity risks for what price and in what areas. In any event, the challenge is a general one. Citizens who in a democracy must oversee the deployment of common resources for health care, as well as consumers who need to purchase health care, must both become better informed in order to aid institutions in responding to the moral interest of health care consumers.[6]

Given sufficient consumer interest, one has every reason to believe that health care corporations will respond by providing services with attention and concern for the interests of patients. In fact, the inroads made by multinational corporations and other for-profit undertakings into the health care markets of countries with well-established national health services have been accomplished in part through attention to the interests of patients. Consider, for example, the following description of a for-profit clinic established in Stockholm.

> It is located close to many business offices, in contrast to the public primary health care centers built in the residential neighborhoods. It remains open late, until 7 p.m., and on Saturdays. Its personnel have received special training in consumer relations and it guarantees a shorter wait than the public hospital outpatient clinics. It provides quick, courteous, efficient care in a convenient location. While it is not equipped to deal with serious problems requiring elaborate hospital technology and sends these to the nearest public hospital, it is a symbol of much that is criticized in the public sector ([14], p. 595).

The point is that the profit motive does not in general appear to be negatively associated with attention to the needs and rights of patients. If anything, quite to the contrary. This is as one would expect, since humanism is a very marketable commodity. To put the matter in less tendentious terms, people tend to prefer one institution over another, if that institution invests time and energy to provide kindness and attention.

One might still object that humane service for hire is by definition insincere, for it is provided *according to duty* but not *from duty*, to repeat the Kantian distinction ([10], IV, 397). Perhaps that is often the case.

However, one can indeed buy loyalty and moral commitment. One might think of the Marines or the Secret Service, where in exchange for salary and certain benefits one commits one's life and in many circumstances one's entire person. An individual can make a morally binding agreement to commit his all, body and soul, for a price. One must grant that such total devotion is rarely part of for-profit health care undertakings and tends in the for-profit sector to characterize only a few areas such as the hiring of personal bodyguards (but it might also include certain surgical subspecialties if their members stand at increased risk of contracting AIDS). But still, one must note that corporations have morale development sessions that share with Marine boot camp the goal of fashioning common commitments to goals. Also, in circumstances where one is given status and remuneration for services rendered, employees often see the bond of employee and customer as being more than a financial one. Somewhat on the analogy of the pleasure bond, which Masters and Johnson have argued provides a basis for the development of enduring commitment and trust [13], so too, the profit bond can come to constitute a basis for true commitment, if not friendship. Though in the beginning one may act only *according to duty*, after acculturation within a well-managed health-care corporation, one may come to act *from duty*.

One cannot simply determine from the origins of an institution's capital and resources, nor from the goals of those who supply and manage the capital and resources, what the goals and demeanor of special groups of various employees will be. After all, to make a profit, one must often attend to issues other than profit-making. Complex institutions like the human body must support numerous functions in different organs that have a significance in their own right (e.g., the brain is more than simply a support organ for reproduction, though a simplistic understanding of evolution may so construe it). Since corporations must, as a matter of fact, be more responsive to the wishes of their customers than governments must to the wishes of their citizens (e.g., governments often have dictatorial control over their citizens and always have police control over them), one can have a well-founded assumption that, in general, for-profit institutions will be better caretakers of the rights and sensibilities of their customers than will governmentally-run institutions. Indeed, the Inquisition in the Middle Ages and the bureaucracies of the Soviet Union and of the United States, as well as the management of health-care corporations, have made assertions regarding their commitments to the best interests and well-being of those in their care. The advantage of for-profit institutions is that the moral presuppositions of the market as the origin of peaceably acquired resources and authority are generally more congenial to the rights and well-being of customers than the claim of divine authority or governmental sovereignty.

To know that a health care institution is a for-profit institution is to know something about how it has acquired and will continue to acquire its capital resources. It means that the corporation has gained its resources (1) by the free consent of individuals who hope to receive a return on capital, rather than (2) by confiscation of funds through taxes or (3) by charitable gifts, often given with a view towards a tax deduction. Moreover, unless a corporation has been given a monopoly by some variety of state enforced restraint of trade, a corporation will continue to have customers only insofar as it satisfies them.

## VII. SOME CONCLUSIONS

This essay began by underscoring the significance of Hegel. Hegel is helpful in reminding us of the importance of institutions that must span various moral communities. Hegel is also helpful in reminding us of the necessity of bringing our abstract moral concerns together with the content of the moral life. That an institution can span various moral communities and freely acquire the capital and the energy of individuals in the service of goals important to those communities and to the state, and can be understood as solving the problem of integrating respect of individual rights with a general commitment to beneficence while recognizing the limits of authority and of reason, is enough to justify that institution. The creatures of the free market, including for-profit corporations, help us solve the problem of understanding free cooperation in the pursuit of important human goals, given the limits of reason and governmental authority. Here I return to my recasting of Hegel in the service of better understanding what it would mean to cooperate peaceably and within the constraints of quasi-Kantian rights across diverse moral communities. This involves the challenge of understanding social structure in a large-scale, peaceable secular pluralist society. The market, driven by interests in profit, is part of meeting this challenge.

It still remains appropriate to evaluate the motives of both individual employees and particular institutions. Individuals are praiseworthy or blameworthy in terms of the reasons for their actions, though they are dangerous or useful in terms of their consequences. The same can be said of institutions. If those who own or manage a for-profit institution are committed to goals other than the morally neutral goal of profit, they attract special attention for either blame or praise, and such judgments tend properly to mark the institution as a whole, since it is the undertaking of those who own or manage it. Here concerns with abstract rights and commitments to beneficence can reach out and direct our judgments regarding social institutions. But aside from requiring that the rights of individuals be respected and that corporations not be malevolent, there is very little that may be said in condemnation regarding what motives should move corporations from the point of view of a secular state. Without a morally canonical ranking of social desiderata such as equality, liberty, prosperity and security, little can be said definitively regarding consequences. Concrete judgments in these matters require a particular concrete vision of morality, and most large-scale for-profit institutions are precisely those sorts of institutions that span and transcend particular moral communities and become part of the lives of citizens in secular pluralist states without endorsing a particular moral hierarchy beyond an interest of profit.

If anything, one should be at home when institutions clearly and honestly pursue profit, because such a pursuit is compatible with respecting the forbearance rights of individuals and is not hostile to interests in beneficence. One must simply recall an obvious truism, one underscored by David Hume and Adam Smith, that profit is not our only motive for action. We are moved as well by sympathy for our fellow-men and by our perceived duties to others. Since for-profit institutions do not do everything, one will need to seek other social devices for realizing our other important moral goals.

*Baylor College of Medicine*
*Houston, Texas 77030, U.S.A.*

## NOTES

[1] References to the works of Immanuel Kant will provide the volume and page in the standard edition of the *Königliche Preussische Akademie der Wissenschaft*, in addition to a reference to a translation in the case of direct quotations.

[2] As Hegel put it, "Reason is as cunning as it is powerful. Cunning may be said to lie in the inter-mediative action which, while it permits the objects to follow their own bent and act upon one another till they waste away, and does not itself directly

interfere in the process, is nevertheless only working out its own aims" ([9], Sec. 209, *Zusatz*, p. 350).

3 The central role played by freedom as a side constraint does not lead to strong property rights in land and natural resources, as Nozick suggests. Both states and individuals will suffer from equal difficulties in establishing exhaustive rights of this sort, leading to an argument on behalf of an internationally collected land tax, the proceeds of which should be paid out as an international negative income tax ([5], pp. 127–145).

4 To underscore how the profit motive inspires corporations and their employees to treat customers with attention and respect, it should be observed that the treatment of customers by for-profit corporations compares very favorably with the treatment of citizens by their governments. The worst treatment visited on customers by the worst of for-profit corporations compares very favorably with the worst treatment visited on citizens by the worst of governments.

5 "Ethics as a commodity" is an offensive phrase only from an aristocratic or egalitarian perspective. "Ethics as a commodity" means that people will often freely transfer their resources to others if that will purchase a change in behavior or a restructuring of institutions in ways that better meet their interests in having their rights respected and in receiving care that is attentive and humane. I have developed this point, and others in this article, elsewhere with Michael Rie and must acknowledge the substantial influence of that article on this essay [6].

6 The Oregon Plan serves as an excellent exemplar insofar as one considers it an endeavor to democratize choices regarding the deployment of communally owned resources for the indigent, recognizing the need to respect the existence of a for-profit tier that provides goods and services beyond those available through the governmentally supported tier of care.

## BIBLIOGRAPHY

1. Aaron, H.J. and Schwartz, W.B.: 1984, *The Painful Prescription,* Brookings Institution, Washington, D.C.
2. Aristotle: 1941, *Politics*, trans. B. Jowett, in R. McKeon (ed.), *The Basic Works of Aristotle*, Random House, New York, pp. 927–1111.
3. Cronin, D.A.: 1958, *The Moral Law in Regard to the Ordinary and Extraordinary Means of Conserving Life*, Typispontificus Universitatis Gregorianis, Rome.
4. Department of Health and Human Services: 1985, 'Child Abuse and Neglect Prevention and Treatment Program; Final Rule', *Federal Register* **50** (April 15, 1985), 14878–14901.
5. Engelhardt, H.T. Jr.: 1986, *The Foundations of Bioethics*, Oxford University Press, New York.
6. Engelhardt, H.T. Jr. and Rie, Michael: 1988, 'Morality for the Medical Industrial Complex', *New England Journal of Medicine* **319**, 1086–1089.
7. Hartmann, K.: 'The Profit Motive in Kant and Hegel', in this volume, pp. 307–325.

8. Hegel, G.W.F.: 1965, *Hegel's Philosophy of Right*, trans. T.M. Knox, Clarendon Press, Oxford.
9. Hegel, G.W.F.: 1965, *The Logic of Hegel*, trans. William Wallace, Oxford University Press, Oxford.
10. Kant, I.: 1976, *Foundations of the Metaphysics of Morals*, trans. L.W. Beck, Bobbs-Merrill, Indianapolis.
11. Lazenby, H.C. and Letsch, S.W.: 1990, 'National Health Expenditures, 1989', *Health Care Financing Review* **12**, 1–25.
12. Mandeville, Bernard de: 1962, 'The Grumbling Hive' [1705], in I. Primer (ed.) *The Fable of the Bees*, Capricorn, New York, 11. 71–72, 202–203.
13. Masters, W.H. and Johnson, V.E.: 1974, *The Pleasure Bond*, Little, Brown and Company, Boston.
14. Rosenthal, M.M.: 1986, 'Beyond Equity: Swedish Health Policy and the Private Sector', *The Milbank Quarterly* **64**, 592–614.
15. Schieber, G.J. and Poullier, J.-P.: 1991, 'International Health Spending', *Health Affairs* **10**, 106–116.

GEORGE P. KHUSHF

# RIGHTS, PUBLIC POLICY, AND THE STATE*

## INTRODUCTION

Baruch Brody has argued that rights language is not needed to justify a health care program and it is not helpful in addressing the crucial questions of what type and how much health care should be provided ([5]; see also [31], p. 149; [26], p. 210). Since the primary function of the debate on the right to health care is to address public policy concerns, Brody concludes that rights language does not serve its primary function and it should be abandoned in favor of a more fruitful language ([5], p. 113). In this essay I accept Brody's contention that rights language has not proven very helpful in the past.[1] But I believe this fact is only accidental and not inherent in rights language. In fact I think that rights language is especially useful because it allows us to appreciate the complex interaction of moral and nonmoral concerns that are involved in deliberations on public policy. The purpose of this essay is to develop rights language in a way that makes its relevance clear. I do this by placing the debate on the right to health care in the broad context of the justification of public policy and the nature of the state.

## I. MORAL AND LEGAL RIGHTS TO HEALTH CARE

To say there *is* a right to health care can mean two things:

1. There *is* an actual entitlement to health care, enforceable by law.[2]
2. There *should be* an actual entitlement to health care, enforceable by law.[3]

The distinction between moral and legal right has been used to avoid the equivocation that results when these two meanings are not properly distinguished.[4] If the meanings directly parallel the above definitions, then:



*T.J. Bole III and W.B. Bondeson (eds.), Rights to Health Care*, 355–374.

1. legal right = there is an actual entitlement to health care which is enforceable by the law
2. moral right = there should be a legal right.

When the terms are so defined, then the argument "there is a moral right therefore there should be a legal right" is irrefutable because it rests simply upon an analysis of the terms. But this would be misleading. First, some argue that there should be a legal right even though there is no moral right. Buchanan's argument in this volume is a good example of this. If "moral right" simply meant "there should be a legal right", then such argumentation would be absurd. Second, it is not necessarily true that if there is a moral right there should be a legal right. We shall take up this point shortly.

James Childress provides some clarification when he defines a right as a "justified claim". In this case

> Legal rights are claims that are justified by legal principles and rules, and constitutional rights are claims justified by the Constitution. Likewise, a moral right is a morally justified claim, that is, a claim validated by moral principles and rules ([7], p. 133).

Although this definition is an important step toward clarification, it still allows for some ambiguity, because the nature of the justification is not fully clear. The problem arises from the fact that there is a two-fold justification, and these two can easily be confused.[5] The right is a justified claim. But the right is also a claim that is in need of justification. When asking about whether or not a right is justified – and this is the question that is asked in public policy debate – one is actually asking about the justification of a justification.

The ambiguity in Childress' formulation of the right can be seen when we focus on the formula he gives: "$x$ has a right to $y$ from $z$" ([7], p. 133–134). The distinction between moral and legal right cannot be clearly delineated until this formula is expanded to read "$x$ has a *claim* to $y$ from $z$ that is, or can be, *justified* by $w$". Here Childress' definition of "right" (=justified claim) replaces the word "right" in his formula. If "$w$" is a moral system of rules, then one calls the right a "moral right"; if a system of legal rules (the laws of the state), then one calls the right a "legal right".[6] The question of the two-fold justification can now be developed as follows:

1. How does one justify the claim of *x* to *y*; i.e. what is *w*?
2. How does one justify the claim that "*x* has a claim to *y* that is justified by *w*"?

If one is sick and is only concerned with whether or not one can call upon the law to justify one's claim to health care, then the second question of justification is largely irrelevant. But if one is a legislator concerned with whether or not to vote in favor of a particular policy that provides legal entitlement to health care, then the second concern is very important.

In the case of a legal right there is an important difference between the justification of the claim and the justification of the justified claim. One could make the same distinction between the justification of the claim and that of the right in the case of a moral right. But in this case the distinction would be artificial. A moral right means "*x* has a claim to *y* which can be justified by an appeal to moral systems of rules". In this case the only way to justify the right is to justify the claim by way of moral argument. And if one morally justifies the claim then one has given the necessary and sufficient condition for justifying the right. Thus, in the case of a legal right there is a difference between the justification of the claim and the justification of the right. Argument is not sufficient to establish a legal right; a legislative decision is needed ([2], p. 119; [3], p. 58). In the case of a moral right no such difference exists. In a moral right the justification of justification is the justification. If *x* can justify a claim to *y* on the basis of moral principles, then that justification can also be taken as a sufficient justification of the claim "*x* has a moral right to *y*". This means that a moral right is simply a moral justification, i.e., a claim that *x should* have *y*. A legal right is a claim that *x does* have *y*, and can call upon the law to enforce that claim.[7]

The relevance of the moral/legal distinction to policy debate can now be developed as follows: a legal right is an entitlement to a benefit that can be justified (and thus enforced) by an appeal to the laws of a state. In policy debate one asks how and if a legal right is justified.[8] This asks about the justification of a legal justification. A moral right is an entitlement which appeals to moral principles rather than legal ones. Thus, as Beauchamp and Faden note, moral rights "exist independently of and can form a basis for justifying or criticizing legal rights" ([2], p. 119). To say that there *is* a moral right is to say that *x* can make a claim that he or she *should* be given health care. But does this mean that because *x* should be given health care, there should be a legal right corresponding to the moral right?

To show that *x* should have *y* or that *z* should provide *x* with *y* is *not* sufficient for showing that there should be a legal right for *x* to have *y*. Only certain types of moral considerations are such that they justify legal enforcement. Thus "*z* should give *y* to *x*" does not necessarily mean that *z* should be compelled by the law to give *y* to *x*. In order to account for this a distinction can be made in ethical theory between strong and weak obligations ([3], pp. 59–60). A strong obligation is one that should be enforced by the law. A weak obligation is still just as much a moral obligation but it is one that should not be a legal obligation. To avoid transgressing the law one needs only obey strong obligations. But to avoid transgressing a moral system one must obey both strong and weak obligations. The word "right" has been used by some (e.g., Beauchamp) to speak of the claim that is correlative to a strong obligation only. Given such a framework, a moral right would indeed imply that there should be a legal right, because a moral right would be the moral claim that should be legally enforceable. Thus there would be a distinction between the statement "*z* should give *y* to *x*" and the statement "*x* has a right to *y* from *z*". Then, by definition, a moral right would mean that there should be a legal right. There is, however, the following qualification: a moral right says that there is moral justification to say there should be legal justification of a given claim by *x* to *y*. But this does not yet imply that there should be a legal right.

If a given society agreed upon a given canon of moral principles and values, then, given Beauchamp's distinction between strong and weak obligations, one could argue that if there is a moral right then there should be a legal right. But there is rarely such an agreement.[9] Wherever there is disagreement, the nature of the distinction between strong and weak obligation will depend upon the particular moral perspective used in arguing for the right in question. To say there is a moral right will mean "from such and such a moral perspective, there should be a legal right". A moral right, viewed from the perspective of a policy maker, must thus be taken as the expression of one moral perspective on the issue of whether or not there should be a certain legal right. The question of policy then becomes: to what degree (if at all) should the moral right (expressing a particular moral viewpoint) be taken as an argument for a legal right?

The problem raised here is that of pluralism. Given a society of "moral strangers" – i.e., "individuals who meet in a moral controversy, but without sufficient moral premises to provide a basis for a resolution of their dispute" ([12], p. 48) – how does one balance competing moral claims when considering issues of policy and law? This question must receive some sort of resolution before one can determine whether and to what degree a moral right can be taken as an argument for a legal right.

We now turn to the question of the justification of a legal right. In policy deliberation the makers of policy – i.e. the elected representatives – must ask: what reasons should or do move me/us to implement a legal right? Four answers can be given:

1. Legal justification: If a court concludes that there is a constitutional right or some other compelling legal reason to provide health care, then it can require a legislative body to enact a policy that appropriately meets the legal concerns. Then one could say that there is a legal justification of a legal right.
2. The will of the people: If the people who voted a given representative in office *want* health care, then the legislator enacts the policy (or does what he can to enact it) as a representative of the interest of the people. Here the "argument" is irrelevant; or rather, the argument is very straightforward: the legislator favors a given policy because the people who voted him into office want that policy. In this case one can view the policy as simply an expression of the way the populace wants its communal resources allotted. Note that the people may want the policy for different reasons. Some may want a health care policy out of reasons of self-interest while others may want it for moral reasons. But all this does not matter to the legislator. All he cares about is whether or not the people want it. The reasons do not matter. In this case one could say that there is a justification by popular consent of a legal right.
3. Moral justification: If, for example, an argument is made that the principle of equality or justice requires that there be a legal right, then, if a legislator finds the argument morally compeling, he may vote in favor of the policy for moral reasons. In this case one could say there is a moral justification of a legal right; i.e., there is a moral right.[10]
4. Immoral justification:[11] If, for example, a particular hospital or organization gives a politician a bribe or promises to give large campaign contributions in favor of a particular vote on a particular

> policy that undermines the interests of the politician's constituency, then the politician may choose to vote solely on the basis of his own self-interest rather than the will of the people or some moral motivation.

In the U.S. there is no legal justification for a legal right to health care ([2], p. 120; [8]).[12] I take this to be the meaning of the statement of the President's Commission that there is no right to health care in the U.S.[13] There is no legally compelling reason for instating health care, although there are moral reasons and thus a moral obligation on the part of society to provide health care ([2], p. 120). This means that a policy debate that considers whether or not a legal right *should* be enacted will consider two types of arguments: popular consent and morality. Although the immoral justification will play a role (even a major role) in actual policy deliberation, it should be excluded when the focus is on the way policy decision *should* be made (as opposed to *is* made).

At this point one could ask in a general way about the way in which a legislator – an elected official, representing the people – should make decisions on the basis of moral considerations. Should a legislator always vote on the basis of popular consent (then his status is more emissary than legislator), or are there circumstances where he should have the courage to vote on the basis of moral concerns even though such a vote may be contrary to the will of the people he is supposed to represent?[14] For example, if a governor of a certain Southern state realizes that the popular consent is against certain changes in policy regarding "civil rights" should that governor stand against the populace for moral reasons or should he simply act in a way that represents the interests of his constituency? And then when the populace changes its position? In this essay I shall not attempt to answer this question. I shall simply assume that legislators make decisions on the basis of both moral argument and popular consent. Both are thus relevant in policy debate. There may be circumstances when popular consent can and should be undermined by moral concerns, and there may be circumstances when moral concerns can be outweighed by popular consent. The former case simply states that the will of the majority should not be made absolute. But the latter case is not so clear. Should there be instances where a legislator allows the will of the people to outweigh moral concerns? I think the answer is yes. It arises from the fact that the legislator is a representative of a pluralistic constituency. Values that the representative thinks should be reflected in policy are not necessarily values that should be reflected in policy [14]. One could say that the argument from popular consent and that from moral concern represent two prima facie concerns that must be appropriately integrated by any elected official. This involves a balance of a politician's own moral value system with the moral obligation to represent his constituency. And the nature of this balance will vary, depending upon the moral values of the particular politician.[15]

## II. PUBLIC POLICY AND THE STATE

By developing the relationship between a moral and a legal right, I have thus far attempted to show the role that moral argumentation may have in public policy debate. I shall now broaden the discussion and develop the implications of a theory of the state for the debate on rights and public policy. I shall show that the theoretical question of balance between positive and negative rights is directly relevant to the policy question of how much health care should be provided.

In general terms one can distinguish between two important functions of the state. Because these functions roughly parallel the distinction between positive and negative rights I shall call them the positive and negative function of state.[16] The positive function of the state is to provide the collective agency for the realization of the good.[17] The assumption here is that certain needs can only (or best) be met, and certain goods can only (or best) be realized when non-market, collectively enforced means are used. Of course, in order to realize the good, the state (via its representatives) must know the good. And this assumption becomes problematic in a context where there is no agreement on the good [13].

Contrasted with this approach is the negative function of the state. Here the role of the state is to restrain evil rather than to promote the good.[18] For example, if one speaks of a natural "state of war" where all individuals are constantly under threat of violence, then, when people make a social contract for reasons of self-preservation, the concern is no longer with the promotion of some particular good. In this context, if the people had no threat to their persons, they would prefer to remain in the state of nature rather than give up their liberties to some sort of commonwealth.

When the focus is on the negative function, then the state is a necessary evil rather than a positive good. And the goal is to make the power of the state as weak as possible while still endowing it with sufficient authority for the protection of individual liberty. In this context a right is viewed as a protected area of individual liberty – a place where the state can not impose itself (negative right).[19] On the other hand, if the focus is on the positive function of state, then the state is a good, and a right is a claim an individual has upon the good realized by the state (positive right).[20]

One's understanding of the state and of human concord differs considerably depending upon whether one focuses on the positive or negative function. If the focus is on the positive function then one envisions a society that is in harmony and agreement. The members of society all have a common vision of the good (e.g. some moral vision of society, or some common purpose). Then, with this vision in mind, they gather together, pool their resources, and establish the structures needed to realize this vision of the good. But when the focus is on the negative function, one envisions society as consisting of people who are self-interested, in discord, and who fear the power and corruptability of the state. They do not share any specific common vision and are primarily concerned about their own liberty; each individual seeks to use his own resources to pursue his own particular end. The purpose of collective structures is simply to protect one's liberty from the violence of another or from the state. The role of the state is to provide a context where "moral strangers" can peaceably meet.

Assuming that people do not simply remain alone, pursuing their self-interested ends independently of any and all, what type of concord is possible in a society dominated by the negative function of state? First, it is important to note that one can have the type of community envisioned by the positive function within the context of the negative one; but with this important difference: only those who contract to pool their resources can be compelled to contribute to the communal end. Thus the state does not require non-consenting people to use their resources to further an end that they do not feel important. The result will be many groups with many ends (some even at variance one with another), each group pursuing its ends simultaneously under the umbrella of the state, which in its negative function guards individual liberty. With this approach, one still does not have any relation among people who do not have the same ends. But is there any way for individuals with competing interests to act together in a way that leads to a common good?

Adam Smith has given the classic answer to this question: the *market* provides a neutral place of meeting in which moral strangers can peaceably relate in a way that leads to their mutual benefit ([19], Ch. III). And it does even more than this. Not only does it provide a neutral framework for relation among moral strangers; it also provides a context in which the pursuit of self-interest works to the good of the whole society. The self-interest of one is checked against that of another by way of competition. And the laws of supply and demand channel the pursuit of self-interest in a way that efficiently satisfies the needs and wants of society.

There are thus two ways to promote the common good. It can be promoted directly by way of the positive function of state. Or it can be promoted indirectly by way of the market. Thus even if there is agreement in society that a particular good should be pursued, that does not mean that it should be realized by way of the positive function of state. One can thus identify two important stages in any policy debate: (1) Should a given goal be viewed as a goal of the society that is represented by a given state? (2) If a goal is established as that of a given society, then should/can that goal be left to the market or should it be realized directly by the state?

If the market can realize a particular end as good or better than the state, then the market should be the structure by which that good is realized. The reason for this is as follows: In the positive function the state interferes (e.g. by taxation) with the liberty that each individual has to use his resources in the way he sees fit. There is an inverse proportionality between the positive and negative function of state. If we assume that there is a legitimacy to both the positive and negative functions (and this seems to be a reasonable assumption) then an approach that satisfies the positive and the negative simultaneously will be better than one that only satisfies the positive. If the market can satisfy a given end as well or better than the state, then the market is preferable because it provides a greater satisfaction of the negative function. Thus it is only when a given communal concern cannot be satisfied by the market that one will turn to the positive function of the state.

The implication of these theoretical reflections for health care policy can be illustrated by the arguments of Buchanan and Friedman in this volume. Buchanan focuses on the positive function of state. He begins with the assumption that:

All of the major traditions of religious ethics and all major philosophical normative ethical theories include general obligations of beneficence or charity – a duty to help others in need ([6], pp. 170–171).

In this way he attempts to identify a vision of the good that is common to all members of society. Health care to the needy is then taken as an instance of this common concern. Arguing that collective provision is the most efficient way to address the moral concern, and that compulsion is needed to avoid the free rider problem, Buchanan contends that the best (perhaps, only) way to satisfy the common moral vision is by way of establishing a legal right to health care.

Friedman challenges Buchanan's claim that a collective discharge of the moral obligation will be the most efficient ([17], p. 304, note 9). Instead, Friedman contends, health care will be more efficiently discharged by way of the free market system. Friedman argues his case by developing an economic model of the distribution carried out by way of public policy. He then compares the efficiency of governmental distribution with that of a market distribution and concludes that the market is on all counts superior.

The argument Buchanan makes for efficiency, however, does not rest on the same assumptions as the argument Friedman advances. The dispute over efficiency depends upon a more fundamental dispute about the nature of society and of human motivation. Buchanan assumes a society of people who act in harmony for the realization of a moral end. He asks about the most efficient way to address the moral concern of providing health care to the needy (*not* the way to provide health care in general). Friedman, on the other hand, assumes people act self-interestedly, in discord, and even immorally.[21] When he develops his "public choice theory" (economic model of government provision and allocation of health care), he assumes that legislators decide almost exclusively on the basis of that which we in the last section characterized as the "immoral" justification of a policy. The concerns of representation (popular consent justification) were only introduced in terms of a self-interested concern with re-election (see note 21). The moral justification of a policy (i.e. the role of moral rights) was completely ignored. Obviously any theory which assumes such motivation on the part of policy makers is going to conclude that it is more efficient to exclude government from the production and distribution of a given marketable good.

Friedman's assumptions about the motivation of policy makers amounts to a challenge of Buchanan's contention that all people recognize an obligation to beneficence. But this challenge is not so much against the contention that people recognize that they *should* be beneficent; it is a challenge against the contention that people *are* beneficent. At this point in the essay, rather than develop the complex dynamic of "is" and "ought" and its implications for the positive and negative function, I shall simply assume that there is some validity to the approaches of both Friedman and Buchanan. Some sort of integration of the positive and negative function is needed.

In the case of health care the concrete balance between the positive and negative function of state will express itself as a balance between a government and market distribution of health care. The question is then: what role should the government play in the distribution of health care and what should be left to the market? In addressing this question, it is important to distinguish between payment and production. Buchanan's argument simply shows that, given the goal of providing health care to the poor, it is more efficient to provide the care by instituting a legal right to health care. This particular goal would remain unrealized if left to the market. But Buchanan says nothing about the most efficient mode of production. If one takes Friedman's argument as one for the efficiency of a free market production of health care, then the positions of Friedman and Buchanan need not be mutually exclusive. Combining them one could advocate a government insurance plan which realizes the positive legal right by way of, e.g., a voucher system that enables the poor to have access to the free market.[22] The result would be a two-tiered system in which checks are put on the amount of insurance and thus the benefits that the government provides the poor. The limit to benefits provided to the poor would be determined by the place the negative function is given in the balance between positive and negative function. The greater the positive right, the greater the taxes required and thus the greater the imposition on individual liberty. The limit on positive benefit to the poor will be determined by the amount of imposition on individual liberty that is considered reasonable within a given society.[23]

As a result of the inverse proportionality between positive and negative rights, any legal right to health care will be a qualified right to health care. At the level of general, theoretical deliberation, the qualification will be developed in terms of the balance of rights: negative rights putting a check on positive rights. The line between the two tiers of health care will be a reflection of this balance. At a more concrete level, the qualification will be developed in terms of the distribution of finite resources. But the latter qualification is only a concrete, practical expression of the former.[24]

The relevance of rights language to public policy can now be summarized as follows: Rights language provides a useful language for developing the moral and non-moral dimensions of the debate on health care policy. The distinction between moral and legal rights provides a language for developing the relevance of moral considerations in ascertaining whether or not a legal entitlement can/should be established. The distinction between positive and negative rights then provides a language to account for two important functions of the state. The balance developed between the two functions is directly relevant to the limit placed on the legal right to health care. Thus, in sum, rights language helps in determining the key questions of policy; namely, how to justify the policy and how to determine how much of a benefit should be given.

*Institute of Religion*
*Houston, Texas, U.S.A.*

## NOTES

* I would like to thank Thomas Bole and H. Tristram Engelhardt for their critical comments and suggestions.

[1] There are some exceptions, for example [9]. Daniels addresses the question of what kind of health care and how much by reflecting on the principle of justice that he uses to establish the moral right to health care.

[2] If one asks a question such as "What kind of a right to health care is there in the U.S.?" then the assumption is that one is concerned with actual entitlements. E.g., in the U.S. the elderly have Medicare. As Beauchamp notes, "Rights to Medicare are, as rights, no different from rights to receive an insurance benefit when required premiums have been paid: Anyone eligible under fixed rules and requirements of eligibility may validly demand due services and goods provided by that program" ([3], p. 58). In this case to say "there *is* a right" means "there *is* an actual entitlement". The right is like an entitlement to property ([3], p. 57). Likewise when Engelhardt talks about creating rather than discovering rights ([10], [14]), he is speaking of creating actual entitlements which would be enforceable by the law. These are just two of many examples that could be given to show that some view rights as actual entitlements.

[3] When Veatch uses the principle of equality to prove that there is a right to health care [32], he does not mean to say that people have an actual entitlement to health care, which is already justified by the law. He intends rather to show that there *should be* an actual entitlement. The same can be said for the U.N. Declaration that there *is* a right to health care ([2], p. 119).

[4] Sometimes the distinction is developed explicitly as in the case of Beauchamp and Faden ([2], pp. 119–120), Childress ([7], p. 133), and Bell ([4], p. 165); and at other times the word "legal right" or "moral right" is used without explicitly developing the meaning, as with the President's Commission ([30], p.32).

[5] The nature of the confusion can be seen in those who decide that because there is no

constitutional or legal justification of the right there is no basis for saying there is or should be a legal right. But when one speaks of "legal right" one does not refer to the justification of the right. One refers to the justification one can use to enforce a given claim which is made for health care.

6 Initially a moral right was related to a legal one as "is" is related to "ought". But now, by defining the two in terms of two systems of rules, they are independent of one another. This approach has the advantage of not begging the question of how moral and legal concerns are to be related. It also has the important advantage of making it possible to raise the question of the relation between law (and policy) and moral argument in terms of rights. To this extent I think it is a more appropriate way to define "moral" and "legal" than that found in the writing of Nora Bell. She states that "the sense of 'right' in "right to health" or "right to health care" is best understood as an institutionalized right or legal right ... What are called moral rights or human rights are simply interests that we have which we feel ought to be made explicit in rules" ([4], p. 165). Recognizing that there are some rule governed practices that we do not want governed by the laws of state, she attempts to expand the meaning of "legal" so that it includes what I have termed moral right. "There is a lawlikeness in certain rule-governed practices that need not be recognized by the law for it still to count as conferring a legal right". Thus she distinguishes moral and legal right in terms of interest and rule-governed behavior. While such an approach points to an important distinction and is not self-contradictory, it has two important weaknesses. First, it uses "legal" in a way that is consistent with a scientific use of "law" but is confusing and misleading in a context where the laws of the state are a major concern. Second, and most importantly, it undermines the most valuable potential contribution of the distinction between moral and legal rights; namely, a discussion of the relation between moral argument and public policy.

7 Frankena points to another side of this important distinction between morality and law: "Morality ... is distinguished from law (with which it overlaps, for example, in forbidding murder) by certain features that it shares with convention, namely, in not being created or changeable by anything like a deliberate legislative, executive, or judicial act, and in having as its sanctions, not physical force or the threat of it but, at most, praise and blame and other such mainly verbal signs of favor and disfavor" ([16], p. 7). In the words of Engelhardt, we could say that legal rights are created (by a legislative, executive, or judicial act) while moral rights are in large part discovered (unless one argues that all morality is created in the same way that laws are created). The discovered moral right is embodied intersubjectively by way of the moral argument. (At this point I simply skirt the issue of whether the embodiment – i.e., the said rule – is the full incarnation of moral principle or simply an expression of a more foundational act.) In the case of moral entitlement "good argument" is both necessary and sufficient for the intersubjective confirmation of a moral right. But in the case of a legal right "a good argument for *X* does not always amount to an entitlement to *X*, and it is not sufficient grounding for a rights claim that one can argue forcefully that one ought to receive a good or service" ([2], p. 119).

8 Viewed in this way, the central concern of policy debate is whether to establish a legal right to health care and, if so, what kind. In this case rights language is very relevant to policy debate. Rights language gives us one type of language, among

others, to speak of the legal and moral concerns. And I know of no other language that can so succinctly and poignantly embody the relevant issues.

9 This lack of agreement has important consequences for the use of moral argument in public policy debate. Agich points to one of these consequences in his response to Buchanan ([1], pp. 193–196).

10 A moral right can thus be viewed as a tool of persuasion in a political context. In considering the persuasive force of a moral right we can appreciate the common "accusation" that rights language is the language of rhetoric. The word "rhetoric" is often used in a pejorative sense to mean language primarily concerned with conviction rather than logic. But if we use rhetoric in the classical sense, then logic is a subset of rhetoric. In the classical sense, it is very accurate to speak of *moral* rights as rhetorical, when considering their role in justifying a legal right. Thus "persuasive power in political negotiation and compromise" should not be set against "theoretical and philosophical" reflection ([1], p. 188; [25], p. 36). Instead one should see that there is a significant overlap between the rhetorical and the philosophical. It is only in the modern period that rhetoric has been divorced from the logical (for a detailed discussion of the relation between the logical and the rhetorical see [28]).

11 That which I here characterize as "immoral justification" could be divided into "immoral" and "amoral" justification. For example, in a utilitarian framework the politician's own happiness should be considered along with the happiness of others. There are several ways in which the politician could allow his own self-interest to play a role in the making of policy. If there are equally convincing justifications of the two sides of an issue, the politician can allow his self-interest to caste the tie-breaking vote. Then he would take a position on policy that, among other things, is in his own personal interest. This approach is perhaps best characterized as amoral rather than immoral or moral. To the extent that other factors were allowed to play a significant role the decision is moral. But to the degree that self-interest plays the deciding role the decision is amoral. (One could also envision a more moral approach. For example, one could imagine the politician making a deal with another politician to trade the vote in question for another vote on a different issue that is clearly justified by morality and representational concerns.) The line between the immoral and the amoral becomes even harder to draw in a case such as the following: In a utilitarian framework the goal is to maximize the overall happiness in the universe. Now let us assume that the net benefit to the constituency of a given policy is negative but the benefit to the politician is so great that it makes the overall net benefit of a policy positive. In this case, if the politician votes on the basis of self-interest, it will not simply be a matter of self-interest serving as a tie breaker. Now the self-interest of the politician will be an independent factor in the calculation, on footing with moral justification and representational justification. Obviously the nature of "moral" and "immoral" will depend on the moral system used to evaluate a given policy decision. In this essay, however, I simply assume that it is morally inappropriate for a politician to allow self-interest to play a role in policy decisions. The remuneration should be the income a politician receives, fame, and the reward of serving the country. I do not think that a politician should use his or her office to further interests that are not simultaneously the interests of the people. For this reason I include both amoral and immoral justifications under the heading of "immoral". In some ways this approach is

an expression of the maxim "duty for duty's sake". In a more complete presentation it would be important to justify this approach to the role-responsibility of the politician. But here I can do no more than acknowledge the problematic as well as the utilitarian alternative.

12 Here I assume that the right to health care is viewed largely as a positive right. In a fuller discussion I would need to more carefully qualify this. There are legitimate senses in which the right can also be viewed as a negative one ([2], pp. 125–126). Then there is a legal justification of the right grounded in the legal protection of life, liberty, and the pursuit of happiness. One could also identify a legal justification of a legal right in the case of the legislative process which established a right such as to treatment under Medicare or to the education benefits for the handicapped (e.g. the hearing impaired [20]). In the case of an already enacted policy one can speak of an a posteriori legal justification of a legal right. But this is not directly relevant in the case of a policy debate on whether or not a given right should be instituted. Then one is only concerned with what may be termed a priori legal justifications such as the constitution or a judicial injunction.

13 It is difficult accurately to pin down the meaning of the President's Commission declaration on rights and obligations to health care. Explicitly the Commission states that there is no moral right to health care ([30], p. 32). But "moral right" means something very different from that which I have defined. This is clear from the Commission's statement that there is a recognized societal obligation to provide health care. As Beauchamp notes, no clear distinction is made between strong and weak obligations ([3], p. 63). The question must then be raised: Toward what distinction is the Commission groping by distinguishing a moral right from a moral obligation? In the context of the discussion (e.g. no Constitutional right, etc.) I think the intent of the Commission is to make a distinction between the moral and legal justification of a legal right. The problem arises due to the two-fold justification of a legal right. The Commission, in intending to say that there is no legal justification of a legal right to health care, was groping for a way to make its case without implying that a legal right to health care is opposed by the law. In the civil rights movement the general claim was that there is legal justification of a legal right. Since there were not yet laws enforcing the right, the distinction between moral and legal right could be used to show that people had a constitutional right to protection that was not yet enforced by the law. "Moral right" meant the right under the constitution (which could also be viewed as a legal right) – i.e. a legal justification of a legal right – and "legal right" meant the policy in which the moral right was recognized and enforced. The President's Commission worked with a similar interpretation of "moral" and "legal" in order to distinguish the "right to health care" from other civil rights – esp. those that played such an important role in the 1960s Civil Rights Movement. It concluded that there was no legal justification that compelled legislation to institute a legal right to health care (e.g. no Constitutional basis). But, nevertheless, there is a moral obligation, and thus moral justification, for instituting such a right. In order to distinguish between a legal justification of a legal right and a moral justification of a legal right, the Commission used the distinction between moral right (meaning legal justification of a legal right) and moral obligation (meaning there is moral justification of a legal right). By "moral obligation" the Commission also sought to go beyond

purely legal concerns and state that there is an obligation on the part of county and community to care for the health of its people.

[14] The conflict between popular consent and moral concerns that may sometimes arise for the politician, as well as its ramifications, is pointedly captured by John F. Kennedy: "In no other occupation but politics is it expected that a man will sacrifice honors, prestige and his chosen career on a single issue. Lawyers, businessmen, teachers, doctors, all face difficult personal decisions involving their integrity – but few, if any, face them in the glare of the spotlights as do those in public office. Few, if any, face the same dread finality of decision that confronts a Senator facing an important call of the roll. He may want more time for his decision – he may believe there is something to be said for both sides – he may feel that a slight amendment could remove all difficulties – but when that roll is called he cannot hide, he cannot equivocate, he cannot delay – and he senses that his constituency, like the Raven in Poe's poem, is perched there on his Senate desk, croaking "Nevermore" as he casts the vote that stakes his political future" ([23], pp. 7–8).

[15] The balance between representational concerns and those of morality also depends upon the view that a given society has of the political system and the nature of democracy. If it is held that the direct participation of the citizens in the determination of policy is the ideal (e.g. the democracy of a Greek city state) then indirect participation by way of the legislator will be viewed as a compromise necessitated by the size and geographical extent of the population. In such a context priority will be given to representational justifications of a given policy. The founding fathers of the American constitution, however, believed that indirect participation was actually advantageous because not all are equally able to make a wise decision in policy matters. They gave priority to the moral sensibilities of the legislator and de-emphasized direct representational justifiction of a policy. Today there is considerably less confidence in the character of politicians and thus a greater weight is given to direct representational concerns. But indirect participation is not simply viewed as a compromise because it is recognized that some are in a better situation than others to make a reasonable decision that is based on all the facts.

[16] For a more detailed discussion of the relation between a given approach to state and positive and negative rights, see [22].

[17] Plato's *Republic* provides a good example of the positive function: "Society originates ... because the individual is not self-sufficient, but has many needs which he can't supply himself" (Republic, 369b). The state is formed "naturally" and as it moves to its more "civilized" form, it distills into three classes: rulers, guardians, and workers. The rulers are situated such that they know the good. On the basis of this knowledge they provide the conditions that are needed for the self-realization of all the members of the state. In this way, the state endows privileges that would otherwise be completely lacking.

[18] This restraining function is well illustrated by Thomas Hobbes. According to him all men are naturally in a state of war with all others. "In such a condition, every man has a right to everything; even to one another's body" ([21], p. 103). Here a "right" (the jus naturale) "is the liberty each man hath, to use his won power, as he will himself, for the preservation of his own nature" (ibid). In such a state of nature, man does not form a society so as to attain unto new privileges and possibilities. Instead he

agrees to a certain limitation of his mights in the interest of self-preservation ([21], p. 125). In giving up his "might to do all things" man becomes "contented with so much liberty against other men, as he would allow other men against himself" ([21], p. 104). John Locke, in many ways the father of the American form of government, further develops Hobbes' approach to the state (see Macpherson's introduction to [24]). Despite significant differences on the description of the state of nature and the role of the "sovereign", Locke sees the primary function of the state in the same way: the state's main role is to restrain evil and thus protect individual liberty.

19 Hollenbach accurately develops negative rights as "defenses of individual liberty. They are immunities from interference by others, claims to 'forbearance on the part of all others.' Rights are the fences around that field where the individual may act, speak, worship, associate or accumulate wealth without restriction by the positive action of either other persons or the state" ([22], p. 14).

20 Hollenbach well illustrates the meaning of positive rights by way of his discussion of the Marxist interpretation: "Rights, therefore, are not claims against society, whether in the political or economic spheres. They are opportunities to participate in the benefits of socialism under the direction of state and party" ([22], p. 23). It is interesting to note that the close association of positive rights with socialism has led to the accusations that the advocates of a right to health care are introducing socialist tendencies ([25], pp. 28, 30). For reasons we cannot develop here, however, it is very problematic to make such sweeping generalizations. The line between positive and negative rights becomes blurred in areas such as health care and education (see note 12). Moreover the appropriate categories for interpreting welfare mechanisms are unclear ([25], Appendix A).

21 To give just one example: "Consider, then, the market for legislation. Individuals perceive that they will be benefitted or harmed by various laws. They therefore offer payments to politicians for supporting some laws and opposing others. The payments may take the form of promises to vote, of cash payments to be used to finance future election campaigns, or of (concealed) contributions to the politician's income. The plitician is seeking to maximize his long run income (plus non-pecuniary benefits, one of which may be "national welfare"), subject to the constraint that he can only sell his support for as long as he can keep getting reelected. Is the outcome of this market efficient?" ([17], pp. 268–269). The answer is very likely to be "no", if one accepts this description of the political process.

22 I have just pointed to one way in which the two-tiered system could be developed. If one challenges Friedman's economic analysis, then one could envision a greater role for the government in the production and regulation of health care. Gavin Mooney has given reasons to think that normal free market mechanisms break down in the case of health care distribution ([27], Ch. 6). As a result of the "agency relationship" the rule of supply and demand is broken. The doctor determines the need of the patient by way of diagnosis and thus determines the demand. But he is also the supplier. And competition is undermined by the fact that the sick patient cannot run around while sick to find a cheeper diagnosis (one that does not determine such a high demand) or cheeper facilities to address the need. Since, as Adam Smith showed, it is competition and the laws of supply and demand that enable a free market to work to the benefit of all, there are reasons to think that the free market distribution of health

care may not work to the benefit of all. One can get around some of these difficulties by speaking of a market for health insurance. But one still has the difficulty of rapidly escalating costs due to the agency relationship. As long as insurance companies can pass on to the purchaser the increased cost there are no significant cost-containment devices at the macro level. To further discuss this theme, however, would take us too far afield from our present concerns.

[23] When the two-tiers are developed as the result of the balance between positive and negative rights one has a slightly different interpretation of government involvement than that developed by H.T. Engelhardt ([11], pp. 360–365; [15]). Engelhardt develops the tiers in terms of the distinction between communal and individual resources. He views negative rights as inviolable. Only communal resources can be used for the government tier. The question of policy is then how to divide up the pie of communal resources. This approach gives an accurate description of what takes place in the U.S. budget debates on Medicare and Medicaid. But it underemphasizes the recognition that, with few exceptions, the communal resources come by way of taxation, and thus are an imposition on individual resources. If one develops the tiers in terms of positive and negative rights, then the status of government supported health care as an imposition on individual liberty can be more explicitly developed. It will make it clearer that there is a significant cost to liberty when the government provides health care and a policy maker should weigh that cost against the benefit of satisfying the communal end.

[24] Halper develops a qualified right to health care by beginning with an unqualified right and then showing how the fact of finite resources forces a qualification [18]. Another approach is seen in the writings of Veatch and Engelhardt. Although they take radically different approaches to the problem, they both develop the qualification in terms of moral/theoretical concerns. Veatch looks at the way rights to liberty balance positive rights like health care ([32], pp. 99–100). Engelhardt looks at the problems involved in justifying a positive right and develops a government tier to health care in terms of the societal use of communcal resources ([11], pp. 360–365). In a policy debate the nature of the qualification would need to bring together the approaches of Halper, and of Veatch and Engelhardt. On one hand it needs to develop the general rational for the qualification. And then it needs to make it concrete in terms of the allotment of finite resources. As Marmor notes, one of the problems with U.S. policy is that it has not worked toward developing a rationale for its concrete allotment ([25], pp. 27–31). This has left the government with a huge commitment but no formulation of its ends. In this essay, by way of a discussion of the balance between the positive and negative function, I have attempted to provide the basis for a rationale that can integrate the important concerns of Veatch and Engelhardt, as well as the more pragmatic, political concerns of Halper.

## BIBLIOGRAPHY

1. Agich, G.J.: 1991, 'Access to Health Care: Charity and Rights', in this volume, pp. 185–198.
2. Beauchamp, T. and Faden, R.: 1979, 'The Right to Health and the Right to Health Care', *The Journal of Medicine and Philosophy* **4**, 118–131.

3. Beauchamp, T.: 1991, 'The Right to Health Care in a Capitalistic Democracy', in this volume, pp. 53–81.
4. Bell, N.: 1979, 'The Scarcity of Medical Resources: Are There Rights to Health Care?', *The Journal of Medicine and Philosophy* **4**, 158–169.
5. Brody, B.A.: 1991, 'Why the Right to Health Care is Not a Useful Concept for Policy Debates', in this volume, pp. 113–131.
6. Buchanan, A.: 1991, 'Rights, Obligations, and the Special Importance of Health Care', in this volume, pp. 169–184.
7. Childress, J.: 1979, 'A Right to Health Care?', *The Journal of Medicine and Philosophy* **4**, 132–147.
8. Curran, W.J.: 1989, 'The Constitutional Right to Health Care. Denial in the Court', *New England Journal of Medicine* **320**, 788–789.
9. Daniels, N.: 1991, 'Equal Opportunity and Health Care Rights For the Elderly', in this volume, pp. 201–212.
10. Engelhardt, H.T.: 1979, 'Rights to Health Care: A Critical Appraisal', *The Journal of Medicine and Philosophy* **4**, 113–117.
11. Engelhardt, H.T.: 1986, *The Foundations of Bioethics*, Oxford University Press, New York.
12. Engelhardt, H.T.: 1989, 'Applied Philosophy in the Post-Modern Age: An Augury', *Journal of Social Philosophy* **20** (1–2), 42–48.
13. Engelhardt, H.T.: 1989, 'Pluralism and the Good', *Hastings Center Report* (19:5), 33–34.
14. Engelhardt, H.T.: 1991, 'Rights to Health Care: Created, Not Discovered', in this volume, pp. 103–111.
15. Engelhardt, H.T.: 1991, 'Virtue For Hire: Some Reflections on Free Choice and the Profit-Motive in the Delivery of Health Care', in this volume, pp. 327–353.
16. Frankena, W.: 1973, *Ethics*, Prentice-Hall, Inc., Englewood Cliffs, New Jersey.
17. Friedman, D.: 1991, 'Should Medicine be a Commodity? An Economist's Perspective' in this volume, pp. 259–305.
18. Halper, T.: 1991, 'Rights, Reforms, and the Health Care Crisis: Problems and Prospects', in this volume, pp. 135–168.
19. Heilbroner, R.: 1964, *The Worldly Philosophers*, Simon and Schuster, New York.
20. Herbison, P.J.: 1986, 'Legal Rights of Hearing-Impaired Children: A Guide For Advocates', *Health and Social Work* **11**, 310–307.
21. Hobbes, T.: 1962, *Leviathan*, Collier Books, New York.
22. Hollenbach, D.: 1979, *Claims in Conflict: Retrieving and Renewing the Catholic Human Rights Tradition*, Paulist Press, New York.
23. Kennedy, J.F.: 1955, *Profiles in Courage*, Harper & Row, New York.
24. Locke, J.: 1980, *Second Treatise of Government*, Hackett Publishing Company, Inc., Indianapolis, Indiana.
25. Marmor, T.: 1991, 'The Right to Health Care: Reflections on its History and Politics', in this volume, pp. 23–49.
26. McCullough, L.: 1979, 'Rights, Health Care, and Public Policy', *The Journal of Medicine and Philosophy* **4**, 204–215.

27. Mooney, G.: 1986, *Economics, Medicine, and Health Care*, Harvester Wheatsheaf, Brighton, Sussex.
28. Ong, W.: 1958, *Ramus: Method, and the Decay of Dialogue*, Harvard University Press, Cambridge, Massachusetts.
29. Plato: 1974, D. Lee (tran.), *The Republic*, Penguin Books, New York.
30. President's Commission for the Study of Ethical Problems in Medicine: 1983, *Securing Access to Health Care*, Government Printing Office, Washington, D.C.
31. Siegler, M.: 1979, 'A Right to Health Care: Ambiguity, Professional Responsibility, and Patient Liberty', *The Journal of Medicine and Philosophy* **4**, 148–157.
32. Veatch, R.M.: 1991, 'Justice and the Right to Health Care: An Egalitarian Account', in this volume, pp. 83–102.

# CONTRIBUTORS

George J. Agich, Ph.D., Associate Professor, Department of Medical Humanities, Southern Illinois University, School of Medicine, Springfield, Illinois 62708, U.S.A.

Tom L. Beauchamp, Ph.D., Professor, Department of Philosophy and Kennedy Institute of Ethics, Georgetown University, Washington, D.C. 20057, U.S.A.

Thomas J. Bole, III, Ph.D., Program Associate, Biomedical and Health Care Ethics Program, University of Oklahoma Health Sciences Center, Oklahoma City, Oklahoma 73190, U.S.A.

William B. Bondeson, Professor of Philosophy and Medicine, Co-Director Program in Health Care and Human Values, University of Missouri-Columbia, Columbia, Missouri 65212, U.S.A.

Baruch A. Brody, Ph.D., Leon Jawarsky Professor of Medical Ethics, and Director, Center for Ethics, Medicine, and Public Issues, Baylor College of Medicine, Houston, Texas 77030; also Professor of Philosophy, Rice University, Houston, Texas 77001, U.S.A.

Allen E. Buchanan, Ph.D., Professor, Department of Philosophy, University of Arizona, Tucson, Arizona 85721, U.S.A.

Norman Daniels, Ph.D., Professor, Department of Philosophy, Tufts University, Boston, Massachusetts 02155, U.S.A.

H. Tristram Engelhardt, Jr., Ph.D., M.D., Professor, Departments of Medicine and Community Medicine, and Member, Center for Ethics, Medicine, and Public Issues, Baylor College of Medicine, Houston, Texas 77030; also Professor of Philosophy, Rice University, Houston, Texas 77001, U.S.A.

David Friedman, Ph.D., John M. Olin Visiting Fellow in Law and Economics, University of Chicago Law School, Chicago, Illinois 60637, U.S.A.

Thomas Halper, Ph.D., Department of Political Science, Baruch College and Graduate School, City University of New York, New York, New York 10010, U.S.A.

Klaus Hartmann, Emeritus Professor of Philosophy, Tübingen University, Tübingen, Germany.



*T.J. Bole III and W.B. Bondeson (eds.), Rights to Health Care, 375–376.*

George P. Khushf, M.A., Research Associate, Institute of Religion, Houston, Texas 77030, U.S.A.

Theodore R. Marmor, Ph.D., Professor of Public Management and Political Science, Yale University School of Organization and Management and Department of Political Science, New Haven, Connecticut 06520–7382, U.S.A.

Nancy K. Rhoden, late Professor of Law, School of Law, University of North Carolina, Chapel Hill, North Carolina 27599, U.S.A.

Hans-Martin Sass, Ph.D., Professor, Kennedy Institute of Ethics, Georgetown University, Washington, D.C., 20057, U.S.A.

Robert M. Veatch, Ph.D., Professor of Medical Ethics, Kennedy Institute of Ethics, Georgetown University, Washington, D.C. 20057, U.S.A.

# INDEX





*T.J. Bole III and W.B. Bondeson (eds.), Rights to Health Care*, 377–380.

# Philosophy and Medicine

1. *H. Tristram Engelhardt, Jr. and S.F. Spicker (eds.): Evaluation and Explanation in the Biomedical Sciences.* 1975 ISBN 90-277-0553-4
2. S.F. Spicker and H. Tristram Engelhardt, Jr. (eds.): *Philosophical Dimensions of the Neuro-Medical Sciences.* 1976 ISBN 90-277-0672-7
3. S.F. Spicker and H. Tristram Engelhardt, Jr. (eds.): *Philosophical Medical Ethics: Its Nature and Significance.* 1977 ISBN 90-277-0772-3
4. H. Tristram Engelhardt, Jr. and S.F. Spicker (eds.): *Mental Health: Philosophical Perspectives.* 1978 ISBN 90-277-0828-2
5. B.A. Brody and H. Tristram Engelhardt, Jr. (eds.): *Mental Illness.* Law and Public Policy. 1980 ISBN 90-277-1057-0
6. H. Tristram Engelhardt, Jr., S.F. Spicker and B. Towers (eds.): *Clinical Judgment: A Critical Appraisal.* 1979 ISBN 90-277-0952-1
7. S.F. Spicker (ed.): *Organism, Medicine, and Metaphysics.* Essays in Honor of Hans Jonas on His 75th Birthday. 1978 ISBN 90-277-0823-1
8. E.E. Shelp (ed.): *Justice and Health Care.* 1981 ISBN 90-277-1207-7; Pb 90-277-1251-4
9. S.F. Spicker, J.M. Healey, Jr. and H. Tristram Engelhardt, Jr. (eds.): *The Law-Medicine Relation: A Philosophical Exploration.* 1981 ISBN 90-277-1217-4
10. W.B. Bondeson, H. Tristram Engelhardt, Jr., S.F. Spicker and J.M. White, Jr. (eds.): *New Knowledge in the Biomedical Sciences.* Some Moral Implications of Its Acquisition, Possession, and Use. 1982 ISBN 90-277-1319-7
11. E.E. Shelp (ed.): *Beneficence and Health Care.* 1982 ISBN 90-277-1377-4
12. G.J. Agich (ed.): *Responsibility in Health Care.* 1982 ISBN 90-277-1417-7
13. W.B. Bondeson, H. Tristram Engelhardt, Jr., S.F. Spicker and D.H. Winship: *Abortion and the Status of the Fetus.* 2nd printing, 1984 ISBN 90-277-1493-2
14. E.E. Shelp (ed.): *The Clinical Encounter.* The Moral Fabric of the Patient-Physician Relationship. 1983 ISBN 90-277-1593-9
15. L. Kopelman and J.C. Moskop (eds.): *Ethics and Mental Retardation.* 1984 ISBN 90-277-1630-7
16. L. Nordenfelt and B.I.B. Lindahl (eds.): *Health, Disease, and Causal Explanations in Medicine.* 1984 ISBN 90-277-1660-9
17. E.E. Shelp (ed.): *Virtue and Medicine.* Explorations in the Character of Medicine. 1985 ISBN 90-277-1808-3
18. P. Carrick: *Medical Ethics in Antiquity.* Philosophical Perspectives on Abortion and Euthanasia. 1985 ISBN 90-277-1825-3; Pb 90-277-1915-2
19. J.C. Moskop and L. Kopelman (eds.): *Ethics and Critical Care Medicine.* 1985 ISBN 90-277-1820-2
20. E.E. Shelp (ed.): *Theology and Bioethics.* Exploring the Foundations and Frontiers. 1985 ISBN 90-277-1857-1
21. G.J. Agich and C.E. Begley (eds.): *The Price of Health.* 1986 ISBN 90-277-2285-4
22. E.E. Shelp (ed.): *Sexuality and Medicine.*
    Vol. I: Conceptual Roots. 1987 ISBN 90-277-2290-0; Pb 90-277-2386-9

23. E.E. Shelp (ed.): *Sexuality and Medicine.* Vol. II: Ethical Viewpoints in Transition. 1987 ISBN 1-55608-013-1; Pb 1-55608-016-6
24. R.C. McMillan, H. Tristram Engelhardt, Jr., and S.F. Spicker (eds.): *Euthanasia and the Newborn.* Conflicts Regarding Saving Lives. 1987 ISBN 90-277-2299-4; Pb 1-55608-039-5
25. S.F. Spicker, S.R. Ingman and I.R. Lawson (eds.): *Ethical Dimensions of Geriatric Care.* Value Conflicts for the 21th Century. 1987 ISBN 1-55608-027-1
26. L. Nordenfelt: *On the Nature of Health.* An Action- Theoretic Approach. 1987 ISBN 1-55608-032-8
27. S.F. Spicker, W.B. Bondeson and H. Tristram Engelhardt, Jr. (eds.): *The Contraceptive Ethos.* Reproductive Rights and Responsibilities. 1987 ISBN 1-55608-035-2
28. S.F. Spicker, I. Alon, A. de Vries and H. Tristram Engelhardt, Jr. (eds.): *The Use of Human Beings in Research.* With Special Reference to Clinical Trials. 1988 ISBN 1-55608-043-3
29. N.M.P. King, L.R. Churchill and A.W. Cross (eds.): *The Physician as Captain of the Ship.* A Critical Reappraisal. 1988 ISBN 1-55608-044-1
30. H.-M. Sass and R.U. Massey (eds.): *Health Care Systems.* Moral Conflicts in European and American Public Policy. 1988 ISBN 1-55608-045-X
31. R.M. Zaner (ed.): *Death: Beyond Whole-Brain Criteria.* 1988 ISBN 1-55608-053-0
32. B.A. Brody (ed.): *Moral Theory and Moral Judgments in Medical Ethics.* 1988 ISBN 1-55608-060-3
33. L.M. Kopelman and J.C. Moskop (eds.): *Children and Health Care.* Moral and Social Issues. 1989 ISBN 1-55608-078-6
34. E.D. Pellegrino, J.P. Langan and J. Collins Harvey (eds.): *Catholic Perspectives on Medical Morals.* Foundational Issues. 1989 ISBN 1-55608-083-2
35. B.A. Brody (ed.): *Suicide and Euthanasia.* Historical and Contemporary Themes. 1989 ISBN 0-7923-0106-4
36. H.A.M.J. ten Have, G.K. Kimsma and S.F. Spicker (eds.): *The Growth of Medical Knowledge.* 1990 ISBN 0-7923-0736-4
37. I. Löwy (ed.): *The Polish School of Philosophy of Medicine.* From Tytus Chałubiński (1820–1889) to Ludwik Fleck (1896–1961). 1990 ISBN 0-7923-0958-8
38. T.J. Bole III and W.B. Bondeson: *Rights to Health Care.* 1991 ISBN 0-7923-1137-X
39. M.A.G. Cutter and E.E. Shelp (eds.): *Competency.* A Study of Informal Competency Determinations in Primary Care. 1991 (in prep.) ISBN 0-7923-1304-6

GPSR Compliance

*The European Union's (EU) General Product Safety Regulation (GPSR) is a set of rules that requires consumer products to be safe and our obligations to ensure this.*

*If you have any concerns about our products, you can contact us on ProductSafety@springernature.com*

In case Publisher is established outside the EU, the EU authorized representative is:

Springer Nature Customer Service Center GmbH
Europaplatz 3
69115 Heidelberg, Germany

**Batch number: 10755813**

Printed by Printforce, the Netherlands